MORAINE VALLEY COMMUNITY COLLEGE LIBRARY
PALOS HILLS, ILLINOIS 60465

[WITHDRAWN]

THE SERVICE CONSULTANT

Principles of Service Management and Ownership

THE SERVICE CONSULTANT

Principles of Service Management and Ownership

2ND EDITION

Ron Garner, Ph.D.
Pennsylvania College of Technology
Williamsport, Pennsylvania

C. William Garner, D.Ed.
Rutgers University
New Brunswick New Jersey

Australia • Brazil • Mexico • Singapore • United Kingdom • United States

The Service Consultant: Principles of Service Management and Ownership, Second Edition
Ronald A. Garner and C. William Garner

Vice President, Editorial: Dave Garza

Director of Learning Solutions: Sandy Clark

Executive Editor: Dave Boelio

Senior Product Development Manager: Larry Main

Senior Product Manager: Matthew Thouin

Editorial Assistant: Courtney Troeger

Vice President, Marketing: Jennifer Baker

Marketing Director: Deborah S. Yarnell

Marketing Manager: Erin Brennan

Production Director: Wendy Troeger

Production Manager: Mark Bernard

Content Project Management and Art Direction: PreMediaGlobal

Cover Image: © Shutterstock / RoyStudio

© 2014, 2005 Delmar, Cengage Learning

ALL RIGHTS RESERVED. No part of this work covered by the copyright herein may be reproduced, transmitted, stored, or used in any form or by any means graphic, electronic, or mechanical, including but not limited to photocopying, recording, scanning, digitizing, taping, Web distribution, information networks, or information storage and retrieval systems, except as permitted under Section 107 or 108 of the 1976 United States Copyright Act, without the prior written permission of the publisher.

> For product information and technology assistance, contact us at
> **Cengage Learning Customer & Sales Support, 1-800-354-9706**
> For permission to use material from this text or product,
> submit all requests online at **www.cengage.com/permissions**.
> Further permissions questions can be e-mailed to
> **permissionrequest@cengage.com**

Library of Congress Control Number: 2013933574

ISBN-13: 978-1-133-61235-3

Cengage Learning
5 Maxwell Drive
Clifton Park, NY 12065-2919
USA

Cengage Learning is a leading provider of customized learning solutions with office locations around the globe, including Singapore, the United Kingdom, Australia, Mexico, Brazil, and Japan. Locate your local office at: **international.cengage.com/region**

Cengage Learning products are represented in Canada by Nelson Education, Ltd.

To learn more about Cengage Learning, visit **www.cengage.com**

Purchase any of our products at your local college store or at our preferred online store **www.cengagebrain.com**

Notice to the Reader
Publisher does not warrant or guarantee any of the products described herein or perform any independent analysis in connection with any of the product information contained herein. Publisher does not assume, and expressly disclaims, any obligation to obtain and include information other than that provided to it by the manufacturer. The reader is expressly warned to consider and adopt all safety precautions that might be indicated by the activities described herein and to avoid all potential hazards. By following the instructions contained herein, the reader willingly assumes all risks in connection with such instructions. The publisher makes no representations or warranties of any kind, including but not limited to, the warranties of fitness for particular purpose or merchantability, nor are any such representations implied with respect to the material set forth herein, and the publisher takes no responsibility with respect to such material. The publisher shall not be liable for any special, consequential, or exemplary damages resulting, in whole or part, from the readers' use of, or reliance upon, this material.

Printed in the United States of America
1 2 3 4 5 6 7 17 16 15 14 13

CONTENTS

Preface   vii
About the Authors   xiv
Acknowledgment   xvi

PART I   Service Facilities and the Service Consultant

Chapter 1   Types of Automotive Service Facilities and Industry Overview   1

Chapter 2   The Role of the Service Consultant and Formatted Operating Systems   27

Chapter 3   The Team Approach: The Service System and Shop Production   47

Chapter 4   Computerized Service Systems and Customer Recordkeeping   69

Chapter 5   Working with Warranties, Service Contracts, Service Bulletins, and Campaigns/Recalls   95

PART II   Communications: Customer Relations, the Service System, and Industry Business Practices

Chapter 6   Personal Communications: From the Greeting to the Presentation of the Invoice   113

Chapter 7   Working out Service Details with Customers and Industry Business Practices   151

Chapter 8   Closing a Sale and Suggesting Additional Work   171

PART III   Internal Communication, Employee Relations and Shop Production

Chapter 9   Writing for the Technician: Communicating Technical Details   188

Chapter 10   Workflow: Production Capacity and Scheduling   202

Chapter 11   Workflow: Monitoring Repair Progress   223

PART IV Service Management Principles, Marketing, and Ownership

Chapter 12   Customer Satisfaction and Marketing   253

Chapter 13   The Assistant Service Manager's Authority and Responsibilities   283

Chapter 14   Management of Operations, Contracts, and Insurance   303

Chapter 15   Service Management: Tracking Efficiency and Improving Effectiveness   327

Chapter 16   Ownership of a Service Facility from Start-up to Expansion   351

Glossary   385

Index   389

PREFACE

This edition of the textbook is devoted to those people who wish to work in an automotive facility as a Service Consultant but also aspire to become an Assistant Service Manager and perhaps even a Manager or an Owner in the future. The authors recognize that the position of Service Consultant and the Assistant Service Manager must be examined from different perspectives and discuss these differences in a fourth part that is devoted to management principles and practices along with establishment of a service facility from an owner's perspective.

Updated Content and Authors' Conceptual Framework

The significant changes over the previous edition include a conceptual framework that was developed to teach automotive management principles to aspiring students who want to become service managers. This edition presents a service system concept and combined it with more discussion on communication throughout the book. The service system is defined as an orderly arrangement of procedures that are linked together to form a process to be followed by employees and used by management to control shop production, customer service, and business operations. This model concept is important to the service consultant's job performance and as a means to effectively process customer requests.

As a teaching tool that aids students in understanding the service system approach, the textbook; makes available activities, examples, and worksheets for students to follow. The service system provides not only a framework for discussion in the classroom but also reinforces management concepts, communication, and information technology with computer application. It teaches the students to understand "what is happening" when work is processed. While the service system approach is a major improvement, the book stays true to the information students must know to be a good service consultant. This means the book wants the student to know "what is to be done" and "how the industry is structured and functions."

General Overview

This book is divided into four parts and 16 chapters to make instructional lesson planning easier. The first part examines the automotive service industry and the duties that the service consultant performs. The perspective is both as an employee and as a member of a team working within a service system. The second part is devoted to customer communication and industry business practices. The focus of the chapters in the third part is to describe the internal communications and how the service team works together to execute the service systems.

Finally, the fourth part discusses management and owner topics that relate to the job of the service consultant and Assistant Service Manager. When the title "Service Consultant/Assistant Service Manager" is combined in the fourth part of the book as a single job title, it is to recognize the overlapping tasks of the two positions. Further it is meant to show how the job is much broader in scope to assist management than a position; such as a receptionist, clerk, customer service representative, or a sales representative; that merely serves customers. The fourth part then concludes with a detail description regarding the steps to be followed when starting a new service facility business.

Second-Edition Changes

Chapter titles may appear to be similar to those in the previous edition; however, each chapter has had significant changes. These changes have greatly expanded the content to explain customer service as a system as well as to integrate more on information technology, business principles, and management topics into the chapters. The authors, as instructors, have been careful to assure the upgrades explain the concepts in a way that students can relate to and understand them. In other cases, especially in the final parts of the book, the chapter titles have changed to better identify the new content that includes management and ownership.

The greatest change since the previous edition has been that information technology and communications is no longer a phone line and a computer to record customer information. This edition focuses on various electronic devices, Internet, and electronic communication mediums as well as the integration of computer programs at every level of service facility operations. This means that the authors explain communications and recordkeeping so that students understand them relative to a seamless service system. Then the authors discuss the different ways that computers are integrated into the service systems (Customer Service, Shop Production, and Business Operations systems) and used at different service facilities to improve interactions between the service consultant, customers, technicians, parts specialists, and even the hotline assistance.

Detailed Overview

PART I: Service Facilities and the Service Consultant

Chapters 1 to 5 present an introduction to the service consultants' career beginning with an overview of the service industry that includes warranties and licenses followed by an in-depth description of the service consultant's tasks and duties. This part also explains how the team approach is connected to the three systems that make up the service system, which are integrated into every chapter. The application of each chapter's content is to explain the role of the service consultant who acts as a pivot working across the three systems of service facilities that range from small garages to large dealerships that use computerized systems to handle every aspect of customers service, shop production, and business operations.

PART II: Communications: Customer Relations, the Service System, and Industry Business Practices

Chapters 6 to 8 take a more in-depth approach to the service consultant's position as a pivot by looking at customer communications as a system from the greeting to the presentation of the invoice. This part pays close attention to the details of these interactions to connect them to the first part of the textbook. Topics in the discussion of the additional tasks and duties include working in teams, service facility size, warranty contracts, parts suppliers, computers, and the details about the three systems in a service facility.

PART III: Internal Communications, Employee Relations and Shop Production

Chapters 9 to 11 move the discussion of the service consultant's position to a broader perspective by looking at internal communication to assure workflow is on track to meet the customer's "promise time." The authors talk in terms of the importance of a seamless system that occurs when the three service systems (Customer Service, Shop Production, and Business Operations) overlap to make the service consultant central to directing information much like a pivot. More specifically, the chapters stress the need for the service consultant to effectively communicate with customers and technicians and to monitor workflow using a Repair Order Tracking Sheet.

PART IV: Service Management Principles, Marketing, and Ownership

Chapters 12 to 16 connect business principles to the automotive service industry for students aspiring to enter the industry as an assistant service manager (ASM) and then move into management or perhaps ownership. The topics are applied to the auto repair industry and include marketing, management, types of ownerships, legal contracts, efficiency and effectiveness, and profitability. The content is designed to help a student, who starts his or her career as a service consultant, gain the knowledge needed to become a functional part in a management team and not a spectator. This introduction is a basic prerequisite that is required to eventually be in a service manager's position as well as an owner of a facility. This final part also leads into the authors' second book in the series, *Managing Automotive Businesses: Strategic Planning, Personnel, and Finance*.

Alignment with Automotive Industry Standards

Chapter objectives are aligned with the ASE Task List for a Service Consultant. However, at the same time, the authors recognize that the tasks and specific duties assigned to a Service Consultant/Assistant Service Manager will vary depending on whether the employer is an independent automotive service facility, a dealer, a franchise, a chain operation, or a fleet operation. Therefore, the book covers numerous tasks, duties, and procedures that *may* be performed by a Service Consultant/Assistant Service Manager.

Admittedly, some service consultants are required to perform a wide range of tasks, while other service consultants are given a limited number of tasks. The effort to provide an in-depth review of as many tasks as possible is intended to prepare the student for a variety of possible Service Consultant/Assistant Service Manager Positions. Further, this approach presents the information needed to move from one type of service facility to another as well as from an entry-level position to one that affords greater responsibility, such as a manager or an owner.

The book includes extensive reviews of how computers are used to manage a service facility and communicate with the public. From personal experience, the authors know the importance of using a computer to run a service facility. However, computer programs and levels of networking can be quite different from each other and the actual computer entries and commands must be learned from the software manufacturer. This book, therefore, thoroughly covers the application of computers at different types of service facilities relative to the tasks to be performed by the Service Consultant/Assistant Service Manager.

While it is unrealistic to discuss a specific service system method or computer program, it is further impossible to describe all the features that can include various communication devices, networks, computer programs, and applications. Therefore, the textbook approaches information technology and computer applications relative to the service system, management, and communication from a generic approach. This is to help promote student comprehension so each has an advantage when being trained to use an employer's specific computer system and procedures.

Specifically, the book emphasizes the integration of computers, devises, and features relative to the customer service system, shop production system, and business operations systems. When appropriate, the textbook mentions specific options available to the service staff and includes the popular use of email, social networks, texting, and so on. It is recognized that the current methods will change just as much in the next 10 years as it has since the first edition was written. What the next 10 years will bring to the service industry is exciting to think about.

Finally, throughout the book, the importance of the customer is stressed. In some cases, automobile owners come to a facility with a clear-cut knowledge of what has to be done to their automobile. In other cases, the owners depend on the Service Consultant/Assistant Service Manager to assist them. From the authors' experiences, dependence on the Service Consultant/Assistant Service Manager for solid advice is a growing need. Thus, the book's approach to customer relations is that the Service Consultant/Assistant Service Manager should offer customers professional assistance when taking care of them. As part of this responsibility the Service Consultant/Assistant Service Manager must keep track of the shop production to assure the customer's demands are met. To illustrate this process the second edition has created service system models and expanded the details of the original Repair Order Tracking Sheet to present it as part of a shop production system. The authors also created a customer service system within the broader service system to illustrate the stages, documents (estimate, repair order, invoice), and interactions that a service consultant performs. This detail is based upon the request of the many first-edition readers and suggestions of graduates of Dr. R. Garner's who served as book advisors. They work in the industry and had the Repair Order Tracking System presented to them in class as a teaching tool.

Special Features

New boxed features present Case Studies, Career Focus, including Career Profiles, and Additional Discussion sections to help and inspire students' to master theory. For example, the additional discussion sections are meant to extend beyond the basic text information for students who want more details. In addition, some topics the book has included are about women in the industry (WRN network input), service facility insurance/accidents, warranty administration, how the manufacturer determines labor times, court proceedings to defend an invoice, business contracts and leases, unqualified technicians performing work, management authority as related to technicians, team leaders, service consultants, and the ASM. OSHA and other government regulations with examples of industry problems throughout the book in addition to other essential information. Further, marketing and communication has been extended to include a comprehensive explanation of social networks, websites, email, browser and other advertising opportunities, text communications, and other types of communication, such as Twitter. All are related to the service facility and making money from satisfied customers.

Career Focus and Profiles

Most chapters start off with a Career Focus feature that speaks directly to the new generation of students. These are intended to help focus students' attention on how the content in the chapter is related to their future success. Plus, they make chapter readings more relevant and personable for the student and helps inspire them to think about their future as a professional in the industry. To further encourage students, career profiles of Factory Representatives, Team Leaders, Field Service Engineers, Technical Trainers, and Technical Hotline Employees have been added among other professional positions and the qualifications to obtain those positions.

Instructor Resources

Instructor Resources have been expanded to include PowerPoint lecture presentations and other teaching tools, such as the Repair Order Tracking Sheet. To measure student outcomes, ExamView text questions for each chapter are available to test student knowledge. There are also in-depth activities found at the end of every chapter that instructors can use for discussion in class. As time permits, instructors can create student "individual or group" projects from the activities at the end of the chapter.

The authors have been thoughtful to present opportunities for students to do research and work with the school's library staff as required by some accreditation bodies. Furthermore, there are activities that an instructor can use to have students visit job sites and perform field observations as well as interact with industry professionals. Therefore, the instructor resources at the end of each chapter go beyond multiple-choice questions and short answers to include more opportunity for interaction and discussion between students, instructors, school library staff, and industry professionals.

Book Overview: To the Student

To help you, the student, understand how a service facility works, the service system approach has been added as a tool to help you learn. The service system provides not only a way for you to understand how to communicate but also how to work with a service team that uses information technology with computer applications. The authors want to help you be aware of "what is happening" when work is processed. While the service system approach is a major improvement in this edition of the book, it keeps to the basics of what you need to know to be a good service consultant. This means the book wants to teach you "what you are supposed to do" and "the principles that the industry is built upon."

Specifically, chapters 1 through 3 focus on the different types of service facilities, the tasks and duties of the Service Consultant and the service team involved in the processing of work within a service system. Chapters 4 through 8 move to the customer process from the initial contact to the presentation of the invoice. Chapters 9 through 12 fill in the details of processing and tracking workflow as well as the actions and procedures described in previous chapters. Chapters 13 to 15 focus on the Assistant Service Manager duties and differences from those of a Service Consultant while they also delve deeper into management topics. The final chapter explains service facility ownership from start up to expansion so you can appreciate the difficulties of owning a service facility and understand how a new business must be managed.

Finally, the final chapters in the book aim to prepare you to study the next book in the series *Managing Automotive Businesses: Strategic Planning, Personnel, and Finance*. Therefore, these chapters are intended to add greater details about general operations and the work environment.

Since the chapter's content is aligned with tasks identified by Automotive Service Excellence (ASE) for the Automotive Service Consultant, each chapter begins with a list of learning objectives that you need to know. Many of these objectives directly reference an ASE task, and then the chapters end with a set of exercises related to the learning objectives. In addition, at the end of each chapter are suggested activities that when completed can offer you some additional insights into the operations of different service facilities. Hopefully, these insights can offer you some practices, issues, and concerns you may think about in the future and discuss with other professionals in the field. The authors hope that you enjoy reading the book and think you will get a lot out of it that will help you in the future.

From the Authors

The content in this book is largely based on our experiences in the automotive industry at large and small dealers, as well as owners/operators of automotive service facilities and other non-industry businesses. This is combined with our studies of automotive technology, business, and vocational education in addition to our work as teachers, automotive trainers, and consultants. Starting an automotive service facility in an empty building with two old lifts and a compressor, we gained invaluable insights into the trials and tribulations associated with creating a business from the ground up and this is reflected in the final chapter. As owners/operators, we did everything: working as the Service Consultant/Assistant Service Manager, cashier, bookkeeper/accountant, manager, technician, and even custodian. Consequently, most of what we talk about in this book we have done. We try to present as honest a picture as we can and hope it will benefit those of you who will enter the automobile industry. If you like being around cars and people, you will enjoy the job of a Service Consultant/Assistant Service Manager.

ABOUT THE AUTHORS

Ron Garner, Ph.D. is a Professor, in charge of the B.S. degree in Automotive Technology Management (http://www.pct.edu/schools/tt/automotive/), at The Pennsylvania State University special mission campus, Pennsylvania College of Technology in Williamsport. The program is accredited by the Association of Technology, Management, and Applied Engineering (ATMAE) and is a 2 + 2 program that is available both on campus and by distant learning. In that capacity for over 15 years, he teaches automotive business courses and advanced automotive technology courses, directs senior thesis projects, and teaches technical training courses that include the Pennsylvania Enhanced Emissions program. In addition to business consulting and authorship of other publications, including the second book in this series, he is a partner in charge of financial operations for an investment company that specializes in owning and managing residential apartment buildings.

Ron completed an A.A.S. degree in Ford ASSET automotive technology from Lehigh County Community College, a B.S. in Vocational Education (Teacher Certification), a second B.S. in Accounting, and an M.S. in Vocational Education with an emphasis in School Administration and Leadership from The Pennsylvania State University. He earned his doctorate (Ph.D.) in Workforce Education and Development at The Pennsylvania State University with an emphasis in Training and Post Secondary School Administration. An A.S.E.-certified Master Automobile technician he was a technician for several years starting at a Roll Royce/Bentley specialty shop near Princeton, N.J., before working as a flat-rate dealership technician at the Fred Bean's Family of Dealerships, outside Philadelphia. He went on to own and manage automobile repair facilities and entered teaching as a Toyota T-Ten automotive instructor. His email address is: rgarner@pct.edu

C. William Garner, D.Ed., served as an airborne sonar technician in the U.S. Navy from 1959 to 1963. In the Navy he attended basic electricity-electronics and airborne sonar school and served as an aircrewman in a helicopter squadron. After the Navy he earned a bachelor's degree in business education, a master's degree in higher education administration, and a doctorate in vocational education (accounting, retailing, business law) at The Pennsylvania State University. After working at Penn State as an administrator for 10 years, he took an appointment with Southern Illinois University at Carbondale as an Assistant Professor of Occupational Education and Site Administrator at March Air Force Base in California. His next appointment was with the University of Louisville as an Assistant Professor of Vocational Education and Coordinator for Educational Programs at Fort Knox.

In 1978 he received an appointment as an Associate Professor of Vocational Education and later in Education Administration at Rutgers University. He served as Chair of the Department of Urban Education, Chair of the Graduate Department of Vocational Education, Director of Continuing Education for the Graduate School of Education, Program Director for Education Administration, Executive Director of the Vocational Education Resource Center, and an Acting Dean. Currently, Bill is a Professor Emeritus of Education Administration at Rutgers University.

Bill has written two books: *Accounting and Budgeting in Public and Nonprofit Organizations: A Manager's Guide* and *Education Finance for School Leaders: Strategic Planning and Administration* plus numerous research, scholarly and professional articles, and monographs. As a consultant, he has worked and continues to advise a variety of profit, nonprofit, and governmental organizations through CWG Associates and has served on numerous boards. With his son, Ron, he owned and operated an auto repair business and has written two books on automotive management. His email address is: cwandkgarner@comcast.net

ACKNOWLEDGMENT

Over three dozen industry professionals have come together to provide input to the book because they feel customer service, management, and business principles are important topics for schools to teach. These professionals make up an impressive list with almost every manufacturer represented as well as input from U.S. Government fleet administrators, small shop owners, large dealership managers, service chain operators, parts store owners and managers, and even faculty from other universities, a court judge, vehicle engineers, and organizations such as AAA. They have taken the time to supply input or review content for the new edition of the book. This along with the authors' industry and teaching experience has resulted in a fresh and updated content with a new title *The Service Consultant: Principles of Service Management and Ownership*.

Special Thanks

Appreciation is acknowledged to all of graduates of the Automotive Technology Management program from The Pennsylvania College of Technology, most of whom hold jobs presented in this textbook and have provided feedback over the years to improve the concepts and how they were presented in class. This edition of the textbook reflects those improvements. Appreciation is specifically extended to these industry professionals who took the time to assist the authors:

- Dr. Joel Anstrom, *Ph.D. Director, Hybrid and Hydrogen Vehicle Research Laboratory at the Pennsylvania Transportation Institute, The Pennsylvania State University, College of Engineering*
- Jonathan Bastian, *Director of Retail Operations, Bastian Tire and Auto Service Centers (regional chain)*
- Ryan Black, *Toyota Training Developer, Toyota Motor Sales*
- Gary Best, *Owner, Best Auto Sales*
- Randall Bounds, *Field Service Engineer, Ford Motor Company–San Francisco Region*
- Celeste Briggs, *Director Women's Retail Network, Dealer Development, General Motors Corporation.*
- Alan Buck, *Library Faculty Associate Professor, Pennsylvania College of Technology*
- Mike Byrne, *Fixed Operations Director, Ed Morse Sawgrass Auto Mall (GM dealer)*
- James G Carn, *Magisterial District Judge, Williamsport, Pennsylvania*
- Dan Crisman, *Owner, 3D glass LLC*

Tim Cleary, *Pennsylvania Transportation Institute, the Pennsylvania State University, College of Engineering*

Anthony J. Cox, *Store Manager, Advance Auto Parts*

Bill Cragin, *AC Delco Marketing Area Manager, Northeast Region Pa/NJ/De*

Eric Dane, *Service Manager, Bob McCormick Ford Dealership*

Kris Doyle, *Penn-NY Group Coordinator for AAA Approved Auto Repair*

Pat English, *Associate Professor, Automotive Service Technology, Ferris State University*

Joseph Fosko, *Manager of AAA Approved Auto Repair, AAA (East Penn Region)*

Charlie Frye, *AAA Approved Auto Repair (East Penn Region)*

Ken Haas, *Owner, Pine Mountain Auto Repair*

John Hiney, *Insurance Agent, Davis Insurance Inc.*

Jeremiah Jones, *District Sales and Service Manager, AC Delco*

Jason Kline, *Customer Service Manager, Hoffman Ford Dealership (Harrisburg)*

John Kouroupas, *Customer Experience Specialist, Ford Motor Company*

Pat McCormick, *Technical Operations Coordinator, Ford Motor Company*

John McDonald, *Automotive Commodity Specialist (Fleet Operations) U.S. Government General Services Administration (Washington D.C.)*

David Paterno, *President, Sageminder Systems Inc.*

Lori Paterno, *Owner, Sageminder Systems Inc*

Barney Penton, *Owner, Penton Automotive*

Dave Plessinger, *Owner, Economy Auto Parts (Carquest Franchise)*

Anthony Piccari, *Service Operations Consultant, Infiniti USA East Region*

Kathleen Ponto, *Program Manager, Advantage ™–Technical Resourcing (Technical Hotline; Ford Motor Company)*

Ben Pyle, *Regional Vehicle Field Sales Manager, Toyota Motor Sales; (Central Atlantic Toyota Distributers)*

Lem Rowland, *Owner, Action Towing*

Carlo Santora, *Technical Training Instructor, Chrysler Group LLC*

Dr. Dorothy Chappel, *Ph.D., Assistant Superintendent, Jersey Shore School District*

James Simon, *District Manager, Toyota Motor Sales (Central Atlantic Region)*

Al Steinbacher, *Parts Department Manager, Sweitzer's Performance Parts and Car Sales*

Donald Stewart, *Service Manager, MacIntyre Chevy, Buick, GMC, Cadillac Dealership*

With special thanks to all of the Service and Parts department employees at *MacIntyre Chevy, Buick, GMC, Cadillac Dealership for their cooperation*

John Thompson, *Service Manager, Fred Beans Family of Dealerships (Dodge, Chrysler, Jeep dealership)*

Dr. Richard Walter, *Ph.D., Associate Professor Workforce Education and Development, Pennsylvania State University, College of Education*

Colin Williamson, *Dean, School of Transportation Technology, Pennsylvania College of Technology*

Dan Zell, *Owner and Manager, Stouts Pro Auto*

CHAPTER 1

TYPES OF AUTOMOTIVE SERVICE FACILITIES AND INDUSTRY OVERVIEW

OBJECTIVES

Upon reading this chapter, you should be able to:

- *Describe the different types of automobile service facilities.*
- *Diagram an organizational structure for the different types of automobile service facilities.*
- *Explain the importance of the guidelines that service facility operations must follow in different states.*
- *Identify diversity issues within the industry and improvement initiatives.*
- *Identify the major financial measures that have a direct impact on the profit or loss of a service facility.*

CAREER FOCUS

If you like cars, people, being the center of the action, and want to be regarded as an expert, then you need to start your career as a service consultant. Being a consultant means you are an expert and they pay you "big bucks" for your advice. The better you are, the more you make. As they recognize your strength, promotion into management positions comes next, perhaps ownership opportunities.

Not just anyone can walk in off the street and become a priceless member of a repair superstore, dealership, franchise, or a successful owner of a service facility without serious preparation. Good looks, great smile, and an award-winning personality cannot make it alone. As professors, we (the authors) have personally prepared very successful automotive service consultants who are admired by their customers, valued by the owners, and looked up to by the technicians. Because we stay in touch with many of them, we can testify to the success and personal satisfaction achieved in their career. It can be a great career and not a job.

To be number one, in this high-technology business, you must understand the industry and business. To help you with this understanding, some of our personal contacts, as listed in the thank-you section, have taken the time to suggest specific content that you will read throughout this book. They are industry experts who want you to enter the industry and be the best because when things get tough in this industry, you have to be the best. Nothing has been held back; it is how we teach it in class, and to get you started, this chapter tells you about the industry.

Introduction

A clear understanding of the terms used in a book is critical. For that reason, this chapter begins by defining two important terms used throughout the book. First, the term *service* means the maintenance, repairs, and diagnosis of an automobile. Second, an **automobile service facility** is a for-profit business that services automobiles.

The purpose of this chapter is to overview the automotive service industry. To point out the differences within the industry, we examine different types of auto service facilities. First, not all automobile service facilities are the same. While the job of the service consultant is similar at most service facilities, there are important differences. Being aware of the business differences is critical for service consultants and students to understand the industry.

Second, automobile service facilities have different organizational structures. An **organizational structure** is the managerial chain of command, and an **organizational diagram** presents the relationship between the different positions at an automobile service facility. A position that is higher in the diagram is responsible for the supervision and evaluation of the position beneath it. Knowing to whom the service consultant reports is critical.

In addition, service consultants must be knowledgeable about the federal and state laws that regulate the operation of their service facility, the people who work in the facility, and the automobiles serviced. This chapter provides a discussion of the regulations found in many states.

Finally, this book does not expect the service consultant to be an accountant but does present financial concepts for the management of a profitable service facility. Therefore, the service consultant must have an awareness of the *bottom line*, meaning the generation of a profit for the facility. Regardless of the effectiveness of the person serving as consultant, a service facility cannot stay in business if it does not make a profit. This chapter presents an introduction to basic financial measures that are important to the service consultant and must be understood as future chapters are studied. For those interested in management or ownership, this first chapter prepares the reader to understand the final section in the text on management and ownership.

Types of Automotive Facilities by System or Product

An automobile service facility sells the labor of a trained automotive technician to one customer at a time. The objective of an automobile service facility is to solve a customer's automotive problems. To achieve this objective the service facility may have to perform a diagnosis to identify the problem, repair or replace a part, and/or perform maintenance to keep the automobile in top condition.

In most automobile service facilities, the service consultant is the person who works directly with the customer and technicians. The service consultant arranges for the service (diagnosis, repair, maintenance) of the automobile. As a result, service consultants must have the ability to communicate with other people and have a thorough knowledge of the types of repairs and maintenance offered by their service facility.

Because the automobile is the most complex machine a person owns, some service facilities specialize in certain services. One of the mistakes a service consultant in a specialty service facility can make is to accept work that is not done by his or her service facility technicians.

Automobile Systems

One way to classify the types of automobile service facilities is by the automobile system or systems that the facility's technicians service. Nine different automobile systems are listed by the National Institute for Automotive Service Excellence (ASE) for automotive and light truck. The areas where certification may be obtained are:

- Engine Repair
- Automatic Transmission/Transaxle
- Manual Drive Train & Axles
- Suspension & Steering
- Brakes
- Electrical/Electronic Systems
- Heating & Air Conditioning
- Engine Performance
- Light Vehicle Diesel Engines

For the collision repair industry, the ASE identifies:

- Painting & Refinishing
- Non-Structural Analysis & Damage Repair
- Structural Analysis & Damage Repair
- Mechanical & Electrical Components

There are also other ASE certification areas that can be helpful in classifying the systems a service facility services. These certification test series include:

- Alternative Fuels
- Parts Machinist
- Medium Heavy Truck
- Truck Equipment
- School and Transit Bus
- Advanced Engine Performance and Electronic Diesel Engine Diagnosis
- Parts Specialists
- Under car exhaust

Naturally, when a facility services one of these automobile systems, it must employ automobile technicians with the expertise needed to diagnose, repair, and maintain the system on the various makes and models of automobiles owned by customers.

Some service facilities may work on all or most of the systems listed above for the automotive and light truck area. Such a facility must have the space to house the different equipment, inventory, and tools needed to work on multiple systems as well as the equipment and space needed by the service consultants. In the service facility that works on multiple systems, the service consultants must be thoroughly familiar with all of the systems. In addition, they must be knowledgeable about the parts used to make the repairs; the specific expertise of each of the technicians; and the means to maintain accurate records of customers, services, and parts.

Some automobile service facilities repair and maintain one system, such as transmissions or brakes. Such a service facility may be referred to as a **system-specific service facility.** The facility, equipment, and tools used by a transmission specialty shop are limited to transmission diagnosis, repairs, overhaul, and maintenance.

Naturally, a facility that limits its work to one or two systems must have technicians and service consultants who are thoroughly familiar with the specific system or systems being serviced. They must understand the system's repair procedures, the parts needed to make the repairs, and the related maintenance procedures.

Product-Specific Service Facilities

Another type of service facility diagnoses, repairs, and maintains a particular make and model of automobile. This service facility is referred to as a product-specific service facility. For example, a facility might specialize in the diagnosis, repair, and maintenance of Volkswagens.

The most common product-specific service facility is an automobile dealership that sells new automobiles. Although it specializes in the models it sells, its service facility may provide similar repairs and maintenance for other makes of automobiles. At a dealership, the service consultant must be thoroughly informed about the different models sold as well as the abilities, factory certification, and sometimes preferences of the technicians who perform the work.

Facility Ownership

Automobile service facilities differ in ownership. For example, one person, several people, or a corporation may own a service facility. A **proprietorship** means that one person owns a service facility, while a partnership means two or more people are the owners. If an owner or the owners wish, they may incorporate the service facility. A **corporation** may be owned by one or more persons.

A person's ownership in a corporation depends on the amount of money the person invests. People who invest in a corporation by purchasing shares of stock are called the stockholders. Their percentage of corporate ownership depends on the number of stock shares they purchase. Because a corporation becomes a legal entity, service facilities are incorporated for legal and financial benefits.

In all of the types of ownerships (proprietorship, partnership, and corporation), the money invested by the owners or stockholders is used to prepare the service facility to conduct business: for example, the purchase of land, buildings, tools, equipment, inventory, and furniture.

In a new service facility, invested money is usually needed to run the business for a short period until it can earn a profit. The profit earned by the service facility goes to the owners or stockholders, and the amount of the profit an investor receives depends on the amount the person invested. From a business standpoint, the service facility may be classified as a:

- Privately owned service facility
- Franchise
- Chain
- Dealership
- Fleet service department

Obviously, the method of ownership influences the job activities of the service consultant and the number and type of people employed by the service facility. In addition, the type of ownership may influence the management structure of the facility. For example, who owns the facility and what role does the owner play in its day-to-day operations?

Privately Owned Service Facility

A privately owned facility may provide services to all automobile systems or limit its services to one system or product. The owners may change the systems to be repaired, services offered, or products sold at their own discretion.

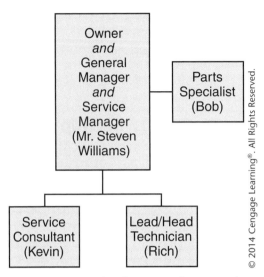

FIGURE 1-1 An organizational structure for a proprietorship.

Figure 1-1 presents an organizational structure for a private proprietorship owned by Mr. Williams. As shown in this diagram, Mr. Williams is in the top box and serves as the general manager and service manager of the service facility. The service consultant (Kevin), the lead technician (Rich), and the parts specialist (Bob) report directly to the owner-general/service manager.

In other words, Mr. Williams is responsible for the hiring, direct supervision, and evaluation of the work of the service consultant, the technician, and the parts specialist. The service consultant and lead technician work directly with each other on a daily basis, but neither is subordinate to the other. Likewise, the service consultant and lead technician interact directly with the parts specialist on a daily basis. In other words, the diagram represents a chain of command and not a diagram for communications or working relationships.

If more technicians are employed, they would be under the supervision of the lead technician, and if people were hired in the parts department, they would report to the parts specialist. Each would supervise and assign work to his or her subordinates, but decisions to hire or terminate them would be made by the owner unless the owner delegates this authority to the service consultant, lead technician, or parts specialist.

Corporate-Owned Facilities

Assume that Mr. Williams decided to change his business from a proprietorship to a corporation and sell stock in his business to investors. Figure 1-2 presents a diagram for a corporate-owned service facility. Note that the name in the top box is *stockholders* as opposed to an owner or owners. Williams now reports to the stockholders and also serves as the general manager.

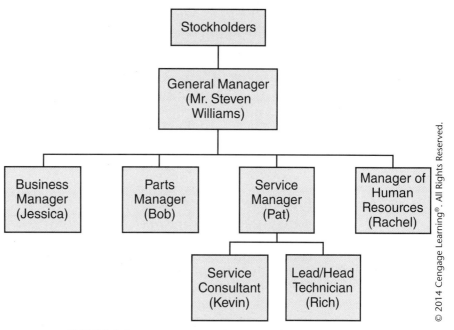

FIGURE 1-2 An organizational structure for a corporation.

In Figure 1-2, Mr. Williams is the immediate supervisor of the four managers. As the general manager, Mr. Williams relies on the service manager (Pat), business manager (Jessica), manager of human resources (Rachel), and parts manager (Bob) to assist him with the operations of the business. This assistance includes working with the day-to-day operations of the service facility, overseeing contracts with and deliveries to parts stores, recruiting new employees, working with the local bank on the checking account and loans, ensuring availability of telephone service, approving computer software purchases, and many other activities.

The four managers work as a team under the general manager. Also, Figure 1-2 shows that the service manager supervises the service consultant and the lead technician. Because Figure 1-2 exhibits the managerial chain of command and not the interactions between employees, the lead technician or service consultant orders parts through the parts manager, and service orders are exchanged between the service consultant and lead technician. **Vendor invoices** are received by the business manager, who works with the parts manager to verify the delivery of parts. The manager of human resources and business manager work together on the payroll.

In Figure 1-2 the service manager has direct supervision of the service consultant and the lead technician. This implies that Mr. Williams has given the service manager the authority to hire, supervise, and evaluate the performances of the service consultant and lead technician.

Further, the service manager works directly with the human resources manager to recruit new employees and with the business manager to get them on the payroll. The service manager reports his actions to the general manager or presents a recommendation to the general manager for approval. Whether or not the service manager takes action and reports or makes a recommendation for approval is up to the general manager. At the same time, if there are personnel in the parts and other departments, the assumption from Figure 1-2 is that, while the managers are responsible for the day-to-day operations, the general manager is personally involved in the hiring, supervision, and evaluation of employees.

Franchise Ownerships

Because the startup and operation of a service facility can be expensive and complicated, some people who wish to own a service facility may choose to purchase a franchise. A **franchise** permits an owner to use a nationally recognized name and receive some assistance, such as training, advertising, and consulting, for start-up operations.

People wishing to start a franchise usually have to submit a business plan when applying to purchase the rights to open the service facility. They must also have the money to purchase the franchise, buy the necessary equipment and supplies, and obtain the building space to house the operation. The national franchise corporation that owns the name of the franchise (such as Midas Muffler, AAMCO, or MAACO, among others) has to approve the request presented by the people making application as well as all local arrangements, such as the location and building.

In return for the purchase of a franchise, the national franchise corporation typically requires a percentage of the gross sales received by the local service facilities. In addition, it usually has the right to conduct periodic inspections of the franchise's operations and records and may close a local service facility as warranted under the contract.

In most cases, the national franchise corporation requires that the local owner and service consultant attend its training programs and learn to use its forms to record all local services and receipts. The training of service consultants may require that they read written scripts when talking to customers on the phone or communicating with a customer in person.

An organizational diagram of a franchise is shown in Figure 1-3. Note that in this diagram the national franchise corporation is in a supervisory position similar to the stockholders in Figure 1-2. As in Figure 1-2, Mr. Steven Williams is the local owner, general manager, and service manager, with Kevin working as the service consultant and Rich as the lead technician. In a larger franchise, the corporation may prefer that a person other than the owner serve as the service manager and may choose to hire multiple managers such as those shown in Figure 1-2 (a business manager, several technicians, a parts manager or specialist, and a human resources manager).

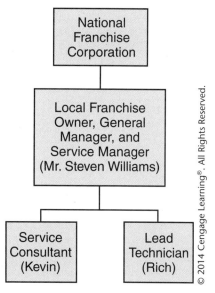

FIGURE 1-3 The organizational structure of a franchise service facility.

Chain Ownerships

A service facility that is one of several facilities owned by a corporation is called a **chain service facility** (such as Sears, Roebuck & Co., Wal-Mart, Pep Boys, and many others). The chain service facilities sell the same services and use the same procedures regardless of location.

Each service facility has a service manager, at least one service consultant, and several technicians. As shown in Figure 1-4, a service manager usually reports to the store manager, who in turn reports to a manager in a national corporate office. The owners of the corporation's chain of service facilities are the corporate stockholders, who provide the money needed to buy the tools, equipment, and building that houses the local service facility.

Automobile Dealership

An automobile dealership sells new and used automobiles as well as the parts needed to repair and maintain them. The dealership service department is considered a **specialty service facility** because its primary objective is to diagnose, repair, and maintain the automobiles sold by the dealership. Actually, a dealership is really a franchise of an automobile manufacturer since the manufacturer approves the dealership, can close it if warranted, and makes money from the selling of the automobiles and replacement parts (called a product franchise).

Figure 1-5 shows a diagram with the owners/stockholders at the top. One or more people may own a dealership; however, it is usually incorporated. Under the owners/stockholders is a general manager, president,

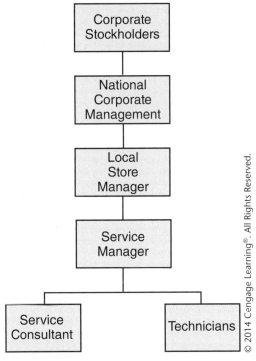

FIGURE 1-4 The organizational structure of a chain-owned service facility.

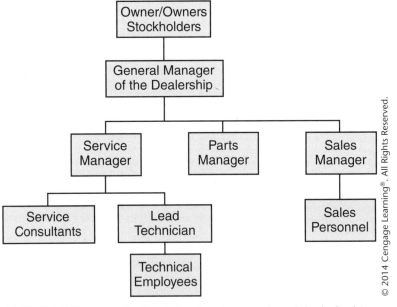

FIGURE 1-5 The organizational structure for an automobile dealership.

or vice president who is responsible for a variety of departments or divisions in the dealership.

In Figure 1-5, three departments are shown, although a dealership typically contains more departments and managers such as a business manager, a manager for the maintenance of the buildings and grounds, and so on. With respect to the service facility, a service manager is shown as the person in charge of the service operations. Reporting to the service manager are the service consultants and the lead technician.

It is important to note again that Figure 1-5 is an example that shows how an organizational structure may appear at an automobile dealership. The service consultants, service managers, and technicians may be in different positions in a dealership diagram, but the point is that the individuals occupying these positions must know to whom they report. The employees must know who is assigned to supervise and evaluate their performance.

Fleet Service Departments

The government, colleges, and some companies may own or lease a fleet of vehicles. The constant need for service requires a service department. The fleet service department is part of the government, college, or company and employs a fleet manager and automobile technicians just as they employ other workers. The job of the technician is to diagnose, repair, and maintain the fleet of automobiles (see Figure 1-6). The job of the fleet manager is to keep the vehicles in the fleet in top condition with the cost of servicing the fleet within an assigned budget.

Unlike the other service facilities, the fleet service department does not service customer vehicles but rather the vehicles owned by the

FIGURE 1-6 Fleet parking garage.

company and driven by its employees. Even though there are no retail customers, every employee is like a repeat "customer" to the fleet service department because they are fellow employees who sometimes feel an "ownership" of the vehicle assigned to them. In other words, the employee assigned these vehicles cannot do their job well if the vehicles are not in good condition.

The Fleet Manager

Very small fleets with perhaps 1 to 10 vehicles may have a manager whose extra duty is to oversee fleet operation as well as other non-fleet-related operations. Slightly larger fleets of up to perhaps 20 vehicles will have a shop foreman who is in charge of management as well as working on the vehicles. In larger fleets, a fleet manager whose job is to oversee the entire fleet operation may be used. Duties of the fleet manager, regardless of size, include acquisition of fuel, new vehicle purchasing, and aftermarket fitting of equipment to the vehicle as well as accident management, and disposal of old vehicles. It is a *"cradle* to grave" oversight of each fleet vehicle with some complicated decisions to be made, such as whether to do work "in house" or contract it to other service facilities. Budget pressures and market competition, meaning to perform the work at a lower price—makes the fleet manager's job challenging. Other pressures include how to handle the fleets' needs when the vehicle cannot return to the service department easily as well as outsourcing of fleet management services to another company.

Automotive Industry Jobs

Up to this point, the automotive service industry as a business has been presented. To make it function, the automotive service industry needs professionals with the skills, knowledge, and attitudes to help customers. Jobs in the industry are as diverse as the people who hold the jobs and include positions from ownership and management to automotive technicians, parts specialists, and service consultants. There are even jobs at the vehicle manufacturer level that include customer/technical hotlines, technical trainers, and factory representatives, among other career paths. The automotive industry is vast, and jobs are located in almost every town and city and include dealerships, independent service facilities, parts stores, as well as vehicle manufacturer regional offices. This book will profile these jobs in future chapters. Each job is interesting and enjoyable with opportunities that lead to career advancement. Perhaps one will appeal to you.

Women as Professionals in the Automotive Industry

While the majority of the industry is staffed by men, women are also in these professional positions and more are needed. Examples of women who own, manage, and staff the many automotive jobs can be found throughout the industry. Their effectiveness is unquestionable, but overcoming the barriers to succeed can be challenging. To help increase the number of women in the industry requires making them aware of the automotive industry's career opportunities and encouraging them to become leaders. To help with this initiative, there are organizations such as The General Motors Women's Retail Network (WRN). WRN has been working since 2001 to support women dealers and increase the number of women involved in automotive retail. WRN is an Original Equipment Manufacturer (OEM) program dedicated to providing resources to women dealers that includes training and networking opportunities. To help with the initiative, WRN offers scholarships to women enrolled in automotive training programs.

Diversity Initiatives

WRN is an example of an initiative that is needed along with others to help the automotive industry attract a variety of people into automotive professions. The leadership that a diverse array of professionals brings ultimately helps the automotive industry better serve our society's transportation needs. WRN is a start! More organizations need to encourage women, as well as other people with diverse backgrounds, to consider a profession in the automotive industry.

Federal Regulations That Influence Service Facility Operation

A service consultant must be aware of certain laws that affect the operation of the service facility. While service consultants are not directly responsible for monitoring these various licenses, they must understand these licenses when consulting with customers and assisting management. For service consultants who will become an assistant service manager (ASM), manager, or owner (discussed in section four of this textbook), these laws must be thoroughly understood; otherwise, significant fines and loss of business can result when they are not followed.

Vehicle Emission Programs

The Environmental Protection Agency (EPA) requires regular automobile emissions inspections in some states under the Clean Air Act. While emission programs vary from state to state, they can be generalized as centralized or decentralized emission inspection programs. Centralized emission inspection programs require customers to drive their vehicles to a state-operated emission testing station to be inspected. If the automobile fails its emissions test, it cannot be repaired at the state-operated emission station but at an automobile service facility. State regulations may also require that the technician and/or the automobile service facility be licensed to perform the repair and may allow for re-inspection at the service facility.

In some states the emission inspection program is decentralized. This means an automobile service facility performs the emission inspection and performs the repairs to pass the inspection. Often, decentralized emission programs require that both the automobile technician and the automobile service facility have a state-issued Emission Inspection license and own state-approved emission testing equipment.

While the emission program is based on federal law, the state administers the program and negotiates the regulations with the EPA, a federal agency. Therefore, the emission license for the service facility and/or the technician is issued by the state and is only valid in the state issued. The actual regulations for a specific state's federal emission program can get very complicated and are beyond the scope of this textbook. In fact, California has more stringent emission regulations (referred to as SMOG inspection) under CARB (California Air Resource Board) than the EPA mandates under the Clean Air Act. Regardless of the specific program details, there are some common requirements, such as following the required emission inspection procedures, to assure a legal inspection is performed. Also, it is a serious offense for a technician to disable vehicle emission components or permanently remove them. Therefore, in all states, failure to follow inspection guidelines will result in significant fines for the technician and the service facility owner.

Air Conditioning and EPA Section 609 Regulations

Another area of regulation is automobile air-conditioning equipment. More specifically, automobile technicians must hold an EPA-approved license to reclaim, recycle, and recharge automobile air-conditioning systems. This certification license is required under the Clean Air Act Section 609. To obtain the federally mandated certification license to service automobile air conditioning, an automobile technician must obtain training and pass an EPA-approved test. This training can be obtained from companies that offer approved training courses.

At the same time, under the EPA's Clean Air Act Section 609 regulations, the automobile service facility does not need to be licensed. However, the automobile service facility must register with the EPA to assure them that it owns the required air-conditioning equipment that meets the standards set by the Society of Automotive Engineers (SAE).

EPA and Waste

Automotive operations generate a significant amount of waste from all of the chemicals and oils used in the repair process. Commonly, the service consultant at the direction of management or owners will oversee the removal of the waste and obtain the proper forms from haulers. The federal and state regulations are not difficult to understand but failure to comply with EPA and the state's procedures can result in an investigation as well as fines and any clean-up costs.

To illustrate the seriousness of these regulations, a service facility had 500 gallons of waste oil, a barrel of used solvent, and two barrels of used antifreeze removed by a hauler. The truck that the hauler used was in an accident and the fluids spilled into a creek. The service facility was held responsible for clean-up costs and fines because waste fluids were their responsibility until deposed of in accordance with government regulations. In this case, the hauler was not licensed to haul waste fluids, he was uninsured, and the required government forms the hauler and service facility were to fill out and file were not completed. The fines and clean-up costs that the garage had to pay were very expensive. The situation illustrated would have had the same results had the hauler removed and dumped the waste fluids illegally with or without the service facility's knowledge because the waste belongs to the service facility until disposed of properly.

The best course of action is to study the regulations and make sure the hauler is reputable and meets all government regulations. The Service Consultant, ASM, or Manager must obtain a copy of any government-required waste hauler permits and manifests for the service facility files. In addition to other documents that might need to be obtained, depending on state requirements, a copy of the hauler's "certificate of insurance" that insures the hauler for accidental waste material spills is suggested.

An easy place to begin to understand the regulations and forms that must be filed when the hauler picks up waste materials is to visit the EPA website (http://www.epa.gov). Then open the link Wastes, which is listed under Popular Topics on the home page (or type in "waste" in the search box), and read it. The site allows the user to click on his or her state to find any additional state regulations. Manifest form examples are available to study as are links to handle other "materials" the shop may use, such as "scrap tires" (just type scrap tires into the EPA search box).

The worst thing for a service consultant is to not know the laws or procedures and in the absence of the manager, give permission to a waste handler to remove waste. For example, a local truck driver removed used tires for local garages at a reasonable cost. He provided invoices to each garage and claimed to be an authorized hauler who was taking the tires to a recycling center. Months later, it was discovered that the hauler had dumped the used tires on land that had to be cleaned up by the government. The hauler was not to be found and clearly not a legitimate business. Each garage was identified by the invoices the hauler left behind and held liable for a portion of the total clean-up cost in addition to government fines. Therefore, it is best to know the law!

OSHA—Department of Labor's Occupational Safety and Health Administration Service

Facilities must comply with Federal OSHA standards to assure workers remain healthy and avoid them being hurt in the workplace. While there are more regulations than can be covered in this textbook, a few that a manager may delegate to a service consultant will be covered. Specifically, material safety data sheets (MSDSs), fire extinguishers, as well as first aid kits and personal safety will be discussed. While service consultants are not expected to be experts in these areas, they are expected to know where the equipment/information is located and how to access OSHA website information.

The first OSHA regulation to be discussed is fire suppression. A fire can start in a shop and the service facility must have the proper fire extinguisher to put out the flames. Placement of the equipment, the correct extinguisher, as well as the annual inspection and labeling guidelines are beyond this discussion. However, the information is readily available on the OSHA website (http://www.osha.gov) and in the search bar type "fire extinguisher."

Another area of interest is first aid. A lucky shop has an employee who is a volunteer fireman or EMT. If the shop does not have anyone with emergency responder training, at least someone should have first aid/CPR training. OSHA compliance demands that a first aid kit be in the shop and properly equipped to Standard 1910.266 App A. (Go to http://www.osha.gov and type in *first aid* kit into the search box to find out more information.)

Material Safety Data Sheets (MSDS) are required for every chemical used by employees (see Figure 1-7). The sheets contain product information such as health effects from exposure, how to treat someone who has been injured by the chemical, characteristics of the product, and contact

FIGURE 1-7 The MSDS for this chemical is found in this box that is accessible to every employee.

information of the manufacturer. As new chemicals are introduced into the shop, even if the same product but manufactured by a different company, a new MSDS must be obtained from the vendor. If the service facility is the vendor, the employee must be able to provide the user with an MSDS upon request. More details can be found at http://www.osha.gov and type *"MSDS"* into the search box.

Service consultants need to wear proper work shoes (to avoid slip and fall injuries) and as soon as they enter the shop, they should wear eye protection as well (see Figure 1-8). While personal safety is covered on the OSHA website, the idea behind safety is to create a safe work environment rather than rely exclusively on protective equipment. Note that technicians are working on broken vehicles that may not function as intended. Accidental acceleration, brakes that don't work, parts that fly off, fluids that squirt out, and engines that start when in gear and lurch forward are not uncommon. Therefore, the service consultant must be on the lookout and have their "head on a swivel" when they are in the shop area and even the parking lot. This means look around and don't have tunnel vision because accidents happen fast and they can be deadly.

As new technology is introduced, new tools are needed and new protective devises will be required. For example, high-voltage systems as shown in Figure 1-9 are becoming more common on vehicles and the equipment shown is required to work on these systems safely.

FIGURE 1-8 Service consultants who enter the shop without safety glasses risk getting something in their eyes and may have to use the eye wash station, as demonstrated.

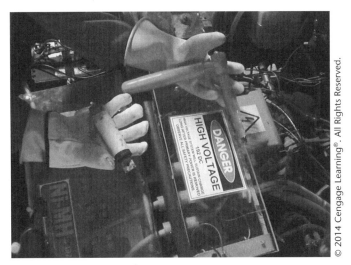

FIGURE 1-9 This high-voltage vehicle prototype can be deadly to work around without proper protective devices, tools, and knowledge about how to use them correctly. Shown are a high-voltage fuse, a torque wrench protected with a high-voltage covering, high-voltage rubber gloves with a protective glove over the right-hand rubber glove to lift sharp and/or heavy parts without ripping the glove.

State Guidelines for Service Facility Operations

Some states require that automobile service facilities obtain a state-issued license to sell services to customers. Service consultants must be thoroughly informed about these legal requirements. For example, California's Bureau of Automotive Repair must issue a license to an automobile service facility before it can sell automobile repairs to customers. In addition, in some states and cities, a technician may have to obtain an automobile technician's license before he or she can provide automobile services to a customer. An example of this regulation is Michigan's State Certified Mechanic or Ontario Canada Automotive Service Technician (restricted certified trade regulated by the Apprenticeship and Certification Act) among other states' and city mandates.

In other states, a service facility and its technicians may need a license before either can perform certain types of work. For example, California's Smog, Brake, and Lamp Adjustor Licenses or Maryland's and other states' Automotive Inspector license. In Pennsylvania, automobiles must be inspected every year to make sure they meet state safety regulations. When a state requires vehicle safety inspection, the program typically follows the standards established by the National Highway Traffic Safety Administration's FMVSS (Federal Motor Vehicle Safety Standards). Each vehicle manufacturer must meet the FMVSS standards; otherwise the vehicles they produce cannot be sold in the United States. The FMVSS standards can be found at http://www.nhtsa

.gov/cars/rules/import/fmvss/index.html. A state-required safety inspection program uses the FMVSS standards among other criteria to establish the state inspection regulations or codes.

Most state safety inspection programs check the function and condition of the vehicle components. To illustrate the inspection regulations for automobiles, aspiring students who wish to become managers or owners are encouraged to examine the Pennsylvania State Safety Inspection Regulations to understand the type of regulations a state may require of a service facility. Pennsylvania makes available the entire procedure a service facility must follow for automobile inspection under the subchapter E link found at http://www.dmv.state.pa.us/inspections/pub_45.shtml. While some states do not have a program, states with a safety inspection program find that accidents from vehicle component failures are fewer according to a Pennsylvania Vehicle Safety Inspection Program Effectiveness Study (copy can be found at the Pennsylvania Department of Transportation website [http://www.dot.state.pa.us/]).

For a technician to be qualified to perform a safety inspection, some states require training and a license. To obtain this license in Pennsylvania, technicians must take a safety inspection course and pass a written as well as a hands-on test before they can inspect automobiles. They must also renew their license at regular intervals. In addition, Pennsylvania requires that automobile service facilities obtain a state-issued license to sell Pennsylvania state safety inspections to customers. To become licensed, the service facility must show the state that it meets its requirements. These requirements include the ownership of required tools/equipment, minimum bay size requirements, minimum insurance coverage limits to protect the customer's car, and a signed agreement to maintain records in accordance with state guidelines. Upon approval, the service facility must display a state inspection sign with the station number on it (see Figure 1-10). Each state's requirements will differ; such as Maryland's safety inspection program requires a dedicated bay that will only be used for safety inspection. The Pennsylvania's safety inspection program is discussed in this textbook to illustrate the point that to perform certain repairs, a license is required. In this example, both the technician and the service facility must be licensed to perform safety inspections.

Industry Partners and Voluntary Certification

In some states, a license is not required for either automobile technicians or automobile service facilities. To show a level of competence many service facilities, regardless of whether they need a license, choose to participate in industry partnerships and the voluntary certification of their employees. The most popular voluntary certification for automobile technicians as well as service consultants is offered by the ASE. While the ASE also certifies the collision repair industry, another certification/

FIGURE 1-10 A service facility with a Pennsylvania State Inspection sign posted in accordance with state requirements.

training organization is I-Car, which is the Inter-Industry Conference on Auto Collision Repair.

The ASE requires technicians to document at least two years of work experience plus pass an exam to be certified within an automobile system. Specifically, there are nine ASE exams for each subsystem of an automobile and light truck repair. When a technician passes the right combination of eight exams, they are considered an ASE master automobile technician (passing all nine is not required at the time of this writing).

In addition to the ASE's examination for automotive technicians, there are exams for the parts specialists, engine machinist, compressed natural gas technician, heavy truck technicians, bus technicians, and advanced exams for engine performance and diesel electronics technicians. There is also an exam for the service consultant. For details on the different certifications, exams, and to register for an exam, consult http://www.ase.com.

AAA-Approved Auto Repair

AAA is a club with members who as part of their benefits receive assistance about where to take their vehicles for service. AAA directs members to AAA-Approved Auto Repair service facilities that are deemed to be competent for the service of their vehicle. Service facilities that participate in the network often have a service consultant trained to provide courteous service and professional advice to customers with a service manager available for quality control purposes.

AAA also requires that technicians have certifications from the ASE, vehicle manufacturer, or I-Car. The AAA-approved facilities must have the proper tools and equipment as well as maintain a clean and professional facility. In addition, the facility must have a satisfactory community reputation that is monitored by AAA through surveys of customers who used the facility. When there is a dispute between a member and the AAA-approved service facility, AAA has technical experts who understand the problem and can help mediate a solution. The Approved Auto Repair program is highly regarded by both AAA members and participating repair facilities due to the high standards that must be met by everyone involved.

Other Organizations

There are many fine organizations and programs that represent the industry and all cannot be mentioned in this textbook. Some, such as the Chamber of Commerce, are not specific to automotive service. Others are industry specific, such as the Automotive Service Association (ASA). Others are "programs" that help service facilities owners improve their business with information, products, warranties, program benefits, and networking, such as the ACDelco Professional Service Center program, American Car Care, and NAPA Autocare Centers. Basically, these organizations endorse automobile service facilities that meet certain standards.

The Service Facility Business Profit and Loss: The Bottom Line

It is imperative that the service consultant understand that after customers are served, they are charged money for the service. The money collected from a customer (not including any sales tax collected) is the gross sale amount. The total of customer gross sales is the gross sales for the automobile service facility.

As shown in Figure 1-11, the money collected from the customer (gross sale amount) pays the salaries of the automobile technicians and the cost of parts. The remaining balance is called **gross profit**. The gross profit is then used to pay the salaries of the other personnel, including the salaries of the managers and service consultant and the overhead expenses. Overhead expenses include rent, heat, light, telephone, uniforms, insurance, and the other expenses incurred when running a service facility. The balance left over after the expenses are deducted from the gross profit is called the **net profit** (see Figure 1-11).

The owner must pay the taxes of the service facility from the net profit. The balance is the profit after taxes and is the amount the owner may claim for a return on the investment of money and time spent working in the service facility. When the expenses cannot be paid by the gross profit, the negative balance is called the **net loss.** The money needed to cover a loss must come from the owner. If losses persist, then the service facility will be bankrupt and must close.

Chapter 16 will cover the concept of profit and owner return further. At this point, the relationship between income relative to the cost of labor and expenses that creates profit is important. Merely making a profit is not the point but making "enough" profit is necessary for the owners

GROSS LABOR and PARTS SALES
- <LESS> Cost of the technician's labor
- <LESS> Cost of parts to service the customer's automobile

EQUALS: GROSS PROFIT
- <LESS> Overhead expenses (examples)
  - Managers' salaries
  - Service consultants' wages
  - Rent (includes real estate taxes)
  - Heat
  - Electric
  - Insurance and benefits
  - Other expenses

EQUALS: NET PROFIT or LOSS
- <LESS> Taxes on the service facility

EQUALS: OWNER'S INCOME

FIGURE 1-11 The relationship between gross sales and profit.

to keep their money in the business. Without the owners' capital, the business will close. The service consultant's role in the relationship is to make service sales to customers (income). Without sales (and enough of them to cover expenses), there cannot be enough profit.

Review Questions

Multiple Choice

1. An automotive repair business makes its money mainly by:
   A. selling parts to the customer
   B. selling the labor of a trained automotive technician to the customer
   C. selling the support staff's services to the customer
   D. selling the professional image of the repair facility to the customer

2. Whose job is it to work directly with the customer and technicians to arrange for diagnosis, repair, and service?
   A. The technician's
   B. The manager's
   C. The service consultant's
   D. The owner's

3. To "service" means to perform _____ on an automobile.
   A. maintenance
   B. repairs
   C. a diagnosis
   D. all of the above

4. A profit from the service of automobiles means?
   A. The service facility made a sale
   B. Income is greater than the cost of labor and expenses
   C. Expenses were low
   D. Profit is not a concern for the service consultant at an automobile garage

5. An automobile service facility that repairs and maintains one system such as transmissions or brakes is called a/an:
   A. system-specific service facility
   B. product-specific service facility
   C. automobile dealership
   D. automobile garage

6. Which structure has stockholders?
   A. A corporation
   B. A proprietorship
   C. A partnership
   D. An automobile service facility

7. A _____ permits an owner to use a nationally recognized name and receive some assistance, such as training, advertising, and consulting, for start-up operations.
   A. general manager
   B. proprietorship
   C. franchise
   D. corporation

8. A service facility that is one of several facilities owned by a corporation is called a:
   A. chain service facility
   B. specialty service facility
   C. franchise
   D. fleet service department

9. This text classifies a dealership service department as a type of:
   A. chain service facility
   B. specialty service facility
   C. franchise
   D. fleet service department

10. The _____ does not service customer vehicles but rather the vehicles owned by the company.
   A. chain service facility
   B. specialty service facility
   C. franchise
   D. fleet service department

Application Problems

1. Given the following information, calculate the parts sales of the service facility.

MONTHLY INCOME REPORT

GROSS SALES		
Labor	$75,000	
Parts	?	
Total Gross Sales		$140,000

2. Examine the following information, and calculate the total gross sales, total cost of labor, and gross profit.

MONTHLY INCOME REPORT

GROSS SALES		
Labor	$50,000	
Parts	40,000	
Total Gross Sales		$_____
COST OF LABOR AND PARTS		
Technician Labor	$25,000	
Parts	15,000	
Total Cost of Labor and Parts		$_____
GROSS PROFIT		$_____

3. Given the following information, calculate the net profit and the owner's income if the business taxes are 12.5%.

EXAMPLE OF FORM USED TO REPORT MONTHLY INCOME

GROSS SALES		
Labor	$60,000	
Parts	60,000	
Total Gross Sales		$120,000
COST OF LABOR AND PARTS		
Technician Labor	$32,000	
Parts	40,000	
Total Cost of Labor and Parts		72,000
GROSS PROFIT		$48,000
EXPENSES		
Management Salaries	$11,600	
Service Consultants	8,600	
Rent	7,200	
Heat (Oil)	1,400	
Electric	800	
Insurance and Benefits	3,000	
Other Expenses	3,400	
Total Expenses		36,000
NET PROFIT		$ _____
Business Taxes		$ _____
OWNER INCOME		$ _____

Short Answer Questions

1. Describe the different types of automobile service facilities.
2. Visit a service facility and explain that you are a student learning about service facilities. Ask for information that will help you create a diagram of its organizational structure similar to one found in this chapter. Use a program such as MS Word to create the diagram like the one found in this book. At the top of the chart, indicate the service facility name and what type of automobile service facility it is (as discussed in this chapter).
3. During the visit, to answer question 2, ask what types of licenses are required for the service facility to sell automotive services to customers. If a license(s) is required, explain what it is and how a service facility obtains it.
4. An accountant will describe how income, cost of labor, and expenses must be in balance to make the profit the owner requires. Identify and provide examples of what might cause each of these to be "out of balance" and how loss of profit is created.

Activity

At your instructor's discretion, as an individual or a group, complete the following activities.

Activity 1: Work with your librarian to research your state's (or a state your instructor chooses) emission and/or safety inspection programs. Also consult your state's websites for information. Write down your findings. Based on the type of program in your state:

- For decentralized programs, interview local service facility owners, a state official who oversees the administration of the program, or an automotive school that offers the technician certification course.
- For centralized inspection programs, visit the inspection center and obtain information about the inspection process and rules of the program.

After conducting your research, create a report about what a service consultant must know about the program in your state and what should be explained to customers who ask about the specifics of the program.

Activity 2: Use the Web and your school's library resources to find the state and federal regulations and penalties for tampering with emission control devises (perform a Google search such as "EPA anti tampering"). Write a brief description of what was found and what a service facility service consultant must know to avoid scrutiny.

Activity 3: Research your state law (or a state assigned by your instructor) and, if necessary, federal law to determine whether a technician must be certified to perform automotive repairs, such as general repairs, specialty repairs, air conditioning (hint: examine EPA Section 609 regulations), safety or emission inspection. Research whether the service facility is required to have a license to sell automotive repairs to customers or any specific services (such as air conditioning, safety inspection, or emissions). Write a memo to a fictitious service manager about what you found.

Activity 4: Study the EPA and your state's website for information on waste removal (oil, antifreeze, and cleaning solvents) as well as scrap (used) tires. Then use the EPA site to find your state's laws concerning these same topics. Examine the manifests and recordkeeping requirements. Write a memo to the owner about the procedures that must be followed.

Activity 5: Obtain an MSDS for a product from a parts store, school shop, or service facility and read it (also consult the OSHA website for details on MSDS for various products). What does the sheet say about treatment if the product gets on your skin, eyes, or is swallowed? What are the hazards to the workers, and how can they protect themselves when using MSDS? Write a summary about the information found.

Activity 6: Examine the regulations found for the Pennsylvania's State Safety Inspection Regulations subchapter E at http://www.dmv.state.pa.us/inspections/pub_45.shtml

Part 1: A service consultant will likely collect important information for the inspection process. Open the subchapter E link and examine section 175.80 on Inspection Procedure (subsection a1) and determine the requirement for demonstrating vehicle ownership and financial responsibility (car insurance). Specifically, what document must the customer present to show financial responsibility and also what forms of ownership are acceptable. What are some of the rules that must be understood to verify the information?

Part 2: A service consultant will likely collect important information for the inspection process. Open the subchapter E link and examine section 175.80 about window Glazing (subsection a2). What might a service consultant need to share with a customer on the placement of stickers and signs on their windows relative to passing Pennsylvania State Safety Inspection?

Part 3: A service consultant must report information collected by the technician to the customer and needs to know why the parts need to be replaced. Open the subchapter E link and read subsection e1 and e7. What are the minimum tire tread and brake thickness readings that will FAIL Pennsylvania State Safety Inspection?

CHAPTER 2

THE ROLE OF THE SERVICE CONSULTANT AND FORMATTED OPERATING SYSTEMS

OBJECTIVES

Upon reading this chapter, you should be able to:

- *Identify the major tasks of a service consultant.*
- *List the duties of a service consultant.*
- *Explain how the duties and tasks may be assigned in the different types of service facilities.*
- *Explain why the presentation of a professional image is important (Task A.1.14).*

CAREER FOCUS

You are building your reputation as a professional and are the center of attention when customers enter the service facility. Whether they will become a fan of your work will be determined within the first few minutes of their arrival. The customer within moments will size you up as a professional based on your hospitality and appealing demeanor. This means you influence whether customers think you can help provide the automobile services they need or not.

Therefore, you need to look your best and must be seen as a professional that is clean, properly dressed, organized, with a well spoken demeanor. If you act professional, this will help the customer focus on what you say and do. As you execute your game plan and work within the customer service system, even customers who are not happy about coming to the service facility can be turned around as you articulate how you will help them. The comfort and confidence in your professional demeanor will be the first step.

Over time your reputation and fan base of customers who like you will increase because they are satisfied with your performance and depend on your advice. These are called repeat customers and earning their respect is the key to being noticed by your boss and by the owners. Such positive attention is critical to promotions or other personal rewards you will get and leads to opportunities in the future.

Introduction

In order to produce satisfied customers, automobile service facilities must not only employ trained, experienced, and, in some states, licensed automotive technicians, they also need to have one or more professionally trained service consultants. Service Consultant training, some of it required by the manufacturer or franchise, assures that service consultants know how to do their job. It helps them feel positive and confident and able to put their customers at ease as they convey automobile service advice.

However, interacting with customers is only one part of a service consultant's job. The service consultant must perform a variety of other tasks to keep the work moving at the shop. This is important because a service facility must have enough customers flowing in and out each day to make a profit. This part of the service consultant's job can be complicated and therefore requires experience as well as training on a service system that works to become proficient. This proficiency cannot be gained overnight. To become proficient, service consultants must first have a clear understanding of the service system, task, and duties to be performed each day.

To conceptualize the larger "service system," it must be understood that it is actually made up of three separate systems:

- the Customer Service System
- the Shop Production System
- the Business Operations System.

The service consultant has duties in all three systems. In the Customer Service System, the service consultant and the customer work together. A repair order is created in the Customer Service System that the service consultant transfers to the Shop Production System. The Shop Production System is made up of a service team (team leader, technician, and parts specialist). The duty of the service consultant is to track the progress of the repair order using the Repair Order Tracking Sheet. Within the Shop Production System's operations, the service consultant will work as part of the service team. The duties performed include keeping the customer informed of any additional work needed and delays or problems in the repair process, as well as provide general information on progress. The Customer Service System and Shop Production System's end when the invoice is created by the service consultant. Then the business operations system starts with the collection of the money from the customer and recording the transaction. The service consultant starts the Business Operations System and may have duties within the system, such as collecting the money from the customer.

The purpose of this chapter is to provide an overview of the major tasks and duties of a service consultant and how they might differ among the various types of service facilities. In addition, because customer relations are often the key to the success of service consultants, this chapter gives special attention to their image (see Figure 2-1).

FIGURE 2-1 A service consultant at a customer station.

Major Tasks of Service Consultants

Job tasks are major work assignments given to employees and should appear in a job description. Each task has assigned duties, which describe the job requirements in detail an employee performs. Some jobs are quite complex, so a job task may have to be broken down into duties with related sub-duties. In this chapter the breakdown of the tasks is limited to duties and procedures.

ASE incorporates the tasks and duties of service consultants into the major skills and knowledge expectations of the position. These skills and knowledge areas are as follows:

- Communication skills
    - Customer relations
    - Internal relations
- Product knowledge
    - Engine systems
    - Drivetrain systems
    - Chassis systems
    - Body systems
    - Service and maintenance intervals
    - Warranties, service contracts, service bulletins, and campaigns/recalls
    - Vehicle identification
- Sales skills
- Shop operations

The tasks and duties discussed in this chapter focus on the operation of the shop. Communication and sales skills that are needed to accomplish the tasks and duties are discussed in other chapters of this book.

Service Consultant Tasks

The owner or manager of the service facility ultimately determines the job tasks of the service consultant. Most of these tasks are common for all service consultants, although some owners may prefer to perform some of them, such as opening the business in the morning. Differences, of course, are usually found in the duties to be performed under a task and the related procedures as shown in the examples that follow.

First, the **job description** must list the tasks to be performed by the service consultant. The service consultant performs the duties to accomplish each task on a daily or weekly basis. An example of a job description for a service consultant is shown in Figure 2-2. This description was used in the repair facility owned by the authors, which is referred to as Renrag Auto Repair in this book.

At Renrag each employee was given a copy of his or her job description when hired. The tasks were reviewed with them before they started their job. When the employees' performances were evaluated,

POSITION DESCRIPTION

Service Consultant

1. Opens shop at 7:30 a.m.
2. Greets customers, answers phone calls, provides information, makes appointments, calls customers for approval of work on vehicles, calls customers when jobs are completed, and places follow-up calls after repairs are made.
3. Prepares customer invoices, reviews parts and labor charges with customers, and receives payment from customers (cash, check, or credit card).
4. Makes arrangements for customer shuttle or comfort if waiting at the service facility.
5. Prepares estimates, repair orders (RO), and computerized invoices (IN) and maintains the appointment book and customer status sheet.
6. Communicates with the technician to ensure service work is completed in a timely fashion.
7. Communicates with technician, as needed, to prepare estimates.
8. Assists, as needed, to help order parts, receive parts orders, and check parts invoices to ensure the charges are accurate.
9. Maintains inventory of office supplies, including supplies for computers, photocopier, fax, credit card machine, printers, forms, reports, and other materials, as needed.
10. Makes daily deposit to the bank and prepares daily report for management.
11. Closes the building at the end of the day and checks customer cars in the parking lot.
12. Other duties as assigned by management.

FIGURE 2-2 A job description for a service consultant.

their performance on each job task was reviewed with them. If their job changed, then the job description was also revised and their pay was adjusted accordingly.

Note that Task 12, "Other duties as assigned by management," covers miscellaneous assignments that do not require enough time to make up a separate task. A good rule of thumb is that when a miscellaneous task takes more than 5% of an employee's workweek, it should become a separate task with assigned duties accompanied by a formal evaluation of the employee's performance of the task.

Finally, because Renrag employed an estimator and parts specialist, Task 8 required the service consultant to assist him. This task would not be in the job description if the service consultant calculated the estimates and ordered the parts.

Service Consultant Duties

Each task has duties assigned to it (see Figure 2-3). The duties are the details about how to successfully complete the task. When the service consultant executes the duties properly, a supervisor should be able to

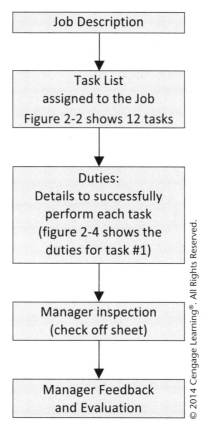

FIGURE 2-3 The relationship of the job description to the tasks and duties.

evaluate the results and provide feedback about whether the service consultant met the job expectations for the task. This list of duties is helpful when training new employees and for evaluating the performance of experienced employees.

To illustrate the relationship, at Renrag Auto Repair, the first task (see Figure 2-2) was to open the shop at 7:30 a.m. The duties, which are prepared for each task (shown in Figure 2-3), are shown in Figure 2-4 for the first task. These duties were listed according to rooms and bays with a list of specific functions to be performed. This detail provided for the opening of the facility and generated a checklist that was used by the service consultant or whoever opened the business to make sure he or she didn't miss anything.

Likewise, duties were outlined for the closing of the facility (see Figure 2-5). Of course, the owners could check the facility after it was closed to evaluate the service consultant's performances. If lights were left on or a thermostat was not set correctly, the owners could make a note and discuss it with the service consultant.

OPENING RENRAG AUTO REPAIR

FRONT OFFICE & WAITING ROOM
1. Enter front side door, turn on lights, and turn off security system.
2. Turn "CLOSED/OPEN" sign around to show "open."
3. Turn on lighted "OPEN" sign.
4. Unlock front door.
5. Turn on computer and printer.
6. Turn on TV security monitor.
7. Switch answering machine to position "A."
8. Turn on photocopy machine.
9. Turn on radio.
10. IF WINTER turn thermostat up to 72°.
11. Enter alignment bay.

ALIGNMENT BAY
1. Turn on lights.
2. IF WINTER turn thermostat up to 65°.
3. Turn on light in bathroom.
4. Go to inside bay.

INSIDE BAY
1. Turn on bay lights and lights to storeroom.
2. Enter compressor room.

COMPRESSOR ROOM
1. Turn on compressor.
2. Enter back bay.

BACK BAY
1. Turn on back bay lights.
2. IN SUMMER open bay door on the left (not facing Arch Street)
3. IN WINTER turn thermostat up to 65°.
4. Enter lube bay.

LUBE BAY
1. Turn on bay lights.
2. IN SUMMER open back bay door.
3. IN WINTER turn thermostat up to 70°.
4. Check for key drop envelopes.
5. Enter front office area.

INTERIOR OFFICE
1. Turn on lights.
2. Uncover computers.
3. Turn on computers and printers.

FRONT OFFICE
1. Listen to phone messages, check parking lot, and proceed to write up repair orders.

PARKING LOT—CUSTOMER VEHICLE INVENTORY
1. Use the **customer automobile inventory sheet** to check all customer cars left overnight in the building or in the lot.
2. Check customer automobiles left overnight for damages and missing parts not noted on the **customer automobile inventory sheet**.

FIGURE 2-4 Task 1 duties: Opening the shop at 7:30 a.m.

CLOSING RENRAG AUTO REPAIR

IF WINTER OR FRIDAY NIGHT—put all company automobiles, including the customer shuttle, in the back bay.

BACK BAY
1. IF WINTER turn thermostat down to 55° (if it is to go below freezing) or OFF if it is to stay above freezing.
2. Close bay doors (make sure motor turns off by looking at spindle in ceiling).
3. Make sure all droplights are off.
4. Check exit door (back wall) and make sure BAR is across the door.
5. Check water faucet (with hose) and make sure it is OFF.
6. Turn off bathroom light.
7. Turn off bay lights by compressor room.
8. Enter compressor room and turn off compressor.
9. Turn off compressor room light.

SIDE LUBE BAY
1. Go to lube bay and close and lock two bay doors.
2. Turn off tire balancing machine.
3. Make sure all droplights are off.
4. IF WINTER turn thermostat down to 55° if waste oil heater is on.
5. Turn off light to lube bay and upstairs light.
6. Go to internal bay lift and turn lights off.
7. Go to bay with alignment machine.

ALIGNMENT BAY
1. Close bay doors.
2. Make sure side exit door is closed.
3. Make sure all droplights are off.
4. Turn off alignment machine.
5. IF WINTER turn thermostat down to 55° (if it is to go below freezing) or OFF if it is to stay above freezing.
6. Turn off light in bathroom—leave door open.
7. Turn off bay lights.
8. Go into front office and enter interior office.

INTERIOR OFFICE
1. Turn off all computers and printers.
2. Cover computers with protective covers.
3. Make sure inspection sticker box is locked.
4. Turn of lights.

FIGURE 2-5 Task 11 duties: Closing the building.

CLOSING RENRAG AUTO REPAIR

OUTSIDE OFFICE & WAITING ROOM
1. Turn off all computers, printer, and TV security monitor.
2. Switch answering machine to position "B."
3. Turn off photocopy machine (leave fax ON).
4. Turn off radio and TV.
5. Turn off lighted OPEN sign.
6. Turn "CLOSED/OPEN" sign around to show "closed."
7. Make sure coffee pot is OFF.
8. IF WINTER turn thermostat down to 55° (if it is to go below freezing) or OFF if it is to stay above freezing.
9. IF SUMMER turn AC off.
10. Arrange furniture and magazines, put cups, etc. in trash.
11. Lock front door.
12. Set alarm, turn off lights, leave and lock door.

PARKING LOT—CUSTOMER VEHICLE INVENTORY
1. Check all vehicles in lot and make sure they are locked.
2. Record all customer automobiles left overnight and placed in the building and left in parking lot on the **customer automobile inventory sheet**.
3. Note any damages or missing parts on inventory sheet.
4. Enter date and time the inventory was taken.

FIGURE 2-5 Task 11 duties: Closing the building. [Continued]

This is feedback about performance relative to the duties required. When service consultants have a list of duties to perform, a check-off sheet should be filled out and submitted to management before they leave the facility at night. This "check and report" method helps assure managers that nothing was missed and holds the service consultant responsible if the duty was not performed, such as a door was left unlocked.

As shown in Figures 2-4 and 2-5, the last set of duties for the opening and closing of the facility required the service consultant to take and then record the customer automobiles on an inventory sheet (see Figure 2-6) for all automobiles left in the building and on the property. If the service consultant did not open the facility, the person who opened it, such as an owner, a manager, or a technician, used the **customer automobile inventory sheet** to check the automobiles left overnight or over the weekend. The customer automobile inventory sheet is important because losing track of a customer's vehicle is terribly embarrassing!

RENRAG AUTO REPAIR

Customer Automobile Inventory Sheet

Date & Time	Customer Automobile			License Number	RO Number
	Year	Make (note location and condition)	Model		

FIGURE 2-6 Customer Automobile Inventory Sheet.

When Employee Duties Overlap

When duties for certain employee tasks overlap, the tasks are combined to create a set of procedures called a system. Employees at Renrag Auto Repair would follow each system to execute the duties that would successfully complete their tasks. There are three basic systems at a service facility where various employee duties overlap. These systems are: Shop Production, Customer Service, and Business Operations. Details that explain how each system functions are discussed throughout this book.

Formatted Operating Systems

A system is a uniform set of procedures that enable employees to know "what they are to do" (practices) and "when they are to do it" (consistency). A system of consistent practices allows management to maintain control of operations and keep the employees as well as the work organized. A system may be referred to as "formatted" when its uniform procedures are documented in a manual so they can be repeatedly duplicated by trained employees. This ensures all customer transactions follow the same process and the system can be duplicated at different business locations when executed by trained employees.

One of the most famous formatted operating systems is McDonalds restaurants. However, there are many successful formatted operating systems that contain detailed instructions, diagrams, tracking mechanisms, and other procedures employees must follow, often written in an "operations manual."

An operations manual is a set of procedures that when executed by a trained employee allows a system to properly function so that the desired outcomes are obtained, for example, the procedures to process repair orders for customer repairs (Shop Production Operations). This process includes interactions between several employees (the service consultant, parts specialist, team leader, and technician).To understand how each employee's duty overlaps with another, a portion of a Repair Order Tracking Sheet is shown in Figure 2-7 for the service consultant and a technician.

Coordination of Employee Duties That Overlap into a System

The repair order tracking sheet in Figure 2-7 shows the second phase of the repair process related to shop production operations. The entire repair order tracking sheet is discussed later in this chapter and in greater detail in chapter 11. Likewise, the service consultant and customer interaction required for completing the estimate, repair order, and invoice is covered in a system called the Customer Service System (discussed in chapter 6).

To illustrate how overlapping duties are coordinated, the job description of each employee is reviewed and the duties are identified where an "overlap" (interaction is required) occurs. To assure the overlap does not cause a problem within a work area or with other work areas, a system is designed. To illustrate this process and how it affects the service consultant job, tasks 6 and 7 (see Figure 2-2) are identified from the service consultant job description where overlap with other employees occurs and interaction between employees is required. As a review:

- Task 6 of the Service Consultant job description: Communicates with the technician to ensure service work is completed in a timely fashion
- Task 7 of the Service Consultant job description: Communicates with the technician, as needed, to prepare estimates.

The Repair Order Tracking Sheet: 2nd PHASE: Initial Work

14	**SECOND PHASE - initial work**	
15	Text message from Tech: TIME started job	
16	TECH Name	
17	TIME Tech reports in person to SC with RO	
18	Additional Work (Hours) recommended?	

(SC) The Service Consult
(Tech) Technician

FIGURE 2-7 Overlapping employee duties are monitored by the service consultant with the Repair Order Tracking Sheet. This helps assure shop production employee duties are coordinated and controlled so customer expectations are met.

In Figure 2-7 are four rows where data is entered. Each row is where an interaction or information is required so the service team (topic of chapter 3) can successfully carry out its current or future duties so the customer repair and the repair order (RO) can continue to be processed. Therefore, after analysis, the overlapping duties of the Service Consultant (SC), Parts Specialist (PS), and Technician (Tech) were identified. The job duties were combined and organized into sequential series of rows to create a system that could be monitored and controlled with a Repair Order Tracking Sheet (such as the portion shown in Figure 2-7).

PRIOR TO ROW 15
DUTY OF TL: Assign work to technicians who are certified to do the job
DUTY OF TECH: Pick up parts for the job from the parts specialist (department)

ROW 15 AND 16:
DUTY OF TECH: Text message to SC that job has started
DUTY OF SC: Record information from the technician's text message (row 15 is the time; row 16 is the tech's name)
DUTY OF TECH: Perform diagnostic or inspection work before any other service work
DUTY OF TECH: Write on the back of the RO the service suggestions as well as any repairs required (cure), the parts needed, and the reason (cause) for the repair on the back of the RO
DUTY OF THE PS: Work with the technician to obtain price and availability of the parts required for additional work (depending on the service facility, the PS may look up the labor times)

ROW 17:
DUTY OF THE SC AND TECH: Meet so the tech can explain the information to the SC
DUTY OF SC: Record the time to the meeting (row 17)

ROW 18:
DUTY OF SC: Look up the additional hours to perform the repairs suggested by the technician and record the hours of additional work in row 18.
DUTY OF SC AND TECH: Review the estimate so the SC can call the customer for authorization

The tracking process of the RO continues past Row 18.

Introduction to the Repair Order Tracking Sheet

Figure 2-7 shows a repair order tracking sheet used to track the initial work (second phase) of a four-phase repair process. The entire sheet is shown in Figure 2-8 and will be referenced throughout this book. This repair order tracking sheet is part of the shop production operations and is useful when monitoring and controlling the repairs as they are processed through a facility, especially on busy days when there is a large number of vehicles being serviced.

	A	B
1	SERVICE CONSULTANT	
2	REPAIR ORDER	
3	TRACKING SHEET	
4	**FIRST PHASE - check in**	
5	Customer Last Name	Garner
6	RO #	X
7	Vehicle year	X
8	Vehicle Make	X
9	Vehicle Model	X
10	Time Promised or Waiting	X
11	Initial hours sold	X
12	Time SC gives RO to TL and PS	X
13		
14	**SECOND PHASE - initial work**	
15	Text message from Tech: TIME started job	9:30
16	TECH Name	
17	TIME Tech reports in person to SC with RO	
18	Additional Work (Hours) recommended?	
19		
20	**THIRD PHASE - additional work**	
21	TIME SC got Customer Approval	
22	Hours approved (change time promised)	
23	Time SC gave updated RO to Tech (or TL)	
24	Time SC placed part order with PS	
25	Estimated delivery time of ordered parts	
26	TIME PS texts (SC+TECH) pick up parts	
27	Text from Tech to SC additional work started	
28		
29	**FOURTH PHASE - Check Out**	
30	Time Tech reports in person to SC with RO	
31	TIME PS supplies parts receipts to SC	
32	TIME final INVOICE has been completed	
33	TIME customer notified that vehicle done	
34	TIME vehicle was Picked up	
35	SC was able to provide Active Delivery?	
36	Thank you note / survey sent to customer?	

FIGURE 2-8 Repair order tracking Sheet with all four phases shown. (SC) is the Service Consultant, (PS) is the Parts Specialist, (TL) is the Team Leader, (Tech) is the Technician this sheet tracks the repair order (RO) and the teams' interactions to reach a desired goal—complete the customers requested repairs by the time promised.

As each of the four phases in the process is completed (Check in, Initial Work, Additional Work, Check Out), the service consultant will enter the information requested, such as time. As a concept, each row monitors the "duty" to be performed and the column is a specific customer job that is being tracked and controlled. For example, Cell B5 is the customer name Garner, and Cell B15 is the time the job was started: 9:30 a.m.

When customers call about the status of a repair on their automobile, the person answering the phone can quickly determine where the vehicle is in the repair process by looking at the repair order tracking sheet. When monitored regularly, problems can be found; such as a repair order that has been stuck in the first phase and has not progressed to the second phase for an unusually "long time.") This will help the service consultant notice the delay and take appropriate action to avoid a potentially upset customer.

Other forms

In addition to the repair order tracking sheet and customer automobile inventory sheet a service consultant might use, there are other forms that may also be needed. For example, Task 10 (Figure 2-2) requires the use of a special form for accounting control. Specifically, Task 10 requires the service consultant to make the bank deposit and prepare a report for the owners. The form in Figure 2-9 is an example of this type of form that required the service consultant to record the money deposited by cash, checks, and credit cards. The information recorded daily on this form was used to provide a picture of the daily, weekly, and monthly operations. For example, the information recorded the number of Lube Oil and Filters changes (LOFs) per day and, when graphed, gave a "picture" of the daily, weekly, and monthly volume of LOFs sold by the business. This information was used to change prices and offer special discounts on the least busy days of the week.

Duties at Different Types of Service Facilities

Different types of service facilities have different tasks, which are important, and possibly use different forms to record processes and daily activities. For example, specialty service facilities such as those that perform a transmission service do not need to train service consultants with extensive product knowledge in other systems, for example, chassis systems. Likewise, service consultants at a luxury brand dealership may have to use different communication skills for customer relations and sales than at a small independent garage whose customers are well known by the owners. For any service facility examined, the systems, related procedures, and forms may differ. However, one element is the same: all are in place to help the service facility personnel meet the customer's expectations.

RENRAG AUTO REPAIR, INC
Receipts for the Day

Date: _____

Cash:		Checks:		Credit Cards	
IN#	Payment	IN#	Payment	IN#	Payment

Total Receipts $ _____

Total Cash $ _____

Total Checks $ _____

Total Cr. Cards $ _____

Total Receipts $ _____

Bank Deposit $ _____

Payment for work done on credit:
Name   IN#   PMT

Work completed today on credit:
Customer   Amount
Name   IN#   Owed

NO. of ROs _____

NO. of LOFs _____

FIGURE 2-9 Task 10: Making daily deposits and preparing daily reports for owners.

Service Consultant Training

Franchise and chain service facilities have predetermined procedures with forms and possibly written scripts for service consultants to follow; therefore, the duties for certain tasks may differ from those of independently owned facilities. For example, to ensure that service consultants at franchise service facilities understand their duties, they often attend training prior to being hired. This often includes training in the business system, customer communication, angry customer resolution, and monitoring production.

Dealership service departments may also require extensive training for service consultants to be certified to serve customers with factory warranties. The manager, an outside consultant, or the automobile manufacturer may conduct this training on line, at a training center, or on site. The focus of the training depends on who conducts it. For example, automobile manufacturer training typically focuses on warranty administration and customer satisfaction, while many outside consultants focus on sales skills, systems, and communications. Service managers who conduct internal training sessions often focus on topics affecting their service facility such as shop operations, communication among employees, and policy procedures. In addition, service manager lunch session meetings may include role-playing and discussion of how to handle tough situations.

Finally, the larger corporate and multi manufacturer auto dealership repair facilities with multiple systems and specialists often relieve the service consultant of some of the tasks commonly performed at smaller service facilities. Specifically, there are typically other employees who handle:

- Customer payments
- Subcontractors such as towing, customer shuttle service, emergency parts pick up and return
- Building maintenance and service facility fleet oversight among others tasks that may be assigned

When it is found that the service consultant cannot wait on paying customers because he or she is busy with other duties, then businesses with larger numbers of employees can easily reassign duties once bottle necks are found. This is not the case at smaller or new service facilities with fewer customers and employees doing multiple jobs. Therefore, the duties assigned to each employee, such as a service consultant, must reflect the tasks needed to ensure effective and efficient daily operations of the business. However, the variety of other factors to be considered when a specific employee's job description, tasks, duties, and system participation is created is beyond this chapter. Consequently, as a service facility grows or adds products, the job descriptions, task, duties, and perhaps systems and procedures must be monitored and possibly changed.

Professional Image

An automobile service facility must present a professional image, and the service consultant must be able to convince customers that the service facility can and wants to help them. The service consultant and other support staff must recognize the customers' presence by smiling and cheerfully saying "Hello" even when the automobile service facility is very busy. If customers are recognized and made to feel comfortable, they are more likely to be patient and wait until they can be helped.

Furthermore, when the service consultant is trained to follow a sequence of steps and project an attitude that conveys interest in solving the customers' problems, they will be more at ease, begin to feel "at home," and be more receptive to the suggestions and recommendations of the service consultant and technicians. Customers also notice the personal appearance of the service consultant and other employees as well as the appearance of the facility. Appearances often give customers a feeling about the care a facility will give to their automobile. To convey this attitude, a service consultant should be clean and neatly dressed. While some odors associated with an automobile, such as oil and gas, cannot be avoided, employees must be sensitive to the odors related to their personal hygiene (some experienced service consultants keep a bottle of mouthwash, deodorant, and after-shave lotion at their stations). Many service consultants and other employees often wear a uniform or shirt with a name or nametag. Likewise, the furnishing and cleanliness of the customer reception area should be constantly monitored (see Figure 2-10). If a facility does not care about the appearance of its employees or business, it implies that the care of a customer's automobile is also not important!

Therefore, a professional image and the right attitude are a fundamental key to the success of a service facility. The service consultant must remember that the service facility is in business to serve the customer. This includes every employee in the service facility, not just the service consultant. Everyone must take the time and be willing to resolve the customers' automotive problems.

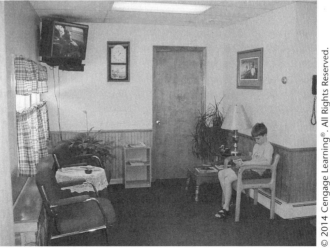

FIGURE 2-10 A neat and clean waiting area.

Review Questions

Multiple Choice

1. A service facility must have enough customers flowing in and out to make a:
   A. sale
   B. business
   C. technician satisfied
   D. profit
2. What organization incorporates the tasks and duties of service consultants into a list of the major skills and knowledge expected of the position?
   A. The Automobile Manufacturers Association
   B. The Service Facility Association
   C. The Automotive Service Excellence
   D. The American Automobile Association
3. What document lists the tasks the service consultant will perform and who creates it?
   A. Duties created by the service consultant
   B. Job (position) description created by the ASE
   C. Job (position) description created by the owner or manager
   D. The repair order tracking sheet created by the technician
4. What form is extremely useful when tracking the automobiles being processed through a facility, especially on busy days when there is a large number of automobiles being checked in and picked up?
   A. Repair order tracking sheet
   B. Customer Automobile Inventory Sheet
   C. ASE job tasks
   D. Job description for a service consultant
5. Customer service can be enhanced by:
   A. a clean and organized facility
   B. the willingness of all employees to help customers
   C. using the repair order tracking sheet to monitor repair progress
   D. all of the above
6. Which of these can have the greatest impact on a customer's decision to do business with you?
   A. Extended business hours
   B. The size of the service facility
   C. The level of trust they feel
   D. Discount pricing

Short Answer Questions

1. What is accomplished in a Shop Production System and how is it different from the Customer Service System?
2. What is the purpose of a Repair Order Tracking Sheet and how is it used by a service consultant?
3. What is the difference between a task and a duty and provide an example based on information found in this chapter for a service consultant.
4. How can a service consultant present a professional image and why is it important?
5. What are the three systems commonly found at a service facility and how can a system help employees when their duties overlap?
6. Given an example of overlapping employee duties and how this is monitored by the service consultant with the Repair Order Tracking Sheet?
7. Why might the duties and tasks assigned to a service consultant be different at different types and sizes of service facilities?

CHAPTER 2 THE ROLE OF THE SERVICE CONSULTANT AND FORMATTED OPERATING SYSTEMS

	A	B	C	D	E	F
1	SERVICE CONSULTANT	PS = Parts Specialist		SC= Service Consultant		
2	REPAIR ORDER	RO= Repair Order		Tech=Technician		
3	TRACKING SHEET			TL = Team Leader		
4	FIRST PHASE - check in	TODAY IS:	Tuesday			
5	Customer Last Name	Jones	Thomas	Grant	Meyers	Sloan
6	RO #	100	101	102	103	104
7	Vehicle year	2012	2011	2013	2015	2012
8	Vehicle Make	GMC	Toyota	Ford	Toyota	Ford
9	Vehicle Model	Truck	Avalon	Taurus	Camry	F250
10	Time Promised or Waiting	waiting	waiting	waiting	waiting	waiting
11	Initial hours sold	X	X	X	X	X
12	Time SC gives RO to TL and PS	X	2:30	X	X	X
13						
14	SECOND PHASE - initial work					
15	Text message from Tech: TIME started job	X		X	X	X
16	TECH Name	X		X	X	X
17	TIME Tech reports in person to SC with RO	3:07		X	X	X
18	Additional Work (Hours) recommended?	1.3		X	X	X
19						
20	THIRD PHASE - additional work					
21	TIME SC got Customer Approval			X	X	X
22	Hours approved (change time promised)			X	X	X
23	Time SC gave updated RO to Tech (or TL)			X	X	X
24	Time SC placed part order with PS			X	X	X
25	Estimated delivery time of ordered parts			3:15	X	X
26	TIME PS texts (SC+TECH) pick up parts				3:05	X
27	Text from Tech to SC additional work started					X
28						
29	FOURTH PHASE - Check Out					
30	Time Tech reports in person to SC with RO					X
31	TIME PS supplies parts receipts to SC					X
32	TIME final INVOICE has been completed					X
33	TIME customer notified that vehicle done					3:01
34	TIME vehicle was Picked up					
35	SC was able to provide Active Delivery?					
36	Thank you note survey sent to customer?					

FIGURE 2-11 Use this Repair Order Tracking Sheet to answer the activity questions.

Activity

Activity: Answer the questions using Figure 2-11 above.

How many vehicles are waiting to move into the second phase?

How many vehicles are in the second phase waiting to move to the third phase?

How many vehicles are in the third phase and working toward moving to the fourth phase?

Howe many vehicles are in the fourth phase?

Which customer is waiting for parts to arrive and what time where they to arrive?

Which vehicle has had the parts arrived and is waiting for work to be started again?

Which RO number is waiting for customer approval? How many hours of work were recommended by the technician? If approval is not granted, what phase will the vehicle go to next?

Which vehicle is done and ready to be picked up? What time was the customer called?

What customer is waiting for the work to start and if it is 3:15 p.m. now, how long have they been waiting?

CHAPTER 3

THE TEAM APPROACH: THE SERVICE SYSTEM AND SHOP PRODUCTION

OBJECTIVES

Upon reading this chapter, you should be able to:

- *Explain why and how a team approach can offer superior customer service.*
- *Present optional team formations and team member assignments in different types of service facility operations.*
- *Identify the major responsibilities of the team leader.*

CAREER FOCUS

You must be able to work on a team and focus on leadership to get ready for your next position—manager or owner. Whether you are a team member or asked to lead a team, as your team works together to reach its goals, you will get noticed. Your recognition is because you understand the team structure and can perform well in it. Over the years some people the authors have managed, taught, and trained to work on a team are now leaders, managers, and owners. Others have preferred to be a "team of one" and cannot work on a team because reaching goals is not about "us" rather the focus is on "I" and "me." You can be a "team of one" and be the best at your job, but you will have a different career path holding different jobs than covered in this chapter. If you can function on a team and can help a team reach its goals, then the information in this chapter will give you the tools that will allow you to proceed into management level positions.

Introduction

The automobile started with very basic mechanical systems when compared with today's computerized systems, such as antilock brakes that were not even considered possible at one time. In addition, the makes and models of automobiles were essentially limited to a few large manufacturers. The operation and repair of the various automobile systems were similar, and repair manuals were rarely needed. What is now known as an automobile service facility was referred to as a repair garage, and service given to customers was generally quite different when compared with today's facility.

One of these repair garages in the "good old days" was owned and run by John. His was an independent repair shop and did not pump gas like many repair garages. John did not talk much and worked alone, except when he could convince his sons to help him. People took their cars to John because of his mechanical ability. He had what was known as a "good ear," meaning he could tune an engine by listening to it run. He was considered honest because he did not put used parts on the cars he repaired and then charge for new parts. His repair charges were not cheap, but he was considered reasonable.

John liked his one-man shop, which meant he was the service consultant, parts specialist, technician, bookkeeper, receptionist, janitor, and groundskeeper, unless he could get one of the sons to mow the grass. When customers wanted John to repair their car, it was expected that they would go into the shop, find John, and wait patiently until he wanted to pay attention. The customers typically had to explain the problem to John while he continued to work on something else. After John agreed to fix the car, the customers would leave it out front and then return, possibly several times, to see if it was done. Calling John did not work very well because he often did not hear the phone when he was working on a car. Eventually, customers would discover that their car was out front again, which meant the repair was finished. John liked to fix only one to five cars a day so he could do bookkeeping at night.

Obviously, John's garage could not survive in today's market without some improvements. For example, aside from greater business expenses, such as heat, light, uniforms, advertising, insurance, and benefits, there are technical advances that require expensive equipment. In addition, the nature of customer service has changed. Customers expect personal service and professional treatment when they take their automobile to a service facility. This, of course, can be taken as a compliment because the image of the repair garage in John's day has shifted to a service facility with trained professionals. Customers expect to be treated differently by professionals who are trained to perform a specific job as compared to John's day.

The tool investment experiences of professionals at a service facility have also changed. John's tools were a basic set of hand tools, a droplight, a jack, and a creeper. This meant that John's personal investment was not very large. In addition, he had few business expenses in addition to his parts bill. Therefore, servicing a large number of customers was not necessary, so two service bays were more than enough. To make a profit, John usually worked six days a week and all invoices were paid by cash or check (credit cards were not in use yet), which made business a lot easier. Today, the investment in setting up and running a service facility is very different. The costs for tools, equipment, and technicians with high levels of expertise have increased dramatically. A fully equipped facility needs multiple bays and a convenient, sometimes expensive, location to conduct the business required to make a profit. Furthermore, there is also the expectation of personal and professional service.

More specifically, in order to earn the money needed to meet the costs of an automobile service facility and remain competitive with respect to the amount charged to customers, a facility must process a set number of customers over a "five-day workweek" to break even. To serve the necessary number of customers per day to make a profit and provide the personal service expected by customers, a facility may need to employ people to perform different jobs, such as a service consultant, a parts specialist, a driver for a customer shuttle, and others. However, just having these different types of employees on the payroll is not enough. They must know their job and be able to perform it effectively. To be efficient, employees must also work together as a team. Certain people can be the best at performing a job independently, but their inability to work with others could mean they are not useful to the business.

The purpose of this chapter is to discuss the composition and use of teams at service facilities and how these teams may differ from one facility to another. Additionally, this chapter explains the importance of leadership. Teams will not function effectively unless someone is willing and able to provide the leadership necessary to get them to work together every day.

The Team Approach within a Service System

Like John's garage, the use of teams at service facilities has changed in recent years. More specifically, a service facility where the technicians worked together as a team is no longer adequate for the larger, more complex facilities. To attract more desirable customers, greater efficiency with expanded services and faster delivery are required. This calls for an approach whereby everyone in the facility is on a team and works together. This means there are team members, each with specific duties. Some of the duties will overlap with each other requiring coordination and cooperation among the members within a service system. A Service System is an orderly arrangement of procedures that are linked together to form a process followed by employees and used by management to control shop production, customer service, and business operations.

Service Consultant, "The Pivot" within the Service System

When customers enter a facility, the services they receive should proceed through a "seamless" Service System, beginning with an introduction by the service consultant and ending when they are handed the keys to their automobile. To understand the team process and responsibilities of each member, the Service System framework must be understood. As mentioned in Chapter 2 and to be discussed throughout the remainder of this text, the Service System framework consists of three separate systems:

- The Customer Service System,
- The Shop Production System, and
- The Business Operations System

The service consultant has a role in all three systems and is therefore a "pivot." Just like a ball joint on a control arm is the pivot that allows the wheel assembly to turn when pushed by the steering arm so the vehicle can change direction, the service consultant has the same function. The service consultant collects and maintains "service information" and, when ready, directs it to be used in the proper Service System (customer service, shop production, or business operations). As a central pivot of the three systems, the service consultant makes sure processes are initiated so the customer's request progresses to completion. Below is a summary of the service consultant's job within each of the three systems.

In the Customer Service System, the service consultant and the customer work together. A repair order is created in the Customer Service System which the service consultant transfers to the Shop Production System.

The Shop Production System is made up of a service team (team leader, technician, and parts specialist). A responsibility of the service consultant is to track the progress of the repair order using the Repair Order Tracking Sheet. Within the Shop Production System's operations, the service consultant will work as part of the service team to keep the customer informed of any additional work needed, delays or problems in the repair process, as well as the general information on the progress of the services being performed.

When the automobile's service has been completed and the invoice has been created by the service consultant, the Customer Service System and Shop Production System ends. The Business Operations System starts with collection of the money from the customer. Then the transaction is recorded for the financial statements, such as the income statement discussed in chapter 1 and in the authors' second book, *"Managing Automotive Businesses: Strategic Planning, Personnel, and Finance"*. The service consultant as a pivot starts the Business Operations System with the delivery of the completed invoice and customer keys to the cashier. However, at some service facilities, the service consultant may have duties within the Business Operations System, such as acting as the cashier by collecting the money from the customer.

Other Customer Expectations

Basically, the point of the Service System is for customers to enter a facility and with minimal effort and confusion and have their service request completed or problem solved in a timely manner. However, the customer may have other expectations as well, and therefore the service consultant, as host, may be asked to oversee or provide services such as:

- Transportation to home or a job while their automobile is at the facility.
- A clean, comfortable, and pleasant place to wait while their automobile is in the shop that has amenities such as coffee.
- Request to get a message to someone to let the person know of customers' problem and where they are at.
- Assistance in obtaining credit for the payment of a large repair estimate.
- Express checkout because the customer needs to leave quickly.
- Ability to schedule follow-up appointments for service and repairs before leaving the service facility.
- Information and brochures about the different quality or grades of parts available, such as tires and brake pads as well as warranty information.
- Restroom facilities, drinking water among other beverages and snacks.
- Provide a referral to a specialty service facility if necessary to get their vehicle repaired.
- Various methods to send messages to the customer updating them on the progress of their service and whether the time promised will be met.

There are a lot of other expectations that a customer may require in addition to just timely and competent service. Obviously, a service consultant as one person cannot provide all of these services to the customers and the team may include support staff, such as courtesy shuttle drivers, custodial staff, and clerical personnel.

Two Primary Missions of a Team

The members of a team cannot just do their job; they must relate it to the team's primary missions—to process the customer's service request and then to perform the services as requested. To do this, first the members must work together to assure the customers are treated as important people. Next, they must effectively perform the customer's requested service. In both cases, customer satisfaction is a responsibility of all of the members on the team. The team members must know their role in achieving the missions and then execute it efficiently without flaw.

Team Profit and Profit per Bay

Profit is important because it is related to production. Greater efficiency leads to greater production and profit.

Profit is important because it means the income was greater than the costs to operate a modern service facility. Therefore, after all of the bills were paid, money was left over. If enough money is not earned, the efficiency of the operations may be the reason. While more service bays should generate more money when properly equipped and staffed, they also cost money to operate (heat, light, electricity) even if unused or underutilized. Therefore, a shop with too many empty bays will drive up the costs and lower the average number of billable hours per bay each day. One method to calculate billable hours per bay each day is to take the total number of dollars earned by the service facility for labor divided by the service facility's labor rate per hour charged to customers divided by the number of bays. For example, $4,000 of labor charges per day ÷ $80 per hour ÷ 10 bays = 5 hours per bay each day.

When Profit per Bay Is Low

The objective of the team is to make sure each bay, assuming each is staffed with a technician, does enough work to maximize the number of billable hours generated in each service bay. The concept is not any different from cylinders in an engine or number of people rowing a boat. Every cylinder has to fire and every rower must put in effort to pull his or her oars. Otherwise the engine will run rough and not produce maximum power, or other rowers will have to do more work to make up for the people not rowing hard enough. For example, if the bays at a facility are generating an average of four to five billable hours of service per day, and the facility is not profitable, it will have to increase the number of billable hours to perhaps six or seven hours each day to stay in business (let us assume each bay has a technician working in it—remember bays without a technician don't make money). To do this, the facility will have to work as a team to determine which bay is not doing enough work and may find they have to increase the number of customers served each day or the number of hours charged to each customer.

Increasing productivity per bay by either method (more customers or higher sales per customer) may require the team of employees to work together more efficiently, not necessarily "harder." If a team does not exist, a team approach using a service system may help the employees work together more effectively. If a team already exists, it may mean there is no system in place for the team to follow, the system has flaws, or the team members do not work well together. To help figure out the problem, a team meeting may help the team members better understand the mission of the business, purpose of the team, and the business systems used to determine if more direct leadership is required. These concepts, with some additional calculations to help with understanding them, are covered in greater detail in Activity 1 found at the end of this chapter.

At some service facilities, an efficient and hardworking team may already exist but may need to be expanded. For example, when the service consultant is too busy answering the phone or collecting money from customers, jobs may not be moving through the system fast enough. In such a case, additional personnel may be needed to permit the service consultant to focus on getting more customers' automobiles into and through the system. This could mean having a cashier, clerk, or receptionist, or cross training another employee to help during high-demand periods. The concept about the growth of the service facility and its team is covered later in this chapter.

The Team Process and the Service System

Remember that both the Customer Service System and the Shop Production System are part of the larger Service System. A Service System is an orderly arrangement of procedures that are linked together to form a process followed by employees and used by management to control shop production, customer service, and business operations. This system is needed when employees' duties overlap (chapter 2) and coordination between them is required.

One application of the team approach is to process the repair orders, described in chapter 2, and use the repair order tracking sheet to monitor progress of the Shop Production System. In order to discuss this from the perspective of a team approach, the entire Service System process is shown in Figure 3-1 from greeting the customer to the presentation of the invoice; this is the Customer Service System. Figure 3-1 includes the Shop Production System at the appropriate place and shows the end of the Customer Service and Shop Production Systems and start of the Business Operations System.

54 THE SERVICE CONSULTANT

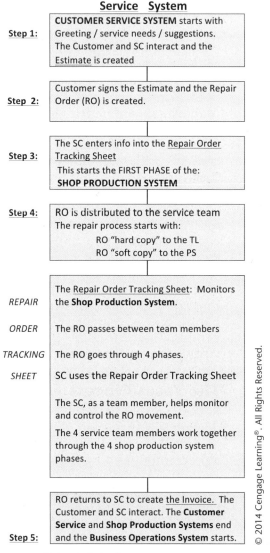

FIGURE 3-1 The Service System.

Notice that the service consultant starts the Service System process and pivots at step 3 (Figure 3-1) and directs information and documents into the Shop Production System. There are four phases of Shop Production and they are monitored by the service consultant using the Repair Order Tracking Sheet. Steps 3 to 5 shown in Figure 3-1 have overlapping functions between the Customer Service System and the Shop Production System as discussed in more detail in later chapters. At this point it is important to know that team members must learn to "keep their eyes on the ball," which, in this case, is the repair order (RO). The Repair Order Tracking Sheet helps the service consultant monitor the RO's progress through the Shop Production System.

The Service Team

The team, shown in Figure 3-1 example, has four members: the service consultant, the parts specialist, the team leader, and the technician. Each team member has to know the service system process and that the RO is monitored when it is "in play" during the repair order tracking process. Specifically, like football, the ball is the RO and has to be passed to the next member on the team. The handoff between team members requires communications to be flawless and prompt (on time), after each team member completes the assigned duties. If this does not occur, the customer's expectations may not be met, such as on-time delivery of all the services they wanted performed. The process of how the RO is handed off between team members is discussed later.

As soon as the service consultant has completed working with the customer in step 1 (this is the Customer Service System and is covered in chapter 6), the estimate is signed making it an RO in step 2. The RO activates the Shop Production System and the RO tracking process starts in step 3 by filling out the Repair Order Tracking Sheet (first phase) with basic customer information and delivery expectation (time promised). In step 4, the paperwork is delivered with the RO's technician copy (hard copy) to the team leader to assign to a technician (second phase of the Repair Order Tracking Sheet). Furthermore, the RO's parts copy (soft copy) goes to the parts specialist. The repair order tracking sheet will follow the RO as it is handed off between members of the team. Naturally, the service consultant must not delay in processing steps 2 to 4 (Figure 3-1); otherwise the entire process would be held up and work would back up. Likewise in step 5, the RO's return to the service consultant from the technician and parts specialist must be prompt to assure the customer's checkout is quick, especially if the customer is waiting at the service facility.

In connecting the Customer Service System and the Shop Production System processes shown in Figure 3-1, the service consultant is the pivot. He or she is expected to monitor and control the processing of the RO as well as the customer requests and paperwork. However, again all members are expected to "keep their eyes on the ROs" in order to keep the work moving efficiently through the facility. Unlike the other members, however, the service consultant plays a major role at the beginning (step 1, initial customer interaction, estimate, and RO), middle (between steps 4 and 5 with the customer approval to perform additional work), and end of the process (step 5, the final invoice is paid, active delivery is provided, and finally follow-up surveys and thank-you note are sent). Consequently, because of their duties and position on the team, service consultants may be viewed as sharing the responsibility of the team leader by default.

The Repair Order Tracking Sheet: A Team Effort

The RO handoff process in the Shop Production System should be a seamless system, meaning it should not be disrupted by having to move to another process and then return at some point to be continued. The repair order tracking sheet with a flowchart of the team members as they process the RO through the service process is shown in Figure 3-2. At the top are the four members of the team: the team leader (TL), the technician (Tech), the service consultant (SC), and the parts specialist (PS). In the flowchart, each block corresponds to the process in the repair order tracking sheet. The passing of the RO is shown by a line with an arrow. The interaction between members throughout the process is indicated by a line. The key processes and the member holding the RO are called out in boxes.

Flow Diagrams on Working Relationships of Team Members

A **flow diagram** is often helpful for team members so that they can "see" how they are to work with each other. Figure 3-3 illustrates the working relationships of a three-member team: the service consultant, the technicians, and the parts specialist. As this diagram exhibits, the owners (or a service manager) are responsible for the team performance and all of the members on the team interact with each other.

To ensure that the process is working, the owners must meet with the team on a weekly basis either before the shop opens or at the end of the day. In addition, on days when either a large number of customers have appointments or a large repair job is taking several days to complete and causing a member(s) to be unable to handle "daily work," the owners or manager may have to assist the service consultant, parts specialist, or technicians.

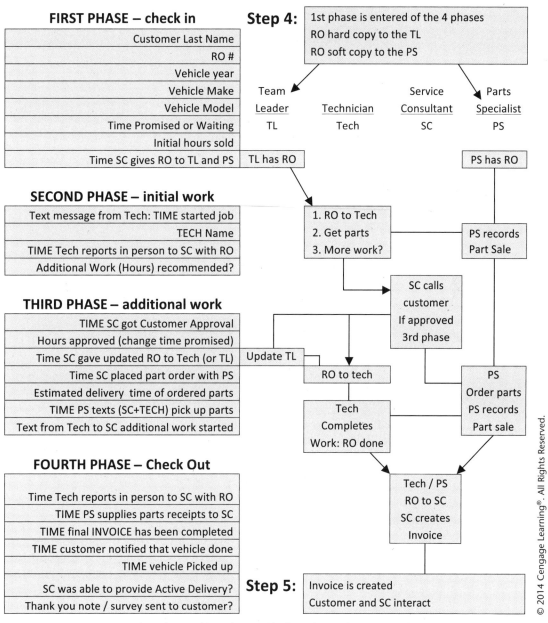

FIGURE 3-2 Repair order tracking sheet with flow chart of team member interactions.

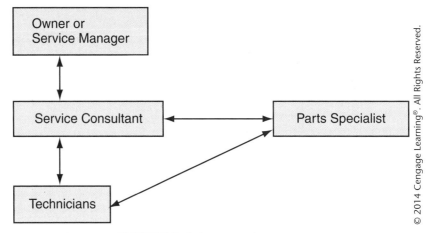

FIGURE 3-3 A three-member team.

Modifying the Team and System

When the volume of customers increases, the number of members on a team has to be expanded or a second team created. The other support personnel that could be added to a service team might include:

- A cashier to receive payments and make the evening deposits.
- A receptionist to answer the phone and make contact with the customers.
- A parts runner to take parts to technicians and go for parts and supplies when needed.
- A customer shuttle driver.

Adding Team Members

From experience, the owners of Renrag knew that if their six-bay garage was not processing 20 to 25 customer vehicles a day, if the appointment book was filled for more than two days, if the bays were not generating at least six billable hours per day, (knowing every team member was working at maximum capacity) a bottleneck existed somewhere. Investigation starts with whether there is adequate number of technicians, then goes to whether the service consultant and parts specialist need some assistance to increase the flow of customers into, through, and out of the facility. The bottleneck must be found and rectified as quickly as possible.

When support staff are added to a team, they must be prepared to perform their job and must understand the team concept. This seemingly minor point cannot be stressed enough. When a part-time or full-time staff member is hired, he or she may not be able to support a team as expected initially, and it may take time before any noticeable change occurs in productivity.

For example, assume that a cashier is added to the team structure in order to provide the service consultant with more time to work with customers and to follow the ROs in the Shop Production System. Figure 3-3 would need to be modified to show how the cashier would be included as a new member of the team. To extend the example, perhaps a technician is assigned to be the lead technician to help assign work to other technicians to further "free up" the service consultant's time. Based on adding a cashier and promoting a technician to be a lead technician, a new diagram (see Figure 3-4) needs to be created to show their working relationships. Notice in Figure 3-4 that the service consultant and parts specialist have a working relationship on a daily basis with the new team members—the cashier and the lead technician.

The Role of a Lead Technician

In Figure 3-4, all work flows through the **lead technician**, who assigns jobs to the other technicians. When a service facility has a lead technician, the team leader's responsibilities are shared between the service consultant and lead technician. The service consultant is technically a part of management and, in that capacity, works with the service manager on issues related to the team's customer relations, warranties, the processing of repair orders, general operations, sales,

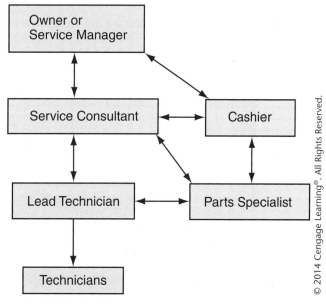

FIGURE 3-4 A four-member team.

volume of work processed, and so on. At the same time, the lead technician assists with coordination of the work within the shop, handing out work to each technician on his or her team, shop operations, quality control, monitoring the condition and use of shop equipment, and so on.

The lead technician is expected to work as a technician in addition to handing out work to other technicians. This is important because jobs must be handed out both to the technicians who can do the job and in a manner that is fair to all technicians. Because of the problems caused by the way jobs are assigned, some facilities have the technicians elect the lead technician. As a result, the lead technician may not be the service facility's best technician but someone the technicians respect and trust to assign the work fairly.

The Flat-Rate Pay System and Assignment of Work

The assignment of jobs by the lead technician is critical when the technicians are paid by the job in a system called **flat rate**. The flat-rate system gives a technician a certain amount of time to do a job. For instance, a job may be assigned a flat-rate time of 1.5 hours. If a technician completes the job in 1.2 hours, he or she is still paid for 1.5 hours of work. Technicians' pay at the end of the week is based on the total amount of time the technicians are given to complete the jobs assigned (not the time it actually takes to complete the jobs) multiplied by their pay rate.

For example, if a technician earned 50 flat-rate hours for the week, possibly taking 38.8 hours to complete all of the jobs, and the pay rate is $10 per flat-rate hour, then the pay would be $500. Therefore, for technicians who are paid by flat rate, how work is assigned is important. If a job takes longer than the time allowed under the flat rate, then the technician will lose time and, ultimately, money. Of course, on occasion, technicians expect to lose time on some jobs. However, when technicians are constantly assigned work that loses money, they will have a morale problem and may even quit their job. Consequently, assigning work fairly to technicians capable of doing the work is critical to the effective and efficient performance of a team.

Multiple Teams at a Service Facility

When the volume of customers at a facility increases to a level in which the bays are consistently producing an above-average number of billable (flat-rate) hours per day, the facility may need to create an additional team. Unlike just adding technicians to a shop and making it bigger causing system issues (typically the ratio is one service consultant for every five technicians), adding teams allows the manager to monitor each team and their performance. After the additional team is set up, the facility should begin to process more customers per day and increase its

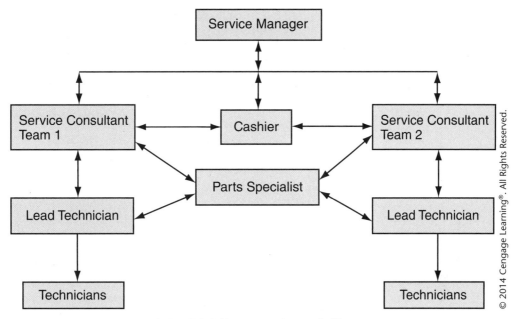

FIGURE 3-5 Two teams in one facility.

total number of billable hours. At the same time, the facility will likely find initially a decrease in the average number of billable hours per bay to the desired average number of hours. This expansion depends on several factors, such as space for more bays and equipment, money to buy the equipment, and availability of technicians and staff to hire.

Figure 3-5 shows the working relationship for two teams. In this diagram the support staff, such as the cashier, works for the service manager, the parts specialist works with both teams, and the technicians have been split to work with one of two lead technicians and one of the two service consultants. If the volume increases to the point where the parts specialist cannot handle the work, a second specialist may be employed to work with each team.

In some cases, the demand for specific services may increase, indicating a need to modify a team's composition. As a result, other options may be considered. For example, at Renrag Repair Garage there was a growing demand for the maintenance of automobiles, specifically quick oil changes, tire rotations, antifreeze fluid changes, and air-conditioning inspections and maintenance, among others. Because of the demand, an option would have been to create a **specialty team** for automobile maintenance while the other team continued to do the diagnosis and repair of automobiles. This arrangement has become popular at many dealers who have fast lube or other light maintenance teams (profit centers which are discussed in later chapters). The working relationships would have been similar to the ones shown in Figure 3-5 with perhaps team 2 exclusively assigned to maintenance services.

Naturally, the addition of a specialty team requires the service processes to be carefully changed because, in some cases, a customer's automobile might be serviced by both teams. For example, one team may change the oil in the automobile, while the other may make a repair to the differential. The arrangement whereby personnel, technicians, and service bays are equipped for the jobs they perform promises to be very efficient. In other words, within the general service facility would be a specialty service facility.

Service Team Formation

In reality, a small independent service facility team is comprised of a service manager, who also serves as the service consultant. In many cases, the service consultant also orders parts and prepares the estimates. If the owner or service manager/consultant has a technical background, he or she may also serve as a technician on occasion. However, when in the technician role, this person should only "pick up the slack" to help the busy technicians. Otherwise the service manager/consultant should oversee and hand out work to the technicians. The relationship structure of the service facility would therefore resemble the diagram shown in Figure 3-3. It is also not uncommon for a part-time or full-time staff assistant to be hired at these facilities to answer the phone, call for parts, receive deliveries, file documents, and so on.

At growing service facilities, more "support staff" are often needed. The diagrams for these working relationships would resemble the one shown in Figure 3-4. Naturally, as more technicians are hired, more service consultants and parts specialists will be needed. A good rule to follow is that, typically, one service consultant can handle between four and six technicians. Important details, such as returning a customer's call, are likely to be missed beyond this number. When additional service consultants are hired, the diagram for the working relationships would resemble Figure 3-5.

In some cases, a larger dealership service department may not have teams and instead have a dispatcher system as shown in Figure 3-6. In this system, the work funnels back and forth from the service consultant to the technicians through a dispatcher. A shop foreman may be employed to "help" the technicians with technical and production problems but is typically not a "production technician." Since there is not a lead technician, as discussed in the service team system, and the service consultant is often disconnected and is too busy to assign work to the technicians, the dispatcher will assign work to the technicians. Each technician reports directly to the service consultant who originally worked with the customers. This arrangement is more common at some chain store-owned service facilities, franchises, and smaller dealerships where there are not enough technicians to form separate teams but there are too many for the service consultant to oversee.

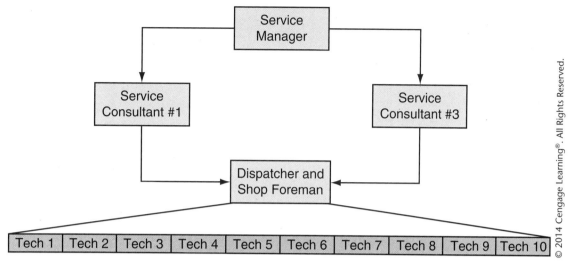

FIGURE 3-6 Dispatcher system.

Since the dispatcher system is more common at service facilities with fewer than a dozen technicians, the system relies on the dispatcher to monitor production in order to maintain control. If that key position fails to perform well, it can cause significant shop production problems. Also since the service consultants will work with all of the technicians in the shop depending on the job assigned to the technician, this can make workflow confusing and organization is the key to maintaining order.

Recognizing the potential problems for a shop that may wish to expand the number of bays and technicians past what a dispatcher can handle or a desire to make the operations more manageable (maintain order), the service manager may elect to divide the shop into smaller work units or teams. This means a service manager oversees several teams which are made up of a service consultant, a lead technician who is a production technician, and three to five technicians.

Lateral Support Groups and the Team System

In this textbook the term *team* is used to refer to a production unit. At some dealerships the team is referred to as a group. The technical difference between a team and a group system of management is mainly how the technicians are paid, not the structure itself. The team system typically averages the technicians' flat-rate hours earned and the pay for each technician is based on the team's average hours. Some feel this produces more cooperation between team members, while others point out that highly efficient technicians are penalized. The group system typically pays each technician the flat-rate hours earned regardless of the other technicians' performance.

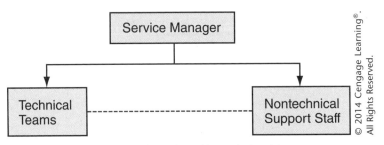

FIGURE 3-7 A formal working relationship.

Finally, at some automobile dealership service departments and many chain store-owned service facilities, it is not uncommon to have more formal working relationships between the technical teams and the nontechnical support staff team. The nontechnical support staff may include service consultants, parts specialists, an assistant to the service manager, a cashier, a warranty clerk, bookkeepers, custodians, automobile detail and lot personnel, and a facility maintenance crew. A formal working relationship between the nontechnical support staff and the technical personnel in the service facility may resemble the diagram shown in Figure 3-7. As shown in this diagram, although the technical and nontechnical people may communicate informally with each other (the dashed line), their formal work communications flow through the service manager.

Leadership Expectations

There are numerous leadership strategies for team leaders to consider. Service consultants should spend some time studying the different leadership styles. In general, a team leader at a service facility has a number of major responsibilities. Several of these are:

- The facility's profit status
- The quality of the service provided
- Customer satisfaction
- Team efficiency
- Employee morale
- The reputation of the service facility

There are many other leadership responsibilities, but the point is that the monitoring and managing of these responsibilities is the function of the team leader. It is not a job that can be learned overnight.

In addition to these responsibilities, the leader of a team has numerous other duties. For example, the team leader is required to set a good example with respect to personal appearance and conduct. Further, leaders cannot expect subordinates to behave in a manner that they do not personally practice. This includes maintaining composure, using appropriate language, interacting with other people, being on time, taking one's job seriously, being honest, respecting others, and so on.

Another responsibility of the team leader is **cross-training** team members. People should learn each other's jobs. Of course, this cannot be done for a job that requires a lot of education and training, but employees should be able to perform other job duties at the facility and cover for each other when someone is absent. For example, at a small independent facility, the service consultant should know how to order parts and prepare an estimate. If a staff member, such as the cashier, is not available, the service consultant, parts specialist, and lead technician should know how to receive a payment from a customer. In this case a set of detailed directions may be needed for the use of the credit card machine.

Another duty for the team leader is team building. There are many recommendations made by experts on team building and the team leader should become familiar with them. The purpose of team building is to have the team members know each other well enough (even if they don't like each other) to work together like a well-oiled machine, meaning with little or no friction.

In addition, proper and clear communication among team members is critical. A team leader should be focused on the ability of team members to communicate. Nothing can cause problems quicker than if people do not understand each other. Communication may include written and verbal exchanges. Handwriting is important when ROs are exchanged. For example, when a part is ordered for an automobile, the year and model of the automobile and the description of the part must be written clearly on the RO. If handwritten messages are not clear, the process may be dramatically delayed. Furthermore, some jobs may require knowledge of a foreign language because some customers and employees may not speak English. For example, in some states, the ability to speak Spanish has become a job requirement at some service facilities.

Team leaders must always be concerned with weaknesses in the process and in team members. This requires them to be observant and to mentally review the activities at the end of the day after the facility closes. When weaknesses are noted, team leaders must correct them immediately. This should be done in a manner that corrects the problem and does not make it worse or create another.

Finally, a more successful leadership strategy is for the leader to treat team members in a respectful manner. Maintaining personal dignity and respecting the dignity of others have been found to be more successful than a top-down, rigid, and controlling style of leadership, especially if people are expected to work as a team with the members supporting each other and the leader.

The Use of Flat-Rate Objectives

When a facility uses **flat-rate objectives**, the team leader, who may be the service consultant or lead technician, must monitor them. A flat-rate objective is the number of flat-rate hours a technician hopes to earn by the end of the week. For example, at the end of the beginning of each quarter each technician meets with management to decide how many flat-rate hours he or she hopes to earn each week. The amount is based on the technician's past performance with input from the lead technician about the type of work the technician is capable of performing. In other words, a paper declaration is not enough since the flat-rate objective must be discussed and be based on a performance review.

Measurement of how close each technician is to his or her objective is provided to the team leader each morning by the service manager or consultant in a daily objective report. The report contains each technician's name as well as the flat-rate objective and the actual amount of time the technician earned the previous day. Figure 3-8 shows that technician 1 did not reach his or her target, while technicians 2 and 3 either reached or exceeded their target. This flat-rate report should be used as a reference when the lead technician assigns work, assuming that the leader is attempting to help each technician reach his or her objective.

The lead technician examines the flat-rate objectives relative to the morning report of the flat-rate hours earned to determine what kind of work to give each technician. If the technician is running behind in the number of hours earned, for example, because of loss of time on a difficult job, the lead technician may give the technician some easier work so he or she can make up for the time lost. When a technician is ahead of the flat-rate objective, he or she may be assigned work that is more difficult as long as the technician has the ability to do the job. More details about teams are covered in section 3 and 4 of this textbook.

Technician		Day 1	Day 2	Day 3	Day 4	Day 5	TOTAL
1	Daily flat-rate total	9.1	3.8	11	13	10.1	47
	Daily flat-rate target	10	10	10	10	10	50
	Ahead or behind	−0.9	−6.2	1	3	0.1	−3
2	Daily flat-rate total	8	8	8	8	8	40
	Daily flat-rate target	8	8	8	8	8	40
	Ahead or behind	0	0	0	0	0	**0**
3	Daily flat-rate total	10	10.3	9.4	9.8	9.9	49.4
	Daily flat-rate target	9.5	9.5	9.5	9.5	9.5	47.5
	Ahead or behind	0.5	0.8	−0.1	0.3	0.4	**1.9**

FIGURE 3-8 A daily labor report for flat-rate objectives.

Review Questions

Multiple Choice

1. _____ must be able to enter a facility with minimal effort and confusion.
   A. Customers
   B. Technicians
   C. Service consultants
   D. Managers

2. According to the textbook, the team's mission is to:
   A. work together to ensure customers are treated as important people and then satisfy them with the service and treatment desired
   B. work together to ensure the efficiency of the facility
   C. both A and B
   D. neither A nor B

3. A typical team has:
   A. a service consultant, technicians, and a manager
   B. a service consultant, a manager, and a parts specialist
   C. a service consultant, technicians, and a parts specialist
   D. a manager, a customer, and a parts specialist

4. At a service facility with a lead technician, the work is assigned to technicians by:
   A. the service consultant
   B. the lead technician
   C. the manager
   D. the customer

5. When technicians are paid by the job, it is called:
   A. the flat-rate system
   B. the hourly pay system
   C. the labor guide system
   D. the technician system

6. Technician A says that a flat-rate objective is the number of flat-rate hours a technician hopes to earn by the end of the week. Technician B says that he or she sets his or her own flat-rate objective and management does not have any input. Who is correct?
   A. A only
   B. B only
   C. both A and B
   D. neither A nor B

Short Answer Questions

1. Explain why and how a team approach to processing repair orders can offer superior customer service.
2. Explain the difference between a lateral support group and the team system.
3. How does a dispatcher system work and what different types of service facility operations might use it?
4. Identify the major responsibilities of the team leader.

Activity

Activity 1: Service Consultant Math Exercises: Calculate the customer labor sales charges based on a labor rate of $80 per hour for this technician.
   A. Diagnose engine misfire (1.0 hour)
   B. Road test and confirm noise (0.3 hour)
   C. Diagnose a no-start, no-crank concern (0.6 hour)
   D. Check computer codes and perform simple circuit checks (1.3 hours)
   E. Perform simple electrical circuit checks (0.5 hour)

1. What is the total amount of the labor sales in dollars?
2. If this work resulted in no additional sales (the job went from the second phase to the fourth phase without any additional work sold) and the technician's wages (cost of labor—refer to chapter 1 if necessary) are $20 per hour (paid hourly for an eight-hour day that is $160 for the day), how much gross profit (sales less cost of labor) did the service facility earn?
3. If the expenses (management wages, taxes, insurance, utilities, etc.) on every bay are $35 per hour, how much does a bay cost each day (a day is eight hours)?
4. Is the gross profit earned for the bay studied enough to cover the expenses? If not, how much is the loss?

Conclusion: With the math completed and the sales understood, discuss from a management standpoint what is the problem and how might this bay's sales be improved so a profit is earned?

Analysis: If this is one of four bays on a team, how much more must the other bays earn to make up for the loss (*hint:* take the loss and divide it by three bays, this is how much more each of the other bays must make to cover the loss)? Obviously this will bring down the team's average in terms of hours sold, as discussed in this chapter.

- Comment on this from a team member perspective and how one member could affect the team as a whole?
- Discuss in your answer if this single bay's problem is encountered by *all* of the team members because the service consultant perhaps cannot close additional work (third phase of the repair order tracking sheet), what will happen to the business? What must be done by management?
- If the problem is with the service system that causes a bottleneck, how might that change the answer to "what must be done by management"?

Activity 2: Calculate the "ahead or behind totals" for F through K. Which days was the technician behind? If the technician is behind due to issues beyond the technician's control (he or she was late for work or not working hard), what might a team leader do to help him or her meet his or her target?

Technician		Day 1	Day 2	Day 3	Day 4	Day 5	TOTAL
1	Daily flat-rate total	9	8	11	7	10	45
	Daily flat-rate target	10	10	10	10	10	50
	Ahead or behind	F	G	H	I	J	K

CHAPTER 4

COMPUTERIZED SERVICE SYSTEMS AND CUSTOMER RECORDKEEPING

OBJECTIVES

Upon reading this chapter, you should be able to:

- *Explain why an automobile service history is important (Task A.1.9).*
- *Define repeat repairs/comebacks (Task D.6).*
- *Describe how an automobile repair history is recorded and stored (Tasks A.1.2, A.1.4, and A.1.10).*
- *Outline the steps followed in a computerized invoice system to open a repair order before the customer comes to the shop (Task A.1.5).*
- *Demonstrate how to obtain and document pertinent automobile information and confirm its accuracy (Task A.1.2).*
- *Explain the importance of first-time, warranty, repeat repair, fleet, and regular customers at a service facility (Task A.1.13).*

Introduction

A person's history of treatment, whether it is with a family physician or at an automobile service facility, is very important to him or her. As a result, people expect professionals, such as a medical doctor, to maintain records of all of their illnesses and examinations as well as the treatment and its results. Likewise, when people are regular customers at an automobile service facility, they expect the service consultant to have records of past services performed on their vehicles. This is expected even if the service consultant is a new employee at the facility and has never met the customers. Currently the expectation has increased to include access to records anytime or anywhere. This means the service facility must be able to post the records to a password-secured website for online review as well as obtain vehicle information online from companies that can check auto histories. Therefore, maintaining a history of services performed on customer automobiles and informing customers about it is important to a service facility that wishes to present a professional image.

In addition, knowing the history of automobiles serviced at a facility is important to the facility itself. For example, at Renrag Auto Repair a customer called and reported that her automobile's cooling system was overheating. She was quite distressed because she had brought her automobile into the shop for the same problem four weeks earlier. The service consultant pulled her name and vehicle up on the computer, verified that the vehicle had been in for a cooling system repair, and asked her to bring her vehicle in right away.

Upon checking the customer's record in the computer, the service consultant learned that the water pump had been leaking and was replaced. Next the service consultant examined the comments on the invoice and the original comments on the technician's worksheet (also called the repair order (RO) or hardcopy). One recommendation recorded on the invoice by the consultant was to replace the thermostat. This was recommended because of the technician's experience that this part had a high failure rate and when stuck closed would cause overheating. In addition, the cost to replace the thermostat when the water pump was replaced was low compared to replacing it later. The written comments on the invoice indicated that the customer was advised of the recommendation but declined to have the thermostat replaced even though it might prevent a future breakdown.

When the customer brought her automobile into the service facility, the technician diagnosed the cooling system and found that the water pump was working properly. Further diagnosis found that the thermostat was broken and stuck closed. When the technician's diagnosis was presented to the customer, she was upset. The consultant then showed her the comments on the invoice at the time the water pump was replaced. She recalled the conversation and knew that she had made the wrong decision. Although she was not happy to pay for the repair because it would have cost her less money a few weeks earlier, she left with a positive attitude about the service facility, the competence of the technician who worked on her car, and the detailed records kept on her car.

There are several points to be recognized in the preceding example. To start, the service consultant had two records to check: the invoice stored in the computer database and the technician's hardcopy notes in the facility's filing cabinet. Without this history, Renrag would have had a customer relations problem and possibly would have had to pay for all or a percentage of the repair charges.

Repeat and First-Time Customers

The Repeat Customer

Service consultants work with different types of customers each day. The most important are the repeat customers and first-time customers who may become repeat customers. Repeat customers have all of the work performed on their automobile at the same facility. They represent a steady income. This means that over time, they spend a considerable amount of money at a facility, and, as a result, influence its sales volume and profit.

Repeat customers, however, are hard to recruit. Often a new service facility has to spend a lot of money for several years to establish a positive reputation in order to build a strong repeat customer base. If repeat customers are not treated properly, however, a service facility can lose them in a short period of time. Service consultants must also realize that repeat customers occasionally "shop around" for better service and lower prices, especially if they think the service facility is not "taking care of them." Therefore, service consultants must not take customers for granted, or they will likely lose them to the competition.

Therefore, one of the major jobs of service consultants is to recruit and retain repeat customers. This requires greeting repeat customers in a personal manner and identifying potential repeat customers when they enter the facility.

For example, assume that a repeat customer, Mrs. McCord, likes to have the tires rotated on her automobile after every other oil change. When Mrs. McCord arrives for an oil change, the service consultant should check her automobile's service history (see Figure 4-1) and determine if it should have the tires rotated. The service consultant should inform Mrs. McCord when her tires were last rotated and then should ask if she wishes to have this rotation service. This shows Mrs. McCord that a record is being maintained on her automobile by the service facility and that service consultant is showing interest to accommodate her preferences. She will most likely be pleased with the special treatment.

The identification of current maintenance needs requires a technician to examine the vehicle. A good inspection of automobiles that results in recommendations can help to retain a customer's business.

FIGURE 4-1 This owner shows how easy it is to access his customer list. In an instant he can click on a customer name and locate past invoices and estimates to provide suggestions so he can give his customers the best service experience possible.

To help identify potential concerns, a maintenance inspection form will help assure the technician check the vehicle thoroughly (see Figure 4-2). Owners, managers, service consultants, and technicians work as a team to identify maintenance inspection items to include on the list. What to include on the inspection list must be given some thought to what types of services the facility sells, the problems commonly found on customers vehicles (such as antifreeze problems), as well as how easy it is to inspect the items. The point is to provide a quality inspection in minimal time to "add value" to the customer's visit. Obviously, the service facility does not want to inspect parts that don't commonly fail or take excessive time to check, such as removal of the dash cluster to inspect burned-out light bulbs. The inspection process must be quick and help the customer better maintain their vehicle, perhaps leading to a service sale.

The First-Time Customer

When new customers are identified by the service consultant, the objective is to convince them that the service facility can help them take care of their automobile. One approach to demonstrating this willingness to provide assistance is for the service consultant to place a flag on the repair order to let the technician know that this is the customer's first visit to the facility.

During the customer's first visit to a facility, the technician should conduct a much more detailed inspection of the general condition of the automobile to identify anything that is wrong with it. This inspection often requires removal of tires to examine the brakes as well as removal of other parts as necessary. At Renrag Auto Repair, the technicians filled

Customer Name _____
Date _____
Vehicle Year _____ Make _____ Model_____
Mileage _____ Engine _____
State Inspection Sticker Date _____
General Exterior Inspection _____
_____

1. Exterior/Interior Lights _____
2. Windshield Wipers/Blades/Washers _____
3. Belts: OK   Cracked/Worn _____
4. Antifreeze:
       Acid Strip Test for pH: OK   Recommend Flush
       Temperature Strip Test: _____ degrees
       Color: OK   Recommend Flush
5. Radiator Hoses: OK   Soft/Cut _____
6. Battery: Date of Purchase _____
7. Transmission Fluid: OK   Topped Off   Burned Odor
8. Air Filter: OK   Recommend Change
9. Fluid Levels: OK   Low
10. Exhaust: OK   Leaks Noted
11. Struts/Shocks/Springs: OK   Leaks/Damages
12. Tires: Pressures at _____ lb
       LF: OK   Feathered   Worn to Wear Strips
       RF: OK   Feathered   Worn to Wear Strips
       LR: OK   Feathered   Worn to Wear Strips
       RR: OK   Feathered   Worn to Wear Strips
13. CV Boots: OK   Worn or Torn
14. Steering Boots: OK   Worn or Torn
15. Oil/Transmission/Power Steering/Brake Fluid Leaks:
       None Noted   Leaks Noted at _____
16. Rear Differential: OK   Leaks Noted
17. Oil: # of Quarts _____
Recommendations: _____
_____

FIGURE 4-2 Maintenance inspection form of basic maintenance items to be checked by the technician.

out the form used for a maintenance inspection of all automobiles first (see Figure 4-2), then went beyond those items to complete a more detailed safety inspection form similar to Pennsylvania State Safety Inspection Requirements (see Figure 4-3). This was appreciated by first-time as well as other customers, such as those who recently purchased a used car, who wanted to know more about their vehicles' condition.

When new customers are offered the inspection, the service consultant might offer them a complimentary service such as a tire rotation since the tires will be removed anyway. This offer shows that the service facility is interested in them and wants to help the customer maintain his or her automobile. After the inspection the service consultant should record the inspection results in the shop database created for the customer and advise the customer of any problems found. All repeat customers should be given a copy of the results of their inspections.

The Vehicle Service History

An automobile service history is a record of the diagnosis, repairs, and maintenance performed on a customer's automobile by the service facility. The records presenting the history are the estimates, repair orders, and invoices that are usually recorded in a computer database. Just as important, though, are the notes of the technician and service consultant on any hardcopies that should be kept on file. The rule to remember is that the three most important duties of a manager are to: (1) document, (2) document, and (3) document.

The sets of estimates, repair orders, and invoices as well as the written documents on customers' automobiles are kept in the **shop management system**. The shop management system consists of procedures to document, store, and retrieve customer information. Service consultants must have a thorough knowledge of the shop management system, whether the system is paper based, fully computerized, or a combination of the two. For example, for a paper system, service consultants must be able to file and retrieve RO hardcopies with handwritten notes and customer approvals (signed and taken over the phone). When the shop uses a computer system, then the computer will store all information in files such as the estimate, technician notes, RO, and final invoice. Some systems will also electronically store other related documents, such as parts receipts (electronic versions), and even repair pictures as well as forms, such as the inspection sheets. When a computerized system is used, the service consultant must know how to back up the records in the computer database on an alternative storage device at the close of each business day. If the computer system "crashes," the cost required to re-create the database is excessive!

Computerized Automobile Records

Customer service histories are often stored in a computer database. In some cases, these records are referred to as service records or service files. To create a record, the service consultant has to enter each customer's

Customer Name _____ Date _____

Vehicle Year _____ Make _____ Model _____

Mileage _____ Engine _____ PA Inspection Date _____

General exterior inspection (checkmark = OK; N = not OK):
- _____ 1. Lights & lenses   _____ 2. Turn signals
- _____ 3. Brake lights   _____ 4. Headlights (hi/low)
- _____ 5. Body (general)   _____ 6. Glass
- _____ 7. Wipers   _____ 8. Ext. mirrors
- _____ 9. Bumpers   _____ 10. Doors
- _____ 11. Gas cap   _____ 12. Hood
- _____ 13. Trunk   _____ 14. Shocks/struts
- _____ 15. Hubcaps/wheels

Under the hood (checkmark = OK; N = not OK):
- _____ 1. WW washer   _____ 2. Belts
- _____ 3. Hoses   _____ 4. Antifreeze Acid Strip Test for pH
- _____ 5. Antifreeze Temperature Strip Test _____ degrees
- _____ 6. Antifreeze color   _____ 7. Antifreeze level
- _____ 8. Air filter   _____ 9. Hood latch
- _____ 10. Master cyl.   _____ 11. Transmission fluid
- _____ 12. Firewall holes

Electrical (checkmark = OK; N = not OK):
- _____ 1. Battery terminals   _____ 2. Battery hold down
- _____ 3. Starter draw test   _____ 4. Battery load test
- _____ 5. Charging system test   _____ 6. Ignition system
- _____ 7. Computer code check

Underside (checkmark = OK; N = not OK):
- _____ 1. Shocks/struts   _____ 2. Springs
- _____ 3. Ball joints   _____ 4. Steering
- _____ 5. Boots-CV/steering   _____ 6. Floor/frame
- _____ 7. Powertrain mounts   _____ 8. Fuel system
- _____ 9. Exhaust
- _____ 10. Tires: RF _____ , LF _____ , RR _____ , LR _____
- _____ 11. Brakes: RF _____ , LF _____ , RR _____ , LR _____
- _____ 12. Rotors/drums RF _____ , LF _____ , RR _____ , LR _____
- _____ 13. Cylinders & calipers   _____ 14. Hoses & lines
- _____ 15. Oil leaks—powertrain   _____ 16. Coolant leaks
- _____ 17. Oil leaks—power steering

FIGURE 4-3 Safety inspection form of items to be checked by the technician.

FIGURE 4-4 This dealership service manager is entering the service information that includes the customer name, telephone numbers, address, and electronic communication information (email, etc.). He will save this before going to the next screen that will have him enter the vehicle information: year, make, model, VIN, license number, and mileage. The information must be accurate; otherwise warranties cannot be processed.

information into the computer database. These entries consist of the customer's name, contact information, and his or her automobile information as shown in Figure 4-4. Often this information must be entered into the database before an estimate or repair order can be generated so that work can be performed on a customer's automobile.

The Customer File and Its Creation

The service consultant begins a file for each customer in the database when the customer initially contacts the service facility. This often occurs when a customer calls to schedule an appointment (see Figure 4-5). The first-time customer's file can be completed (perhaps the VIN is needed) when the customer arrives for the appointment and the vehicle is checked in by the service consultant.

After the customer's information has been entered into the database and saved, many computer programs will permit the service consultant to schedule an appointment, produce an estimate, generate a repair order, and print a final invoice. However, it depends on the program because some will not permit progression into other functions without all of information completed. The reason is the service history files (past invoices and estimates) will be connected to the customer file. If it is incomplete or incorrect, program errors can occur that may cause the service consultant problems. For example, customers often have multiple vehicles and some may be similar years, makes, and models. It looks

FIGURE 4-5 When the customer calls the service facility, this owner (who serves as the service consultant) will look up the customer file (repeat customer) or start a new file (first-time customer).

bad when the service consultant confuses two of the customer's similar vehicles and records information into the wrong file.

Information for Management

Modern computer programs are really amazing because they have a lot of power to search on all automobiles in the database that are of a specific year, make, and model. The programs can indicate how much money a customer has spent at the service facility since his or her first visit. Data about sales averages for the day and technician sales for a time period specified is available at an instance. All of the data ever needed by a manager can be accessed by the program, but the trick is to know what is needed to make a decision. For example, a customer may have an issue with a service that wasn't the service facility's fault but it warrants "goodwill" consideration by the owner. The service consultant may obtain data that indicates the repeat customer has spent $10,000 with the business. As a result, the $75 goodwill gesture is an inexpensive cost to keep the longtime customer satisfied. Specifically, the goodwill gesture can go a long way in assuring future sales from this customer. In addition, the program can search the customer's automobiles by year, make, and model. This is useful when buying inventory or supplies. For example, some oil filters only fit certain automobile engines. If the facility's database shows that no customers have an automobile with a specific type of engine, it does not have to order any of the filters for that engine for the shop inventory. Of course, inventory orders of filters for engines in more popular models can be based on the number of automobiles in the database. The same goes for air and fuel filters, tires, special fluids, belts, batteries, and other supplies that the shop wants to keep on hand.

Service Systems and Operations

The service system is defined as an orderly arrangement of procedures that are linked together to form a process followed by employees and used by management to control shop production, customer service, and business operations. For this textbook, the service facility's three main operations are:

1. Customer service (first and fourth phases of the repair order tracking sheet)
2. Shop production (second and third phases of the repair order tracking sheet as well as technician information systems and equipment)
3. Business operations (banking deposits and payments, payroll, financial statements)

Each operation must have procedures employees must follow to make up what is known as a "system." For instance, a customer service system is a process to serve customers from the service consultant's initial contact with the customer to final invoice and extends to the customer recordkeeping. The customer service system procedures may overlap with the shop production system when it comes to:

- Parts acquisition (cost, markup, and invoicing)
- Customer authorization
- Use of the repair order by the technician that will become an invoice

Once the invoice has been created, the customer service system will overlap with the business operations systems as payment is processed, deposits are made, the sale is recorded to the financial statements, and flat-rate technician pay is calculated. This overlap of the operations such as customer service and shop production systems is seen in the repair order tracking sheet in the second and third phases as discussed in chapter 3 (refer to Figure 3-2) as well as steps 2 to 5 of the service process (refer to Figure 3-1).

Though the three operations are "separate" with perhaps separate personnel, tasks, policies, and procedures that make up the area's operations manual, the service systems employee duties may overlap (discussed in chapter 2). The use of a computer in an operation is merely a program that in conjunction with the system makes the system more efficient and provides data to managers and owners. This is the complexity of describing the computer program(s) used by each operation.

Computers and Service Facilities

Management's use of computer programs at service facilities has changed just like cars have evolved. This edition of the textbook provides three different levels of computer program integration at a service facility as discussed later. They are:

- Fully computerized (paperless) service systems
- Partially computerized service systems
- Non-computerized service systems

FIGURE 4-6 This car dealer needs to have information to run the business. All operations are interconnected with the same computer program that is integrated into each department's operations.

An owner may wish to have a computer perform all of the service systems functions in a single program that is integrated into each operation (fully computerized system, see Figure 4-6). The owner may elect to use one or more computer programs but they are to perform only selected functions in the service system but not others (partially computerized system). For some owners they may even choose not to have any service systems rely on a computer program. When they do, it is often for a single purpose, such as "stand-alone" bookkeeping or shop computers used to fix vehicles (non-computerized service system).

To illustrate the point, the business operations system will be examined. This system includes financial operations, recording income and expense for accounting operations (bookkeeping), payment of invoices, warranty administration, employee time keeping and payroll, and customer payments for service (cashier). At dealerships the business system may also include parts inventory, retail part customer payment, and wholesale parts credit sale accounts.

Furthermore, the business operation system will track new and used car inventory, purchase (trades), and sales, among other business functions. Given the number and complexity of the transactions, dealership owners often wish to integrate the business operations with all of the other operations at the dealership (service, parts, and vehicle sales). They use an advanced computer program that can support each operation and generate significant information for management. The names of program manufacturers include ADP and Reynolds and Reynolds, as well as other names such as Autosoft (ASI). The operation of any specific computer program is beyond the scope of this book and requires training, once employed in the field, to master the various functions of an employer's program. However, the concepts of the systems will be presented.

Shop Production Computers

Shop operations have the most computers and programs at a service facility (see Figures 4-7, 4-8). In some cases, they could be classified as "stand-alone" computers that may perform only one function, such as operating a chassis dynamometer or alignment rack. The shop computer program(s) may or may not be networked to other computers or connected to the Internet. There are also shop computer programs that are merely programs loaded on a PC, such as a high-performance engine parts selection program that estimates power output. Examples of computerized shop production programs are technician repair information, technical websites and related apps, state inspection that reports information to an outside agency, shop equipment that is computerized, scan tools or laptops that interface with vehicle computers, and small stand-alone computers such as oil change label makers. For discussion purposes, computer integration in the customer service operation does not include the shop computer programs or "technical" programs, apps, databases, or websites.

Typically the shop computer programs the owner chooses to purchase depends on the type of work the service facility performs. It is usually driven by the need for a computerized program, apps, database, website, or interface to help the technician get the work done. The computer program(s) an owner chooses to integrate into the service facility's

FIGURE 4-7 A technician uses a computer that is designed to access repair information from the vehicle manufacturer's network. He downloads information onto a scan tool that will reprogram a vehicle powertrain control module (PCM). In the background is the computer used for emission inspection; it is connected to the Department of Transportation database.

FIGURE 4-8 This technician's laptop is wireless and is networked to check for service information. The computer can also be used to interface with a vehicle's computers and can download information to reprogram the vehicle's computers.

customer service and business systems is based on what the owner believes the business needs to operate efficiently.

In choosing a computer program(s) the owner should examine the cost of the computer equipment and related software needed to make the program work properly. The owner should also consider the amount of employee time and training it will require to make the program function as intended with the existing service system. This means there needs to be a customer service and business system in place that works before the computer program, as a tool, is integrated. Therefore, employees need to know the procedures that make the customer service system work in a variety of situations before management plugs in a computer program and hits "enter."

Consequently, the computer program that an owner chooses to use at a service facility will not automatically create a better quality of service for the customer. When service is chaotic to start with and then the owner adds a new computer program to the service system, there can be frustration for everyone and new problems can be created. For service that is effective to begin with, the computer program(s) can help make the service system easier to process the customer's service request so more customers can be served. It also allows faster access to more information for various end users, such as managers and owners among others who are granted access. A "manageable" computer program can take an effective service system and make it more efficient for customers and employees. However, if the service system is not effective to begin with, the computer program as a tool will not necessarily help.

For illustration purposes, three types of businesses are examined to explain how a computer program is integrated into customer service system. Therefore, from chapter 2 one way to understand the duties of a service consultant is to know how much of the customer service system is dependent on a computer program. It should be noted that computer programs neither serve customers nor fix cars; only people do and as a result a shop without a computerized system is not necessarily any better or worse than a shop with a fully integrated and networked computer program. The bottom line is the customer wants his or her car fixed on time, properly the first time, and with everything requested completed. The customer also expects courteous service with a final bill that does not have an error and is within what was quoted on the estimate. Money is the way the customer says "thank you for the good service" and the whole process relies on people while a computer program is a tool that only helps the process.

Non-Computerized Customer Service System

Service facilities without a computerized service system are like cars without any computer controls, such as engines with a carburetor and point-type ignition. The shop does not use a computer program in the customer service system, maybe not even in the business system, but does own shop computers to help with the repair of vehicles. The customer service system processes the customer work order through the use of paper forms from write-up to ordering parts and the generating of the final invoice. This means customer and other service records are also on paper. Technicians who start a business "on the side" usually have a non-computerized service facility because paper systems are inexpensive and are easy to use for a low volume of customers served. Longtime established businesses typically with few employees may continue to use a non-computerized customer service system because there is not a need to improve it. The paper system is adequate to process the number of customers served each day.

Partially Computerized Customer Service System

The partially computerized customer service system uses a "paper system" for portions of the system and a computer program for others. For example, the repair orders given to technicians may be paper, while the invoice for the customer is from a computer program. Commonly, paper documents, with handwritten notes on them, are found on desks and clipboards. Therefore, as employees carry out customer service functions, the transfer of information between employees is at least in part by paper. At some point, someone is typing in handwritten information to enter into the computer program so the customer service system can process the customer, such as creating the final customer invoice.

Most service facilities are at least partially computerized service systems. How much of the service system is computerized depends on the owner and the service facilities needs. For example, if technicians cannot type, repair orders need to be printed out for them to write down information for the service consultant. A service facility with a partially computerized customer service system may not, depending on the owner's desires, be integrated with other operations' computers, such as the business operations accounting program.

Fully Computerized (Paperless) Service System

These service facilities are like a vehicle that operates with many computers multiplexed together so each can share information with one another through a network. Vehicle computers share information to reduce redundant sensors and relay needed information for optimal vehicle operation and control. In the same way a fully computerized service system sends information directly to the shop production and the business operations computer programs. The entire service facility is integrated (see Figure 4-9).

The customer service system prepares computer schedules, creates estimates, looks up part prices and labor times, generates ROs, finalizes invoices, and saves customer files. Relevant information is sent to other operations' computers, such as warranty repairs to be processed. In addition, these programs can send electronic communications directly to customers, such as surveys, repair status, and special offers and can allow a customer to even access their vehicle records through the Internet on some systems. The network of computers can also allow a customer to automatically schedule an appointment through the web instead of calling the service facility. Needless to say, owners and managers can access and monitor a service facility status from any location.

Facilities can choose many computer programs available today to have a fully computerized system. However, some service facilities, as discussed previously, choose the functions they wish to use, such as estimating and invoicing. When this happens and the system integrates

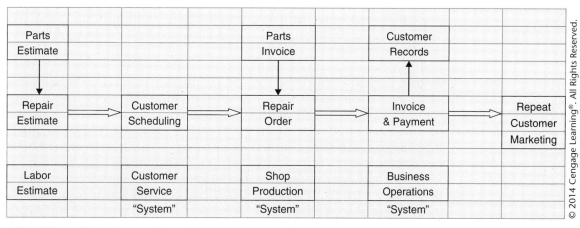

FIGURE 4-9 This diagram shows the customer service operations' major functions and the relationship to shop production and business operations. When the entire system has "paperless transfer of information" and all users can access information, the system is likely a fully computerized service system.

ADDITIONAL DISCUSSION

Paper Files vs. Computer Files

Commonly the partially computerized customer service system assigns a specific task, such as estimate creation and final computerized invoice to the computer program. The reason may be to improve efficiency or just to make the final invoice appear more professional than a handwritten copy. For the scheduling of repair orders, the technician may use paper. The RO may even be printed from the computer for the technician to write notes on it. Then the RO with the technician's notes are typed into the computer by the service consultant and saved to the customer's file.

While the customer file and information may be saved in the computer, there may still be paper records that are filed: such as parts invoices, documents related to the repair, and warranty paperwork. While some partially computerized customer service facilities scan these documents into the computer file or share them by email or other database access, others file the paper copy in a folder. To identify the service facility as a partially computerized service system is not about how much paper is used but rather how the computer program is integrated into the customer service system. It is observable when paper is used to transfer information between end users, particularly to other operations computers, such as the accounting program.

"paper" into the process, the system becomes a partially computerized service system. The commonly used computer programs at dealers were reviewed previously. The fully computerized systems at independent service facilities include Mitchell, Shop Key, and Alldata, among others.

Data Needed to Make Management Decisions

Fully computerized service systems can be wonderful but the devil is in the details. Some work very well and are easy to use to enhance an already effective customer service system. Other programs can be difficult to navigate and require significant training to use properly. While all have a capability to collect data, some give management the specific information that they desire. However, some service managers and owners extract data using the program and then put the information into their own spreadsheet or another company's program to get information that fits their needs. The information available from the fully computerized service program is extensive, but it is useful for management decisions only if it can be analyzed. With too much information, a manager does not know on what data to focus his or her attention. At a busy service facility, there is little time to study so an owner may hire an outside consultant for assistance. In these cases, this book's authors, serving as consultants, have examined service facility operations, collected needed data, applied statistics and other modeling techniques, and discussed the results with owners to help them formulate solution options.

The bottom line is that all the information available is only as good as the ability to use it to make a decision. A manager needs

ADDITIONAL DISCUSSION

Dealership Computer Integration with the Manufacturer

These fully computerized systems are also integrated and networked with all or part of the information contained on a computer network that can be accessed by the vehicle manufacturer and its regional office. This may include data from the dealer's vehicle inventory to warranty information. Also headquarter offices (franchises, chain operations, multiple dealerships owned by the same owner), and other service facilities on the same network can share and update information automatically. Given the two-way nature of the network, dealers can not only submit warranty claims to the manufacturer but they can also order new vehicles, parts, and process requests for technical assistance among other functions.

information to foresee a problem that is arising from the service system and to make plans. But the danger of too much data is referred to as an information overload. Therefore, the final chapters in this book and the author's next book in the series (*Managing Automotive Businesses, Strategic Planning, Personnel, and Finance*) concentrates on management and ownership that may help to shed additional light on this important issue.

Preparation of Computerized Estimates, Repair Orders, and Invoices

To comprehend how the computer program is used to create documents for the service consultant and technicians, an overview of the process is illustrated Figures 4-10 through 4-15. The captions explain what is occurring as the service consultant collects information from the customer (as discussed further in chapter 6) and enters it into the computer program to create the estimate that will become the repair order, and finally the invoice that customer will pay.

After the necessary information is entered into the computer database, an estimate for the cost of a repair or maintenance is prepared, as discussed in the previous chapters and earlier. Most state consumer protection laws require the estimate to include the parts and labor costs plus any additional fees, such as towing and tax. Most computer programs can make the necessary calculations for the cost of the parts, labor, additional supplies, other charges, and taxes. The computer then prints the estimate for the customer's review. Typically, an estimate by a service facility is valid for 90 days, although it may be more or less depending on the parts needed and state law.

After the estimate has been prepared, it can be saved to the customer file or printed for the customer to review. It will not become a repair order until the customer approves of the work to be performed on his or her automobile. Depending on the service system procedure, the estimate may be printed and if it has the information required, such as what

FIGURE 4-10 A Service System that creates an estimate begins when the customer calls or comes into the service facility. The service consultant must either open the customer's file (repeat customer) or create a new customer file (first time customer) and save it to the data base.

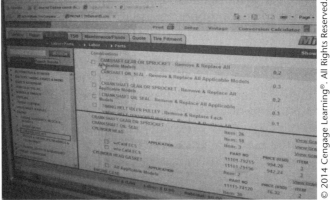

FIGURE 4-11 This screen allows the service consultant to estimate the labor charge by selecting the labor operation and then selecting the parts. The parts selection in many cases provides a part number and a list price (dealer).

was approved with the proper legal disclosure (which can vary by state requirements), it is signed by the customer and becomes a repair order.

If the customer is not available at the facility and approves the estimate over the phone, the service consultant must write the name of the person (typically the owner of the automobile) approving the service on the repair order, what was approved and not approved, and the date and the time of the approval, and must sign or initial it. In both cases, the approved repair order becomes a binding contract between the customer and the service facility.

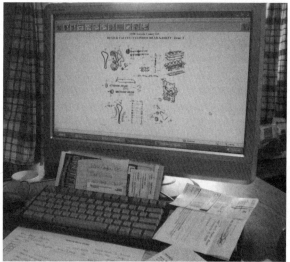

FIGURE 4-12 This screen helps the service consultant identify the parts by showing a diagram of the engine components. This can speed up the estimating process by picking out the parts needed for the repair.

FIGURE 4-13 The service consultant can turn the estimate into a repair order with the click of the mouse after it has been authorized by the customer. It is transferred to the shop either as a paper RO (partially computerized customer service system) or electronic RO where a technician will receive the RO on a shop computer near his or her bay.

After the work is completed, a third document is prepared by the computer. This document is an invoice that lists all of the work performed, the parts used, any other charges for the service, the tax to be paid, and the total amount to be paid by the customer. Depending on

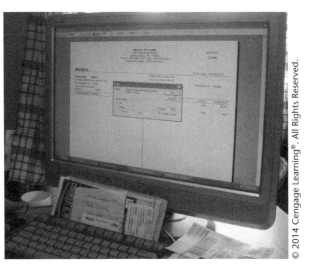

FIGURE 4-14 After the RO has been completed by the technician, an invoice that contains the final charges as well as any service comments or suggestions is created by the service consultant. Next the invoice will be printed out.

FIGURE 4-15 The final invoice will be printed, and signed by the customer acknowledging the repair process and that the repair charge is correct. Payment will be received from the customer and the keys are returned to the customer. When possible, active delivery techniques should be used (covered in chapter 10).

state law, the final invoice may need to be equal to or less than the estimate. If the invoice amount exceeds the amount approved on the work order, the customer must have agreed to the additional charges in writing or in a phone conversation that is documented by the service consultant.

A good customer relations practice is for the final invoice for a repair to be less than what the customer expects to pay. Experienced service consultants know that when the final invoice charge is more than what a customer expects to pay, he or she is likely to become upset and dissatisfied with the services performed regardless of the quality of the work.

Dealership Satisfaction, Warranties, and Computers

Automobile manufacturers and dealers use computer programs extensively to process warranty and monitor operations. They need to make sure that the people who buy new cars from them continue to purchase new automobiles in the future, so customer satisfaction with the repairs and customer service is extremely important. Therefore, dealership service departments perform a large number of manufacturer warranty repairs and free maintenance work, such as oil changes, for customers who buy new automobiles from them. To ensure that their customers are satisfied, the automobile manufacturers survey their new customers about the services they receive (covered in greater detail in chapter 12). If the survey results are not positive, the sale of new automobiles can be affected. The consequences of negative results on a survey at a dealership service department can range from the restructuring and retraining of personnel in the service department to taking disciplinary actions, such as the termination of an employee.

Another reason special attention must be given to warranty customers is because state laws, which are referred to as lemon laws and discussed further in chapter 5, set limits on the services needed by a buyer. Specifically, these laws set a maximum number of visits a person who purchases a new automobile has to make to the dealership service department for the same repair service and/or the number of days a customer's automobile can be out of service before the dealer or manufacturer must buy it back. The service consultant must help the dealer and automobile manufacturer avoid any costly and troublesome buybacks of new automobiles. This is done by closely monitoring the computer files for customer invoices and the number of visits for the same warranty repair and the number of days a customer's automobile is out of service.

Comebacks

Repeat repairs are called **comebacks** or **second attempts**, meaning a customer had a repair performed at the service facility and must return because the same repair must be made again on the automobile. The reason for the return is typically unknown, and just as warranty repair customers are extremely important to a service consultant, so too are repeat repair customers. When a customer returns an automobile because the repair is thought to be unsatisfactory, the service consultant must express an interest in examining and correcting the problem. The customer should be advised that after a thorough diagnosis by a technician, an action plan will be prepared. The service consultant should never propose that a charge for the repair will or will not be made until the automobile is examined.

The key to handling comebacks is for the service consultant to examine the customer's (computer) file and obtain the information about past services conducted on the automobile. By retrieving the past records of an automobile, the consultant can identify when the service was performed and the number of miles that the automobile has been driven since the service was performed. Date and mileage are important because they usually determine whether the customer will be charged for the repair. Next, the service consultant must review the notes and comments about the repair from the computer or hardcopies in the file as well as any related documents such as subcontractor or parts receipts.

Repeat Repair Policy and Consumer Protection Law

There is no standard number of miles traveled or days that have lapsed after the repair was made to determine whether a customer will be charged for a repeat repair. Often the determining factor is the service facility's policy, an automobile manufacturer warranty policy (discussed further in chapter 5), a state consumer protection law, or the parts warranty. Service consultants must be aware that the attorney general in their state must enforce the consumer protection laws assuring customers that they are protected from unfair treatment when their automobile is serviced.

Therefore, the consumer protection laws have guidelines that set forth the conditions related to automobile repairs. To understand the law and the position of the Office of the Attorney General, service consultants must obtain a copy of the consumer protection laws from it or from its website. Upon reading the consumer protection laws, service consultants should work with their managers and facility owners to ensure that their service facility's policies, procedures, advertising, and promotional campaigns do not violate the law. For example, if a service facility states that a customer will receive a "money back guarantee if not satisfied," then the state consumer law will likely define what that means. In Pennsylvania, when the customers are not satisfied, their money must be returned within five days.

Parts Warranty

When a comeback is caused by a defective part, many parts suppliers have a warranty that covers the cost of the new part. In some cases, this warranty may cover the labor cost to replace the part. The parts warranty should state how long and under what conditions a repair will be made free of charge. Therefore, service consultants must know the terms of all of the warranties issued by their service facility and be able to access parts receipts when necessary. Failure to obtain parts receipts to determine the vendor, part number, dates, and any other information may cause delays in processing a warranty claim for a customer.

For example, assume a parts supplier offers a one-year warranty on all parts and labor on remanufactured alternators. This means that neither the service facility nor the customer should have to pay anything if the alternator needs to be replaced within the warranty time period. However, in some cases a warranty may cover only a percentage of the costs (such as a battery warranty) after a period of time has passed or the car has been driven a specified number of miles. In this case, if the service facility has to charge the customer for the amount not covered by the warranty, then he or she must be advised of the difference. If not, the service facility must pay the difference.

Working with Fleets as a Service Contractor

Fleet customers own a few to several dozen automobiles that need regular service, as discussed in chapter 1. A business, such as a construction company, or a government agency, such as a police department, may own a fleet of automobiles serviced by a contracted service facility. The fleet is similar to repeat customers because it provides a considerable percentage of the sales made by a service facility. When providing fleet maintenance services, it is common to have prearranged agreements about when the work is to be performed and detailed files often on a computer. For instance, in many cases certain maintenance tasks are to be automatically performed when an automobile is at or beyond a certain mileage. If the facility also repairs the fleet automobiles, the usual expectation is that their vehicles will receive priority service.

Unlike other customer automobiles, fleet automobiles that need work are either dropped off by a fleet employee, who may or may not be the regular driver of the automobile, or picked up by an employee of the service facility. As a result, when the maintenance or repair of a fleet automobile is needed, the service consultant may not have the opportunity to talk to the driver of the automobile to obtain information about the problem or the performance of the vehicle. Rather, this information may come from the fleet manager or through some other means, such as a note left in the automobile, email, or text message.

Although fleet customers may be desired by a service facility, care must be taken in the service because many fleets require service discounts on the final computerized invoice and special treatment, such as "on-demand" and evening service. They also often take 30 days to pay their invoices and therefore the business system may need to be altered in addition to the service system and the shop production system. This may not be a problem for larger service facilities that have extended hours and extra technician flat-rate hours to sell, and an accounts receivable already established in the business system. However, for smaller service facilities, fleets can cause cash flow problems and operational headaches.

At Renrag Auto Repair, for example, one of the fleet service customers that had over 15 vehicles received a discount on all services and often had to have work conducted in the evening or on weekends. To meet this service schedule, employees had to be paid overtime, assuming they would be available to work, and additional part-time employees had to be hired to shuttle the vehicles to and from the location where the fleet was parked. As a result, even though the fleet provided a lot of business for the facility, the discount demands, overtime pay, and additional costs became too expensive and a loss on the service was incurred.

JOB PROFILE Customer Helpline and Technical Hotlines

A customer helpline is a question-and-answer center for new vehicle owners. Often new owners have a question about their vehicles or want to know more about their warranty coverage. Sometimes they need to express a concern about a repair and whether it should have been covered by the manufacturer's warranty. When the center cannot resolve the problem, the customer helpline representative will create a report that will help other departments assist the customer. The customer helpline is often staffed by professionals who are "good with people" and are usually well trained in company policy and warranty contract details. However, they may not have extensive automotive service backgrounds.

A factory technical hotline is a question-and-answer center for dealership technicians. The center is staffed by automotive repair experts who often have extensive factory training and ASE certifications. They have the same certifications and training as technicians at the dealership and often have worked as technicians. In addition to having technical training and work experience, the hotline employees hold formal degrees such as an AAS degree in automotive technology and sometimes a BS degree. Dealership technicians call the hotline when they have a tough problem and need direction. The technical hotline employees sometimes provide simple solutions overlooked by the technician or use their many resources to provide a more detailed answer that is not readily found in a repair or diagnosis procedure. The resources available to technical hotline professionals include manufacturer databases, advanced diagnostic information that is beyond what appears in the dealership materials, engineering department assistance that includes vehicle design information as well as other technical information. The technical hotline employees also have access to many new vehicles and special equipment that allows them to test vehicles so they can analyze how systems work and better answer the technician's questions.

Dealership technicians have access to factory hotlines but there are also other hotlines that are available to all technicians. These hotlines or networks, some web-based, allow technicians to get information beyond what is available in the vehicle service manual. Often the hotlines have databases of information that can provide technicians with confirmation on a specification or diagnosis as well as special "tricks" and even frequency of similar problems because of the large number of technical calls they track. The professionals who staff these positions are often former technicians with training and certifications, such as ASE. They often have experience working on vehicles, and while they are not in the shop "looking at the problem," they are the next best thing with the information and advice.

Some hotlines charge money, while others come at no charge when other products or business services are purchased by the service facility. A few "hotlines," such as the iATN.net network, are actually networks where technicians who share their experiences about problems and cures can help each other fix vehicles faster and more accurately. These networks require typing in the vehicle information and searching for others who have had similar problems. What is hopefully found is a list of others who had similar problems and can share a diagnosis hint as well as the repair that fixed the problem. The network functions only when everyone participates and updates the information so it is accurate and useful. Therefore, the network often has rules that it asks participants to follow.

In addition, when fleet customers have to have 30 days or more credit, problems can be created depending on the size of the facility and the fleet, and the amount of money the facility must spend before getting paid. For example, most employees are paid weekly and other charges, such as the parts bills, are paid at the end of the month, although some parts must be paid when they are delivered. This means

JOB PROFILE: Repair Authorization or Fleet Hotline Representative

A repair authorization hotline or fleet hotline is a call center that fleets will hire to process repair requests for their vehicles. Specifically the company dispatcher or fleet manager does not directly provide authorization for a service facility to do repairs to the fleet vehicles. Rather a service facility service consultant calls the repair authorization hotline. The hotline records the problem, examines the vehicle history, determines that the repair request is reasonable, and authorizes the repair at the price requested. If the price or repair does not seem reasonable or the diagnostic procedure seems flawed, they will handle the matter for the fleet. This way a fleet manager does not need a technical background to handle fleet repairs nor does a fleet have to maintain a fleet garage with technicians.

The fleet can have vehicles all over the country and be able to maintain and repair them at any service facility. Often the repair authorization hotline is staffed by automotive repair experts who have extensive training and often ASE certifications. Many have worked as a technician themselves.

In some cases these hotlines will arrange for emergency services such as towing and even set up for a service facility appointment at the closest facility to accept the vehicle for immediate diagnosis and repair. A few hotlines have GPS tracking to tell where the vehicle is located and have the ability to get trouble codes to assess the danger in continuing to operate the fleet vehicle. Depending on the technical nature of the hotline, it may be staffed with trained technical personnel.

the facility must be able to finance the maintenance and repair of the fleet vehicles. This may cause a cash flow problem because the facility may have to wait up to 60 days to receive a payment from the fleet for a service. For example, if the fleet has a service at the beginning of the month, the invoice will not be sent until the end of the month (30 days) and then the fleet may take another 30 days to pay. As a result, when making a contract with fleet managers, service consultants should consult with the facility owner or manager before an agreement is reached.

Review Questions

Multiple Choice

1. Service consultant A says that when writing up a second attempt (comeback) or warranty repair order, it is necessary to review any previous repair orders. Service consultant B says that when writing up a second attempt (comeback) or warranty ticket, it is necessary to ask the customer to state the symptoms he or she is experiencing. Who is correct?
   A. A only
   B. B only
   C. Both A and B
   D. Neither A nor B

2. A vehicle in the shop for an oil change shows approximately 59,000 miles on the odometer. What should the service consultant do?
   A. Suggest an appointment for a 60,000-mile maintenance service.
   B. Offer the customer a discount to perform the 60,000-mile service.
   C. Advise that the 60,000-mile service is covered by the manufacturer's warranty.
   D. Provide a ballpark estimate for a 60,000-mile maintenance service.

3. Each of these represents an example of customer information that might be included on a repair order EXCEPT:
   A. an email address
   B. the customer's Social Security number
   C. a cell phone number
   D. the service consultant's name
4. The technician notes on the repair order that the fuel filter appears to be the original one on a vehicle with almost 60,000 miles on it. The item calls for replacement at 30,000 miles. Which of these should the service consultant do?
   A. Estimate a maintenance tune-up, including the fuel filter.
   B. Ask the customer when and if it was replaced.
   C. Tell the customer that it has not been replaced in 60,000 miles.
   D. Leave the item for the 60,000-mile service.
5. A customer calls and states that his or her vehicle has a problem that has had several repair attempts. Which of the following should the service consultant do?
   A. Determine if the dealership or shop has ever worked on the vehicle.
   B. Offer to take the vehicle in immediately.
   C. Ask the customer to provide previous work orders.
   D. Explain that sometimes a problem can take several attempts to resolve.

Short Answer Questions

1. Explain why automobile service history is important to both the service facility and the customer.
2. What are repeat repairs/comebacks and how are they handled by the service consultant? (Task D.6).
3. How is an automobile repair history recorded and stored in a computer program?
4. How are first-time, warranty, repeat repair (comeback), fleet, and repeat customers different?

Activity

Activity 1: Use Google to research the following dealership software companies: ADP, Reynolds and Reynolds, and Autosoft (ASI). Describe the company history and summarize the software products offered. From the information on the website, describe the software features, its basic function, and how it can help the dealership management run the business. Create a report for a fictitious owner.

Activity 2: Use Google to research the following service facility software companies: Mitchell, Shop Key, and Alldata. Describe the customer service and management software, its function, and its benefit to an independent service facility owner. Examine at least one other software program found on line and compare it to the information found. Create a report to an independent service facility owner.

Activity 3: Work with your librarian as well as research your state's website for information on the state's consumer protection law. Find any sections that are related to the automotive industry trade practices and read them carefully. Summarize the main points a service manager must know.

CHAPTER 5

WORKING WITH WARRANTIES, SERVICE CONTRACTS, SERVICE BULLETINS, AND CAMPAIGNS/RECALLS

OBJECTIVES

Upon reading this chapter, you should be able to:

- *Locate and use reference information for warranties, maintenance contracts, and campaign recalls (Task B.6.2).*
- *Define warranty policies and procedures/parameters (Task B.6.1).*
- *Verify the applicability of warranties, maintenance contracts, technical service bulletins, and campaigns/recalls (Task B.6.4).*
- *Explain warranty, maintenance contract, technical service bulletin, and campaign/recall procedures to customers (Task B.6.3).*

> **CAREER FOCUS**
>
> Expanding your career starts by understanding the multitude of warranties and this makes you an irreplaceable professional. There are many different types of warranty contracts and each has its own procedures. Because you know how to navigate them, you know how to get paid! From obtaining approvals to processing, you know it all and can write it without making a mistake. It isn't exciting because it takes time to read the manuals and ask the right questions, but in the end you know the warranty guide better than the people who wrote it. That means other employees, managers, owners, and even customers come to you for answers. You have the information and "know-how" others need!

Introduction

Not all customers at an automobile service facility personally pay for the maintenance or repair of their automobiles; sometimes, payment is covered by a warranty, a maintenance contract, or a recall/campaign repair order. When working with these customers, service consultants are required to know (1) what each type of contract represents, (2) why manufacturers use them, and (3) how to work with them.

Although warranties, maintenance contracts, and recall/campaign repair orders may not make up a large volume of the work conducted at an independent auto service facility, dealership service departments do a considerable amount of this type of work for automobile manufacturers. In both facilities, therefore, the service consultant must know how to take care of the customers' contracts, or they could lose regular customers and the reputation of the facility could be harmed.

Specifically, service consultants must be knowledgeable about the various types of contracts because payments depend on whether the automobile is properly serviced according to the guidelines stated in the agreement. Therefore, the purpose of this chapter is to define, describe the purpose of, and explain the procedures related to the use of warranties, maintenance contracts, and recall/campaign repair orders.

Definitions and References for Warranties, Maintenance Contracts, and Campaigns/Recalls

A **warranty contract** functions much like an insurance policy. A claim is made to a warranty company that will pay for the vehicle repair if the terms outlined within the warranty contract are met. In simple terms, customers buy a warranty contract because they are betting that a covered component may break before the warranty coverage period expires, and the warranty company is betting that the covered component will not break.

A **maintenance contract** (also referred to as a service contract) is often bought by (or in some cases given to) a customer when an automobile is purchased. The contract pays for maintenance services the automobile needs for a specified coverage period. This is like prepaying for the maintenances an automobile will need in the future. When used regularly at the mileage intervals specified in the contract, customers often save money and are assured that their automobiles stay in top running condition.

Finally, **recall/campaign** repairs are often the result of a **National Highway Traffic Safety Administration (NHTSA)** or an **Insurance Institute for Highway Safety (IIHS)** investigation that found a problem with a certain year, make, and model of an automobile. The manufacturer is required under penalty of law to repair the problem for the consumer free of charge.

Eligibility

To determine whether an automobile is eligible for warranty or maintenance contract work, the service consultant should ask customers if they own one. If they do, the service consultant should ask for it.

In some cases, a phone number for contract assistance is available, particularly if the contract was purchased from a source other than a new automobile dealership. When the contract is purchased from a new automobile dealership, it is common for its service consultant to be able to access the information from the automobile manufacturer's database. This is because most new automobile dealerships have a computer that is linked to the manufacturer. Unfortunately, a service consultant at an independent service facility may not have access to this same information; in this case the customer has to contact the dealership or manufacturer's hotline for specific details about the contract.

Recall/campaign information for a specific automobile can be obtained through the manufacturer's database by dealership service consultants and technicians. Automobile owners can sometimes obtain information about their automobile by calling the manufacturer's customer hotline. In some cases, owners who are in the manufacturer's database will be notified by e-mail about a recall on their automobile. Service consultants at independent service facilities can often obtain recall information from a:

- Computerized database (Mitchell-On-Demand and Snap-on's Shop Key system, among others)
- Publication (Automotive News or the Automotive Service Excellence publication *Tech News*).
- Website (the NHTSA at http://www.nhtsa.dot.gov or the Insurance Institute website at http://www.highwaysafety.org).

In addition to recalls/campaigns, automobile manufacturers release repairs for other concerns. Instructions about how to repair these concerns are described in a manufacturer's publication called a **technical service bulletin (TSB)**. Manufacturers release TSBs regularly for dealership service department technicians. Many computer data systems purchased by independent service facilities can provide TSB information to technicians at independent service facilities as well.

Warranty Policies and Procedures

There are four types of automobile warranties that service facilities and the service consultant should recognize and administer. The categories are:

- New automobile warranty contract
- Bumper-to-bumper warranties
- Extended warranty contracts
- Emission warranties

New Automobile Warranty Contracts

A **new automobile manufacturer warranty** contract typically applies to new automobiles sold to the consumer. The warranty provides for the repair of the automobile "free of charge" (in some cases there is a deductible discussed later) provided the automobile is within a predetermined time frame and under a predetermined mileage. This time period is called the **warranty coverage period.** The warranty coverage period may be 12 months from the time of purchase, but it can be as long as several years and 100,000 miles (typically with certain conditions) for some manufacturers.

Customers with automobiles covered by a manufacturer's warranty must take their vehicles to a service facility that is authorized by the manufacturer. This is typically the dealership service department where they purchased their automobile. This does not mean, however, that they must use the same dealership where they bought their automobile. In addition, if a dealer does not have a certified technician who can work on the problem, they will arrange for another dealer to assist.

The predetermined mileage set forth in a new car warranty, for example, may be 12,000 miles or higher, depending on the manufacturer and the state's laws. Sometimes a customer buys a "factory authorized" preowned automobile from a new automobile dealer and the balance of the predetermined mileage may be passed on to the new owner.

Customer Accidents, Modifications, and Abuse

When a customer's automobile is covered by a new automobile warranty, the diagnostic charges, repair charges, and parts are provided free of charge. Sometimes the customers are provided a loaner or rental vehicle. The rules are laid out in the manufacturer's policy and procedure manual. This is assuming the automobile is within the warranty coverage period and the repair is not a result of customer abuse or an accident. For example, one customer drove his automobile at high speed through a flooded street and got water into the engine through the air intake. The connecting rods inside the engine were bent because they could not compress the water. The owner's misuse of his automobile caused the damage. Therefore, the manufacturer did not cover the cost of the repairs, even though the automobile was within the warranty coverage period for a normally covered repair. This would become an insurance matter between the customer and his or her insurance company.

In cases that involve modification, the customer cannot install components that cause a warranty issue. For example, if a truck lift kit causes wheel alignment issues, the repairs are not covered. Another example is stereos and other electronic devices that are incorrectly installed and cause a warranty issue. The repairs or replacements will not be covered. A more recent issue has been powertrain control module (PCM) programmers. Typically these are not an issue unless they cause a warranty concern. For example, a new sports car PCM was flashed and the new parameters allowed the engine to reach higher RPMs than the factory setting. The customer raced the vehicle and the higher speeds allowed the engine to be damaged. The programmer allowed the factory settings to be reinstalled before going to the dealer for a warranty claim. Unfortunately even though the factory settings were in the PCM at the time of dealer diagnosis, a mismatch of computer programming codes left a "finger print" that could be traced. Since the mismatch could not have occurred when the vehicle left the factory, the warranty engine claim was denied. The process used to diagnose and document the issue is relatively simple, but the consequences can leave a customer angry.

Valid Warranty

For a warranty to be valid, the automobile must be under both the mileage and the time interval limits. For example, if an automobile owner with a 12-month, 12,000-mile warranty has a covered repair but the automobile has run over 12,000 miles or he or she has owned it for more than 12 months, the warranty will not pay for the repair. However, in practice, if an automobile is just slightly outside the warranty coverage period limits, many manufacturers often cover some, if not all, of the repair because they know that each customer represents many thousands of dollars in repeat business.

Therefore, when a situation dictates, manufacturer warranties often have procedures a service consultant can follow to request a warranty coverage period waiver for "special circumstances" or for a "goodwill" request. In some cases the manufacturer's warranty policy and procedures may require the customer to pay for the repair until the claim can be reviewed by the factory representative. The claim will either be turned over to the factory representative when he or she visits the dealership or be sent to the manufacturer's warranty audit department for review. If approved, and the customer paid for the repair, he or she will be reimbursed for the payment of the repair. However, the actual process will vary by manufacturer.

Bumper-to-Bumper Warranty

When a manufacturer's warranty covers all of the systems and parts on the automobile, it is referred to as a **"bumper-to-bumper warranty."** This means all of the automobile components are covered for a specific period, such as the first 12 months or 12,000 miles. In recent years some manufacturers have sold automobiles with limited bumper-to-bumper warranties that exclude certain components after a specified mileage or time interval has passed. Therefore, some contracts may cover only selected systems and parts for an additional period, such as the last 12 months or 12,000 miles in a 24-month or 24,000-mile contract.

To be more competitive in the market and to sell more automobiles, some new automobile manufacturers provide very long warranty coverage periods on selected parts. For example, a manufacturer may offer a powertrain warranty that covers engine and transmission parts for 7 years or 70,000 miles. Note that the powertrain warranty is included in the first X months and/or XX,000 miles of the bumper-to-bumper warranty.

In some states, the law may set the minimum number of months and miles a manufacturer must provide a warranty on all new automobiles, for example, 12 months or 12,000 miles. Therefore, a 7-year or 70,000-mile powertrain warranty that covers certain engine and drivetrain (transmission, differential, and axle) failures is like receiving an added 6 years or 58,000 miles past the initial warranty period. If, during this extended warranty period, repairs to the powertrain are needed, the warranty will pay for all, if not most, of the cost. However, if repairs to the brake system are needed in the extended time period, the powertrain warranty will not cover any of the costs.

In addition, new car dealers may have damages caused by the new vehicle hauler before the cars are unloaded at the dealership. Some damages such as a scratch below the clear coat may be covered by a warranty, but a scratch in the clear coat is a claim against the vehicle hauler. Likewise, there are procedures if a vehicle invoice contains options not on the vehicle that was delivered, such as an optional radio. Again the policy and procedures manual will direct how this be handled with the manufacturer; for example, they typically will reissue new paperwork and not send a new radio. Therefore, new car deliveries must be examined carefully as the vehicles are unloaded from the hauler.

Extended "Factory" Warranty

In some cases, customers may elect to extend their warranty coverage beyond what is provided by the manufacturer (for new automobiles) or dealer (for preowned automobiles with a "used-car" warranty). To do this, a customer buys what is known as an extended warranty. When purchased, the extended warranty provides coverage for certain systems and parts for a period that begins at the time the customer purchases the automobile.

For example, assume that Mr. Williamson bought a 36-month or 36,000-mile extended warranty at the time he bought his two-year-old preowned Jaguar with 26,100 miles on it from the Jaguar dealer. As a result, his Jaguar would be covered under the extended warranty for the next 36 months and up to 62,100 miles. The price Mr. Williamson would pay for the extended warranty is determined by the length of the coverage period and the system components covered. The longer the coverage period and the more components and systems covered, the higher the price.

To help control warranty costs, the extended warranty contract will often stipulate where service can be performed. For example, when an extended warranty is purchased from a new automobile dealer, the contract will often require that the warranty repairs be performed by an authorized service facility. In this case, the authorized service facility would be limited to new automobile dealership service departments for the make of automobile the customer owns.

In addition to service facility limitations, extended warranties often have a deductible that the customer must pay toward the warranty repair. The deductible could be as little as $25 but typically is between $50 and $100. Therefore, if Dr. Farmer bought an extended warranty with a $75 deductible and his automobile needed to have a $500 repair performed that was covered by the warranty, he would pay $75 and the warranty company would pay $425.

A contract may also stipulate that the customer pay for noncovered service items, such as maintenance, certain repairs, and possibly the cost of the diagnosis (see Figure 5-1). For example, assume that Mr. Pat McCormick

FIGURE 5-1 After a new PCM is installed, it is programmed.

owns an automobile with an extended warranty with a $25 deductible. Pat's engine was within the extended warranty coverage period when it developed a strange noise upon acceleration. The diagnosis of the problem cost $150 and was not covered by the warranty, while the repair cost of $600 was covered. In this case, Pat's extended warranty would cover the repair but not the diagnosis, and the deductible of $25 would be applied to the repair and not the $150 diagnosis. Pat would therefore owe the service facility $150 and $25 for the diagnosis and the deductible, respectively. The warranty company would then pay the service facility $575 for the repair.

The warranty contract does not cover the cost of a repair if customer neglect is the cause of the problem. For example, if the technician found that Pat's engine noise problem was due to engine sludge caused by infrequent oil changes, Pat would need maintenance records to show that he changed the oil within the time frame recommended by the manufacturer of the automobile. If he does not show that he maintained the automobile as recommended by the automobile manufacturer, the warranty company may refuse to pay the repair.

In addition, warranty companies will not pay for a repair if components not covered by the warranty contract are defective. For example, if Pat's warranty contract states that it does not cover idler pulley bearings that are bolted to the engine and the technician finds that the idler pulley bearing is the cause of the noise, Pat would have to pay for the repair. Furthermore, warranties will not pay for repairs if the problem is caused by system modification. For example, if Pat changed the engine's stock camshaft to a high-performance camshaft and it is determined to be the cause of the engine noise, then Pat's warranty may not cover the cost of the repair.

Extended "Nonfactory" Warranty

Warranty contracts change slightly when customers buy an extended warranty from a used automobile dealer or from a company on the Internet. Although the warranty coverage period and deductible function like the extended warranties sold by new automobile dealers, the warranty often allows the customer to have the repairs done at a service facility he or she selects. At the same time, however, the warranty may set a maximum amount of money it will pay for a system's repair or a maximum labor rate it will pay per hour. When this occurs, the customer must pay for any charges that are beyond the contract's maximum limits. For example, if Pat's repair was based on 10 hours at $60 per hour ($600) and the warranty company's maximum labor rate was $50 per hour ($500), Pat would have to pay the $100 difference plus any deductible.

Emission Warranty

Another type of warranty is the **emission warranty** mandated by the federal government. Emission warranties require new automobile manufacturers to repair problems associated with an automobile's emissions components for a stated period of time.

In practice, emission components are covered under the bumper-to-bumper warranty for the coverage period or at least two years and 24,000 miles, whichever is longer. After the bumper-to-bumper warranty expires (assuming it is the longer period of time), the key emission components, such as the catalytic converter and powertrain control module (PCM), are covered for a time period directed by federal law as of 2012, such as eight years and 80,000 miles.

Customers can obtain emission warranty service when they go to a new automobile dealership service department for the make of the automobile they own. Emission warranty repair is provided free of charge to the customer when the automobile is within the warranty coverage period, and the broken components meet the coverage requirements. Under government regulations, after performing certain emission recalls or repairs such as installing a processor, a sticker is placed under the hood to notify of the component replacement.

Processing Manufacturer Warranty

The biggest concern in processing a warranty is to get paid and avoid mistakes that can lead to a chargeback. A **chargeback** occurs when the dealer is "paid" too much and the manufacturer will reduce a future "payment" to the dealer. Common aspects of processing factory warranty claims are that the customer complaint (problem) must be connected to the cause (defect), which must match the correction (repair). This is called the 3C of warranty documentation.

Most manufacturers monitor the relationship by examining whether the labor operation on the claim is connected to the parts needed for the job. This is not true for every manufacturer, but one reason for a claim to get rejected is when there are mismatches. For example, when the part numbers, such as transmission gaskets, appear on an engine repair claim, the labor operation (repair) does not match the parts and the claim will likely not be paid. Another reason for a claim not to be paid is when the technician who performed the repair does not have the proper training and certifications from the manufacturer. Some manufacturers offer a battery of training programs to prepare technicians to perform warranty repairs. Technicians who have not completed the training may not perform certain warranty repair operations.

The manufacturer also looks at the number of parts required to complete a repair. When the number of parts needed to complete the repair and/or the cost of the parts gets too high, the factory representative assigned to the dealership will want to know the reason. Basically think of it as an allowance; when the dealer gets above the allowance, the factory looks into the problem. For example, a dealership had four "manufacturer-certified" transmission specialists, but the three surrounding dealers had only one. The surrounding dealers could not serve the customer's transmission overhaul needs, so more customers took their transmission warranty repairs to the dealer with four transmission specialists. The dealer was not allowed to turn warranty work away and performed them. Since transmission repairs are expensive, the dealer's total warranty claims in dollars were much higher than the other three surrounding dealers. The dealer was audited to determine why and what needed to be done about the problem.

Reason for Factory Warranty Claim Audits

There are different stages of a warranty claim audit depending on the manufacturer. There are self-audits that allow the dealer to examine their own documents to assure they are correct. There is a warranty review process where documents are examined by the manufacturer through a process that is either at the dealership or information is sent to a regional office, often electronically. Then there is a comprehensive claim audit where a factory representative (various titles) visits the dealership and asks to review warranty claims. The factory representative at his or her discretion may review a single claim, several specific claims, or a random sample of claims made in the past several weeks. The factory representative may also want to see the returned parts that the parts department keeps (see Figures 5-2 and 5-3). The representative examines the parts to make sure they are the correct ones for the vehicle and determine if they are defective when possible. In some cases, the manufacturer will mandate the return of the part for its review. Different manufacturers have different policies about returned parts.

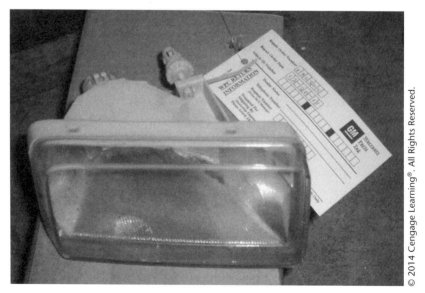

FIGURE 5-2 Defective headlight replaced under warranty. Notice the warranty parts tag that identifies the repair information so it can be identified and retrieved if needed in the future.

FIGURE 5-3 In the parts department is storage for defective parts replaced under warranty.

Generally, the parts department at a dealership is obligated to keep the parts from a warranty repair for a minimum number of days, such as 15, before disposing of them. Although full claim audits are less common since computers can match repair operation codes to parts numbers, they still occur periodically. When submitting claims, factory representatives suggest that the claim is complete with correct vehicle information, part numbers, and proper labor operations codes. If there are unusual problems that require additional diagnosis, or broken bolts that hamper the repair process, there are procedures to be followed that permit additional time, up to a certain limit, to be charged to the manufacturer. When the limit is not enough, the factory representative should be contacted to discuss the problem and possibly authorize additional time.

In a few cases, there is a Technical Service Bulletin (TSB) that provides technicians with technical information on common problems for a certain make of vehicle. When a TSB is involved there may be a special operation number and different repair times with even special parts than found in other factory information bulletins. TSBs provide the most recent information and must be known and followed over any previous factory information. Finally, there might not be a labor time assigned to some repair operations because it is a relatively unusual repair. In these cases, a special warranty code number is used and the labor operation time is when the technician starts the job until it is finished. An example of this type of repair is a wiring harness replacement.

> **CAREER PROFILE**
>
> ## The Warranty Claim Specialist (Warranty Clerk)
>
> Ms. C. has been a warranty specialist for over 20 years. She has been through various factory warranty specialist training programs and has read thousands of pages of warranty policy and procedures manuals. She has worked at different new car dealers and has seen a full "paper only" process to "computer only" processing plus a hybrid of both systems. She has prepared for warranty audits and dealt with various chargeback issues over her career.
>
> In an interview, Ms. C. explained that her job is about being organized, working with management and technicians to get the paperwork in order to submit to the manufacturer for payment. This is usually not difficult, she pointed out, but it can be challenging when there is a claim problem that will not be processed properly. So it can take minutes to submit a claim on a computer to weeks of back and forth trying to get a claim to be paid. Basically, then it is about being patient and doing a good job to avoid mistakes.
>
> Ms. C. noted that to avoid making mistakes with technicians, she had to be sure they prepared a proper summary covering the 3 Cs: complaint (problem), cause (defect), and correction (repair).
>
> She also found it to be important to make sure technicians understood what they can get paid for and what they can't. For example, she pointed out: sometimes flat-rate technicians don't understand that they can't get paid for removing a door panel, diagnosing the problem, and then ordering the part that comes in a week later. They get paid only when the part is installed, which means they must remove the door panel again after the part arrives.
>
> Further, Ms. C. explained that a service consultant can't promise a customer something and expect the warranty will pay for all of it. While most managers understand this when they work with customers, a new service consultant usually does not and can unknowingly make a customer angry because they have to pay money toward the repair. She found her job to be enjoyable but not without its headaches. Her recommendation was that the more service consultants know about warranty procedures, the better they can serve the customer. This means the service consultant must read the manufacturer's warranty policy and procedure manual and know it well!

Verification of Warranty, Maintenance Contract, Technical Service Bulletin, and Campaign/Recalls

When a customer's automobile is covered by a warranty, the following information becomes important for the service consultant to verify before service is provided.

- Is today's date within the warranty's specified time frame limits?
- Is the automobile's current mileage less than the warranty's maximum mileage qualifier?
- Does the contract's (VIN) match the number on the automobile being repaired?
- Is the name on the contract the same as the customer's name?
- Is the system to be repaired covered by the warranty?
- If so, was abuse, modification by the customer, or neglect by the customer the cause of the failure?
- Does the contract specify a deductible that the customer must pay?
- Are diagnostic charges covered by the warranty?

If this information is verified to pertain to the warranty, the service consultant can allow the service facility to perform the repairs under the warranty guidelines.

When a customer's automobile is covered by a maintenance service contract, the following information becomes important for the service consultant to verify.

- Is today's date within the contract's specified time frame limits?
- If so, has the amount of time that has passed since the last maintenance been long enough?
- Is the automobile's current mileage less than the contract's maximum mileage qualifier?
- If so, has the number of miles since the last maintenance been long enough?
- Does the contract's VIN match the number on the automobile being serviced?
- Is the name on the contract the same as the customer's name?
- Is the service to be performed covered by the contract?
- If so, are the parts and fluids needed to service the automobile also covered by the contract?
- Is there an amount of money the customer must pay for the services performed? (This is typically referred to as a copayment rather than a deductible.)

If this information is verified to pertain to the maintenance contract, the service consultant can allow the service facility to perform the maintenance required.

When a customer's automobile has a problem and the service consultant or technician finds that the automobile may be repaired as specified by a manufacturer's TSB, the service consultant must verify the following information.

- Do the make, model, and year of the customer's automobile match the TSB specifications?
- Does the automobile's design (engine size, transmission type, brake system, and other technical information) match the TSB specification?
- Does the TSB "cure" match the customer's complaint?

If this information is confirmed to match that of the TSB, then the TSB procedures may help to repair the customer's vehicle. In some cases, a warranty may cover the repair cost associated with the TSB.

For campaigns/recalls, the service consultant must verify the following information to determine whether the automobile qualifies for a repair:

- Does the customer have a recall notice?
- If not, does the manufacturer's database indicate that the VIN is involved in the recall/campaign?

- Do the vehicle's year, make, model, and design characteristics (engine, transmission, etc.) match the campaign/recall requirements?
- Does the dealership parts department have the parts to perform the recall/campaign?

Warranty Contracts

Warranty contracts are generally governed by the 3 Cs; complaint (problem), cause (defect), and correction (repair). Service consultants must disclose warranty information to customers very carefully. They must never assume that a warranty contract will cover the cost of the repair as it will depend on the cause of the problem. Therefore, service consultants must always prepare customers for this possibility and explain the terms of the contract to them. Service consultants must also never assume that customers have read or understood anything in the warranty contract.

For example, a contract may have a deductible that the customer must pay before the warranty company begins to pay for the repair. Some contracts may even specify a maximum amount of money or maximum labor rate that the warranty company will pay for a certain type of repair. The customer must be told about the terms of the contract and any charges the customer is responsible for, or a dispute with the service facility could occur.

To minimize the chances of a dispute between the service facility and the warranty company or the customer, or both, the service consultant must help the customer understand the warranty contract. The service consultant must also work effectively with warranty company representatives and follow its required procedures. To fully understand the warranty contract terms so each can be explained to the customer, the service consultant should start by carefully reading the entire warranty contract before the repair process starts.

After a repair has been determined to be covered, some warranty companies will outline specific procedures that the customer or service facility must follow. A service consultant's failure to follow the procedures or to have the customer agree to the repairs and the terms of the contract could mean that neither the service facility nor the customer will receive payment for the repairs.

When the exact procedure that is to be followed is not known or outlined in the contract, the warranty company must be called so the details of the transaction can be clarified. For example, a warranty company may require a warranty adjuster to visit the service facility to examine the automobile and approve the repair work before any work begins. Other warranty companies may require telephone authorization prior to a repair. When a warranty company requires authorization, a claim number must be obtained from it before work can begin. This number must be recorded on the repair order and final invoice. Again, if the service facility fails to follow proper procedures, the warranty company will not pay the claim.

Payment in Credits

In terms of payment, non-manufacturer warranty companies, which often allow any service facility to perform the repairs, typically issue a check 30 days after the repair is completed. Because of this delay and contract terms, such as deductibles and maximum payment provisions (maximum labor rate allowances for a given repair), service facilities may require customers to pay all non-manufacturer warranty contract repairs in full before releasing the vehicle. This allows the warranty company to reimburse the customer directly for the amount owed under the terms of the contract. Therefore, the service facility does not have to wait for payment and avoids not being paid in full for repairs because of deductibles or maximum payment provisions.

The payment of automobile repairs made under a manufacturer's warranty and extended warranty contracts is handled differently than a non-manufacturer warranty contract. First, the manufacturer's dealers are typically the only type of service facility permitted to perform the repairs and process the claims. This is because the manufacturer trains dealership employees on how to complete the warranty claim paperwork and process it with the proper codes to identify the automobile, the repair operation, the dealer, and the technician. A service consultant who fails to follow all of a manufacturer's warranty claim procedures will have the warranty claim returned "unpaid."

Manufacturer warranty repair claims are often paid in "credits," not with cash or a check, by the manufacturer to the dealer. This is because under the franchise agreement, the dealership must buy parts from the manufacturer (except in some rare circumstances) and install them on the automobiles it services. Therefore, the dealership buys many of the parts, among other items such as tools, from the manufacturer. Then when a manufacturer owes the dealership service department for warranty services, the credits may be awarded and not money. This means the service department will not receive a check at the end of the warranty repair process, but rather manufacturer credits that can be exchanged for parts that are ultimately sold to customers.

The Lemon Law

The failure to repair a problem to the satisfaction of the customer can result in the customer filing for protection under the **lemon law**. Each state has a lemon law that requires vehicle manufacturers to "buy back" customer automobiles when they are not repaired properly within a time frame set in the state's lemon law. In most states, the law sets a maximum number of days that the automobile may be out of service (for example, 25 days), a maximum number of repair attempts (for example, three) within a given time frame (typically one year), and a maximum number of miles (often 12,000 miles or less) to qualify for lemon law action. Lemon laws are enforced when a customer reports the problem to the state attorney general's office and files the necessary paperwork to start the judicial process. This process may begin with a meeting between the customer and a manufacturer's representative in an effort to resolve the complaint. In other cases, a mediator may be brought in to work with the parties, or an arbitrator may be hired to hear the facts of the case.

Because lemon law actions put the manufacturer, and in some cases the dealer, at risk, the service consultant must first ensure that the customer's complaint is accurately recorded. One manufacturer recommends that the words "customer claim" be used to begin every warranty-related repair complaint. In other words, instead of writing "The vehicle's brakes do not stop the car," the service consultant should write, "The customer claims the vehicle's brakes do not stop the car." Although the difference seems trivial, the first version implies that the service facility verified the problem and will make the repair. The second version implies that the customer has presented a possible problem that must be examined and verified by a trained technician. This is an important difference, because under most states' lemon laws a service facility is not responsible for repairing a problem until it is verified or duplicated by a technician.

In some instances, the defect cannot be found in the automobile's system and the problem cannot be replicated by the customer or technician. Also, the manufacturer's technical assistance hotline may not report any defect specific to that automobile system. When this occurs, the customer's claim cannot be considered a failure under many states' lemon laws. In such a case, service consultants must carefully and accurately document all activities according to the provisions of their state's lemon law. State lemon laws vary in specifics and therefore must be researched for the state where the vehicle was purchased or registered (when a vehicle is purchased in one state but registered in another, the consumer may have a choice about which state to file lemon law action). Details of specific state lemon laws can be found on that state's attorney general website or through a website search engine to find a site such as http://www.carlemon.com.

Service consultants at a dealership must also carefully monitor the number of times and days that a customer's automobile has been out of service. This may easily be overlooked if the customer's automobile has been in for the same repair over an extended period of time and the customer is given another automobile to use temporarily.

For example, assume that a service consultant at a dealership takes a warranty claim on an automobile purchased by Mr. Ronald Taylor within the past 12 months and it had been driven less than 12,000 miles. Mr. Taylor's automobile had been in for repair several times over the past 7 months to the point where the number of days it has been out of service was nearing the maximum number of days permitted by the lemon law. Upon diagnosis, the technician decided to replace several parts and Mr. Taylor was given another car to use temporarily. Instead of ordering the parts needed to repair the automobile overnight, the service consultant decided to include them in the regular weekly order. This caused the automobile to be out of service for the entire week, exceeding the maximum number of days under the state's lemon law statutes. The manufacturer must now "buy back" the vehicle from Mr. Taylor.

If a dealership is at fault for the "buyback," a court can legally require it to pay the manufacturer for the cost of the automobile. In Mr. Taylor's case, the dealership would have to abide by the court order since the service consultant failed to follow the manufacturer's policy to order the parts overnight to avoid lemon law problems. As a consequence, service consultants must keep their state's lemon laws in mind or a serious financial loss to their dealership could occur.

Review Questions

Multiple Choice

1. Which of these is NOT needed to determine applicability of a vehicle's service contract?
   A. Current mileage
   B. Vehicle identification number
   C. In-service date (the date when the vehicle's service contract began)
   D. Production date (the date when the vehicle was built)
2. A customer will receive a letter or notification from the manufacturer for which of these actions?
   A. A TSB release
   B. A vehicle campaign
   C. The end of the vehicle warranty period
   D. A vehicle recall
3. Which of the following DOES NOT describe a purpose of a TSB used by a technician?
   A. A TSB is a document mailed to customers to let them know about a problem with their vehicle.
   B. A TSB outlines how to install a redesigned version of a component.
   C. A TSB may revise a shop manual procedure or provide additional details the technician needs.
   D. A TSB may provide a repair procedure for a pattern of failures found in a vehicle or group of vehicles.
4. There are four types of automobile warranties that service facilities and their service consultants should be able to recognize and administer. Which of the following is NOT one of them?
   A. A new automobile warranty contract
   B. Bumper-to-bumper warranties
   C. Collision repair warranties
   D. Emission warranties
5. Service consultant A says a new automobile manufacturer warranty contract is almost always used by owners who buy the automobile secondhand (used). Service consultant B says the warranty coverage period pertains to a predetermined time frame and a predetermined mileage as provided in the warranty contract. Who is correct?
   A. A only
   B. B only
   C. Both A and B
   D. Neither A nor B

Short Answer Questions

1. What are the different ways a service consultant can locate reference information about a customer's warranty, maintenance contract, and campaign/recall?
2. What is the difference between a warranty policy and a warranty procedure?
3. What are the differences among warranties, maintenance contracts, TSB, and campaigns/recalls?
4. How can a service consultant explain the differences among warranty, maintenance contract, TSB, and campaign/recall procedures to a customer?

Activity

Activity 1: Use your school's library and/or the web to find information about lemon law regulations in your state. Your state's attorney general website should have the information desired or try a website search engine to find a site such as http://www.carlemon.com. Read the lemon law for your state and report the important details a service consultant should know such as the following: maximum days out of service, maximum mileage/time coverage under the law, defects that are covered or perhaps not covered under the law, resolution when the vehicle meets the legal tests to be considered a "lemon."

Activity 2: Conduct a web search for an aftermarket warranty that can be purchased for a used vehicle. Determine the details of the warranty coverage (deductable, mileage restrictions, length of coverage in terms of time and mileage, and any limitations about who can perform the repairs. Study the information to determine how a claim is submitted. Report your findings in a memo to a customer, Ms. C., who asked you to find an aftermarket warranty for her vehicle.

Activity 3: Obtain warranty information for a new vehicle from a web search or new car dealer. Read the warranty information to determine what is covered, how long is the coverage, and if there is a deductible (if so, how much). Write a report and include this information as well as any other information a service consultant might need to know.

CHAPTER 6

PERSONAL COMMUNICATIONS: FROM THE GREETING TO THE PRESENTATION OF THE INVOICE

OBJECTIVES

Upon reading this chapter, you should be able to:

- *Demonstrate the fundamentals of proper telephone skills (Task A.1.1).*
- *Demonstrate how to identify and document customer concerns and requests via electronic communications (Task A.1.3).*
- *Demonstrate appropriate greeting skills (Task A.1.6).*
- *Describe how to obtain and document customer contact information (Task A.1.4).*
- *Identify and document customer concerns and requests (Task A.1.3).*
    - *Address the customer's concerns (Task C.3).*
- *Explain why repair authorization is important (Task A.1.12).*
- *Demonstrate how to open a repair order and confirm accuracy for both computerized and paper repair orders (Task A.1.5).*

CAREER FOCUS

You know the name of the game. It is all about communicating with the customer and getting paid. You understand the entire shop's income depends on your expertise at getting the information needed from customers followed by their approval to execute the repair order contracts. However, you also know there is more because you act as a representative for your business and the person upon whom customers put their trust and confidence. This is where you, as the service consultant, distinguish yourself.

Introduction

Business communications are unlike social communications where two people are engaged in a personal and friendly dialogue. In business communications customers do not want to mess around with people who cannot help them. The customer has a pressing problem and is looking for a service facility that can solve it. While these conversations may be personable (not personal) and friendly, they are about business; meaning there is a monetary exchange for goods and services.

For the sake of clarity, this chapter is set up in three parts. The first part discusses communications, the second explains the customer service system, and third is about the importance of legal and other business issues. Many books have been written on effective communications; however, this chapter is limited to the automotive service industry.

Communications

Communication, whether it is face to face or through some other means, is critical to the survival of a service facility. The service consultant is the employee who is involved in most of these interactions. This means the service consultant must be effective (say and do the right things) and efficient (do them with minimum time or expense). The service consultant is always "on stage" and, as a result, must maintain a proper demeanor; like an actor on stage. People pay to be assisted by someone who is ready to work with them. This means the service consultant must be ready to work before the shop opens at 8:00 a.m. and will likely be helping the customer after the shop closes at 5:00 p.m. The service consultant must greet new customers appropriately, to answer the phone promptly, and to interact through other electronic media in a timely manner. If this is not done quickly and flawlessly, technicians will be delayed in starting work and this will cost the service facility money because time lost is money lost.

Telephone Communications

For many customers the first contact with a service facility is by phone, although email, Internet, or texting may be used and will be covered later. When the service consultant receives a customer's call, regardless of how busy he or she is, the consultant must be aware that the reason for the call is a paying customer wants service. The customer may not be in a pleasant mood and may be anxious, tense, or even angry. The service consultant must "read" cues such as the tone of voice, choice of words, and phrases selected. The reason for different moods is because for many, a trip to the service facility is an inconvenience and means they must spend money for which they will not receive anything "new" in return, such as a new cell phone or ticket to a concert. As a result, the initial phone call or other electronic communication message must be carefully analyzed. A service consultant must think before responding so that communication with the customer is effective. The customer must be served before the service facility can earn his or her money and a customer cannot be served if the initial communication is poor.

To start the telephone communication, the service consultant must have a warm and polite greeting along with a tone of confidence. The customer needs to be put at ease and feel comfortable about the people who are going to work on his or her vehicle. Remember that the customer may have never met the service consultant or been to the facility. Whether the customer is a first-time or repeat customer, the tone is the same but the greetings may vary. In reality there are many different types of phone conversations, and while the basic techniques are the same, the detail of the communication and purpose will differ. This means every communication is unique and a master will be able to handle any communication well.

Basic Telephone Techniques

Since telephone communication is a two-way communication that takes place in "real time" and can't be "delayed until later," like a text or email, the service consultant needs to practice basic telephone courtesy. The customer who calls should receive the service consultant's full attention. The service consultant should not attempt to multitask while listening to a customer on the phone. In addition, keep conversations private. With the sensitivity of the phone devices today, customers can hear background noises, such as another conversation or comments by another person, and this will likely annoy them. Keep telephone customers at ease and focus on an upbeat pleasant tone of voice with a professional feel. Speak clearly and at a pace that is not too fast or slow. Older customers may have hearing limitations plus they may need more time to cognitively process information being heard—so set your pace to match the customer.

Furthermore, since many customers call when the shop opens in the morning, service consultants should avoid putting a customer on hold unless they are currently talking to a customer who is at the counter or on another line. If the service facility is backing up with waiting customers and phone calls, then assistance from a manager, owner, or employee trained on the appropriate use of a phone is an option. In terms of priority, the customer at the facility who is talking to the service consultant is the primary focus, while the person on the phone is secondary.

To be able to put callers on hold without offending them takes practice by saying something such as: "Hello this is "your name" the service consultant. [Let the customer respond and obtain the customer's name]. I am glad you called "customer's name," can I call you back as soon as I am finished or would you like to hold?" After the customer has made a choice, thank the customer for being understanding. When a customer wants to be called back, many telephone systems have caller ID to obtain the phone number of the person who called. If this option is not available, then the phone number must be obtained from the customer.

Finally, when addressing customers, a service consultant should refer to them as Dr., Mr., Mrs., or Ms. When customers wish to be addressed otherwise, such as by their first name, they will request it. Of course, this would not apply to a close friend; however, when addressing a friend while other customers are present, the service consultant should be respectful.

Beyond telephone courtesy, the service consultant must realize that customers typically call an automobile service facility because they have a problem. They are often worried about what is wrong with their vehicle, the cost to fix it, and promises they made to others that they may not be able to keep. However, the service of an automobile is not an option, so service consultants must be sensitive to the customer's concerns and express understanding, especially on the phone. A service consultant must choose words carefully, and complex technical jargon commonly used by technicians must be avoided. Focus on terms that a person without a technical background can understand and encourage customers to ask questions and make comments. Patiently correct any wrong information, misgivings or inaccurate information they may have about the services needed and the service facility among other impressions. At the same time, they must be sensitive to time and not get bogged down in long discussions, especially on topics not relevant to the business call.

Customer Service Scripts

To assure consistent performances, a service facility owner or franchise may provide a script that must be practiced until perfect. In these cases, new service consultants should be given the script for working with customers in person and a second one for answering the phone. An example of a phone script greeting is as follows:

> "Thank you for calling, (name of the business)."
> "This is the service consultant, (state your first and last name)."
> "How may we help you?"

First, the service consultant states the name of the facility to tell customers they dialed the correct phone number. Next, the consultant states his or her first name. Finally, the next question allows the customer to state a variety of concerns or requests from a service concern to a question about the service facility's current Internet provider. Some phone scripts are longer and may contain a "special offer" message. For example, instead of saying, "How may I help you?" the service consultant may ask, "Are you calling to schedule our oil change with free 30-point inspection special for just $____?" The customer will respond and the service consultant will either schedule the appointment or next ask, "How may I help you?" A common practice at some businesses is the use of a recorded message to answer the phone before a person picks up the receiver. The recorded message usually states the name of the business, any specials being offered, and then connects to the person or department being called. However, care should be taken because too much information can overwhelm or irritate a customer with a problem and time limitations. Many customers just want to talk to a service consultant as soon as possible.

A script should also contain a list of responses to common questions or problems, and this is often helpful to newly hired service consultants. The responses used at Renrag Auto Repair were designed to be helpful when the service consultants became flustered and were grasping for a proper reply to either a question or a problem. For example, it is common for a service facility to offer a special. When a special is advertised, a copy of the advertisement must be near the service consultant's telephone. Important information about the sale should be highlighted on the desk copy, such as the date the sale ends, the cost, and any limitations or special considerations. Catching the service consultant off guard, such as not being able to find the copy of the advertisement, can cause embarrassment for both the caller and the service facility.

Other responses commonly provided in a script format or on a desk copy were related to specific services provided, such as a computer diagnosis. A script might state that "our computer diagnosis is $95 and includes up to 1 hour of labor. If additional time is required past the first hour, the charge is $79 per hour. Our policy is to call you when we complete each hour of labor to update you on our progress." A simple written statement, such as this one, often helped service consultants at Renrag do a better job.

Phone Shoppers

In some cases, a person will call for a repair estimate, repair rates, or other price-related information. Caution is recommended when responding to these requests. Customers (and perhaps another shop) often price shop. In other words, the person calling may already know their problem and may want to determine if you will beat an existing estimate or give them your service facility labor rate for comparison. It is best to tell them: "We would really like a chance to help you, but to do good job I need to look at the problem before we discuss the price. Can I schedule you for ____ (provide first day/time, such as 'later today at 3:00 p.m.') or __ (provide a second alternative day/time, such as 'tomorrow at 10:00 a.m.')." Then ask,

"Which one fits your schedule best?" If neither is good for the customer, try to get them to provide helpful information, such as when the customer is off work and then try to schedule an appointment. Finally, the cost of an advertised maintenance package special may be quoted on the phone. If the caller becomes irritated, you can suspect that you are being "phone shopped."

In addition, some callers may attempt to get a "diagnosis and solution to their problem" over the phone. In some cases, they may be trying to fix their own automobile or get a second opinion to compare to a recommendation received from another facility. *Never try to diagnose customers' automobile over the phone.* If they wish to come into the facility for a diagnosis, then schedule the appointment and give directions or call a tow truck. Otherwise, problems can come from answering these requests incorrectly, no matter how well a recommendation is intended.

Face-to-Face Communication

A service consultant must have an honest desire to assist people. This must be shown by presenting a warm welcome when a person comes into the service facility. To illustrate an improper welcome, Jerry was excited to be a new service consultant and rushed out to meet his first customer before she got out of the car. He blurted out, "I am the service consultant. What is wrong with your car?" The customer was startled and asked, "Where is West Third Street?" In this situation, the service consultant was too enthusiastic and he likely caused a potential customer to feel uncomfortable.

In other words, good intentions are not enough for a service consultant to do a good job. How customers are greeted (nonverbal cues and proper body language) and what is said can set up an effective working relationship. To illustrate the importance of nonverbal cues, Tom was a service consultant and was not really interested in the idea of helping people. As a result, when customers entered the service facility, they had to wait or sometimes even search for Tom. Just the sight of a customer caused Tom to get on the phone, start to file papers, or just leave the service desk to go to the bathroom or shop. Tom was not concerned that the service facility technicians counted on him to get the work they needed to earn a living. When customers happened to find Tom, he never greeted them or even said, "Hello." Rather, in a gruff voice he would ask them, "Why is your car here? What's wrong with it?" The customer would describe his or her concern to Tom, who would look at the customer with a blank face. At any opportunity Tom might leave a customer in the middle of a sentence if he could find an excuse.

The descriptions of Jerry and Tom point out extremes in performances. Jerry was too eager and could not establish an effective working relationship with the customers. Tom really did not want to greet or even help customers so he could not transition to a working relationship. Jerry could be taught how to effectively greet customers and to work with them. Tom, unfortunately, does not have the important qualification for a service consultant, which is a desire to help people. In both cases, their nonverbal cues and body language gave negative impressions causing the customer to cast them in an undesirable way.

The Greeting

Greeting a new customer begins with a warm welcome when the person enters the service facility. Although only service consultants typically greet customers, all employees should be taught how to greet and interact with customers or at least recognize them with a smile. The warm greeting to be used by service consultants is fairly simple but requires practice with management mentoring. A greeting should begin with a smile and an extension of the arm to shake the customer's hand (see Figure 6-1). The handshake should be a firm grip that is not too tight or too loose. The arm should then be moved shallowly up and down two or three times before releasing the grip on the customer's hand to terminate it. To know what to say, a new service consultant should practice shaking hands with another person, especially females, by extending the arm and at the same time saying, "Welcome to (the name of the automotive repair business)." "I am the service consultant and my name is __(Paul)_____."

"Your name is _____?" (Wait for an answer then *remember it*).

Next, the service consultant should repeat the name and say:

"Mr./Mrs./Miss/Ms./Dr. (customer's name), how may we help you?"

FIGURE 6-1 A handshake is important to welcome a customer.

When welcoming someone to the facility, the service consultant should maintain good (but not piercing) eye contact. The customer's response will indicate how the service consultant may be of assistance. In most cases, the customer will either request maintenance work or discuss the need for a repair to his or her automobile. This, in turn, will lead the service consultant to create a record for the customer in the customer service computer program database.

Welcoming repeat customers also starts with a handshake and then a greeting such as: "Welcome back to Renrag Auto Repair Mr./Mrs./Miss/Ms./Dr.(customer name), I am Kim the service consultant." When a facility has a large customer base, the service consultant may not remember the customer name. In these cases, the greeting may begin with an apology for not knowing the person's name or may start with "I am glad to see you again, let me look up our record on your automobile in my customer service database. Could I have your first and last name please." Once the record is found, the script would continue, "I found your records Mr. Mrs. Ms. Dr._____, how may we help you today?"

Considerations for Using a Social Network to Communicate with Customers

Social networking developed as a two-way communication medium that connected individuals together, so they could interact much like a service consultant communicates with customers when they enter the service facility. A service facility may want to consider a social networking site for communications. To start, the service facility must determine which social networking site customers are using. Facebook is a successful pioneer of social networking, but something new may develop in the future. A social networking site that has a specific purpose can have an advantage to consumers and businesses, such as one directed toward those interested in drifting or racing and using a chassis dynamometer to improve a vehicle's performance (see Figure 6-2). For a service facility this gives an opportunity to market its business and obtain feedback as discussed further in chapter 12. The point of this discussion is to understand its power to inform customers and help with "mass communication" to a customer base.

There are also sites such as LinkedIn and Twitter that can send out a single message to a group of users. This has a lot of B2B (business-to-business) applications. For example, if a race team has an interesting development that they want to share with fans before a race or an exciting new vehicle that has arrived at a dealership, they can tweet the information to the fans. But unlike a simple text message, this is a one-way communication that allows all fans to receive the same message but they cannot send a message back. Advances in social network communications can be an exciting way to keep customers informed and make B2B operations more effective.

FIGURE 6-2 Race team on the chassis dynamometer getting data and making adjustments before a race. Facebook can help keep fans informed.

Email and Text Messages

Customers often want to use email and text messages as the preferred means of communication. Message etiquette is not part of this chapter, but it is important to choose the proper words so a message is not taken improperly by the customer. Unlike telephone or face-to-face communication, a "positive tone of voice" is not possible. Therefore, other techniques can help, such as showing appealing shop pictures or video clips (links to YouTube perhaps), a nice font choice and color, as well as phrases with a positive tone. Examples of phrases that express a positive tone include "…hope this helps, …" "…glad to do the work, …" "…not a problem, happy to help…" among dozens of others to convey both fondness and desire to help. Phrases such as these could be taken poorly by a customer "…call if you want it scheduled, …" "… I can do it but it is going to be a pain, …" "…you will have to wait; it will be done when it is done, …" "… we're too busy … maybe we can do the work for you next week."

When conveying information to a customer, the message should stick to facts and be short. For example: "Your vehicle is completed and the bill is under the estimate. Thanks for letting us serve you." Avoid putting prices or additional detail in the written communication that might compromise a simple "check-out" transaction. If there is more to tell the customer about his or her vehicle, then the service consultant should call the customer. Pictures and video clips that show the customer's repair could be shared but must be approved by the owner or manager first.

One reason to keep communications short and direct is that they can become a permanent record and used against the service consultant or the business later. Therefore, details and comments regarding cause and cure for a problem should not be discussed over the phone. If they must be done by email or text, then it should be brief with little stated beyond factual information. For example, "Brakes need replaced; I will call at 1:00 p.m. to help you understand the problem. Is that ok?" If the shop has an advertised special for brake replacement, then the advertisement could be attached to the message; otherwise relate the details of the special in the phone call to the customer.

Going beyond basic statements should be avoided, especially if they involve opinion. A service consultant would not want to put in a message, "You are right, that guy didn't know what he was doing when he put on your brakes, what a 'HACK' LOL, we can fix it right for you, just give me the ok." This statement can end up in court for defamation plus it does not define any detail about what will be done by the service facility. Likely this could eventually lead to a gap between the customer's expectation and the garage's actual performance. It would be best to say, "We diagnosed the problem and I will call to explain the repair options, thanks for letting us examine your vehicle."

Further, any statement about a vehicle's design should be avoided such as "This car's tie rod ends always break because of its poor design. We have to replace them all the time." Shops fix vehicles and typically do not have engineers on staff with the credentials to make these statements. Therefore, they must be avoided especially in any electronic communication. These statements only have the potential to make the service facility look bad and potentially expose the business to undesired consequences.

The Customer Service System

Customer Service is a system that interacts with two other systems. The three systems allow the service facility to function. These three systems are:

1. The Customer Service System
2. The Shop Production System (operations)
3. The Business Operations System(s)

The Business Operations Systems typically include payroll (paying employees), bookkeeping (financial statements), payables (paying bills), receivables (customer payments), and so forth. Although a thorough discussion of business operations is beyond the content of this textbook, some of the operations are included in the final section of the textbook and in the second book written by the authors: *Managing Automotive Businesses; Strategic Planning, Personnel, and Finance*. The Shop Production System was introduced and discussed in earlier chapters of this text to explain how to use the repair order tracking sheet to monitor repair progress. Additional details about shop production (operations) as a system that uses the repair order tracking sheet will be covered in part 3 of this textbook (chapters 9 to 11).

Therefore, the following deals with the **Customer Service System** and the service consultant's role in that system. The customer service system (as discussed in chapter 4) is not a computer program although it may use a computer program(s) as a tool. The most important purpose of this system regards how the service consultant carries out nine customer interactions that are to take place from the greeting of customers to the receipt of the payment of their invoice. Within the Customer Service System are three key documents: (1) the estimate, (2) repair order, and (3) invoice. As shown in Figure 6-3, these are the service system documents.

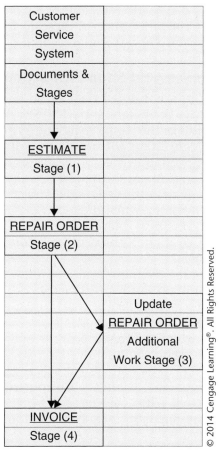

FIGURE 6-3 The Three Customer Service System Documents and Process (estimate, repair order, invoice) as related to the four customer communication "stages" (estimate stage, repair order stage, additional work stage, invoice stage).

To help understand the Customer Service System as a process, there are four customer service system "communication stages" or just "stages" (see Figure 6-3). These "stages" should not be confused with the repair order tracking sheet phases, which track the repair order in the repair process. The customer service stages are:

1. The estimate stage
2. The repair order stage
3. The additional work stage (repair order is updated when additional work is sold)
4. The invoice stage

The term *"stage"* is used because specific actions must be taken so the three service system documents (estimate, repair order, and invoice) are completed properly. This means that the service consultant explains the information to the customer so that the information provided is understood and considered, and authorization is obtained so the document is "legal." The preparation of these documents in each of the four stages is explained in the upcoming sections.

Customer Service and Shop Production Systems Overlap

Shop production is monitored via a Repair Order Tracking Sheet and is used by the service consultant to track the work being done on customer vehicles while it is in the Shop Production System. It is a document used by the service consultant within the overall customer service system to monitor the overlap between the Customer Service System and the Shop Production System. Both systems must work together to meet the customer's goal—a properly fixed vehicle that is on time and at a price agreed. Therefore, when the repair order tracking sheet is used at a service facility, it shares information needed for the service consultant to monitor the overlap of the Customer Service System and the Shop Production System.

To illustrate the overlap, Figure 6-4 shows at check in (first phase) the service consultant in the Customer Service System obtains customer and vehicle information for the repair order stage. They overlap again between the initial work (second phase) and additional work (third phase) phases when the repair order is updated for additional work (see Figure 6-4). Then the final overlap between the Customer Service System and Shop Production System occurs at check out (fourth phase) as shown in Figure 6-4, when the Customer Service System's final invoice is completed. When this does not occur as intended, there can be confusion, errors, and customer dissatisfaction.

The Customer Service System Interactions

In the Customer Service System, filling out the Customer Service System documents is not a goal but is merely a marker that signifies the end of a stage in the process. Within each stage are interactions that the service

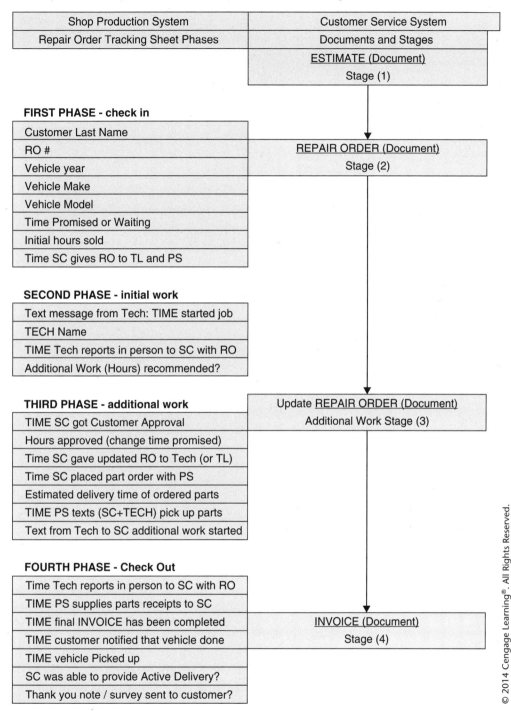

FIGURE 6-4 *Customer Service and Shop Production System* Shown is how the Repair Order Tracking Sheet overlaps with the Customer Service System documents (estimate, repair order, and invoice). Completion of each document as the repair process enters each of the four customer communication "stages" (estimate stage, repair order stage, additional work stage, invoice stage) is a milestone that will help the service consultant meet the customer's expectations.

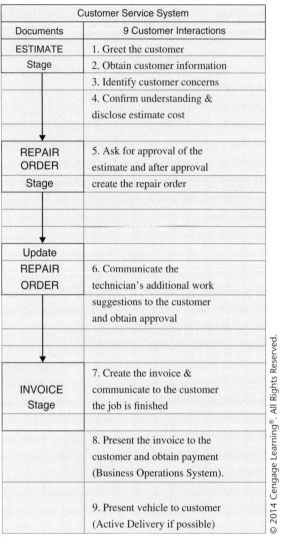

FIGURE 6-5 This is the Customer Service System. On the left column are the four stages and three documents that make up the customer service system. To execute each stage and complete the required document, nine customer interactions are used. The right column shows which of the nine interactions executed by the service consultant and how they align to the Customer Service System stages and documents.

consultant must have with the customer. There are a total of nine interactions in the entire Customer Service System process from the greeting of the customer discussed earlier in this chapter to the presentation of the invoice (see Figure 6-5). It is after the invoice is completed that the Business Operations System begins, which is beyond the scope of this chapter but discussed further in the author's next textbook.

Each interaction between the service consultant and a customer has a purpose that is important to the Shop Production Operation as well as for legal reasons covered later. The following present the nine customer interactions within the four customer service "stages" and the three related documents that are created in each stage. Of the nine customer interactions, the service consultant must master, the first four customer interactions relate to the preparation of the estimate. The fifth and sixth relate to the repair order and any additional work needed. The seventh, eighth, and ninth relate to the completion of the work and the preparation of the invoice. The service consultant must execute each of these interactions for every customer transaction with the exception of the sixth, which may not be needed if additional work is not recommended by the technician.

The Estimate, Repair Order, and Invoice

To understand the nine interactions (Figure 6-5) that the service consultant has with the customer, the information contained in the three customer service documents (estimate, repair order, and invoice) must be fully understood. It is critical to recognize that at the end of the Customer Service System process (Figure 6-3), the final invoice must be completely filled out to be considered a **legal contract** (contracts are covered further in chapter 14). The final invoice is created differently than other legal contracts because it evolves in the last stage from the work of the previous stages and the previous documents.

Specifically, the information in the three service documents is carried over from the estimate stage (estimate service document) to the repair order stage and its update due to additional work sold (repair order document). It finally enters the final invoice stage (invoice document) that ends with the customer paying the invoice and picking up the vehicle. Therefore, in each stage the service consultant must complete each document in detail otherwise the final invoice can have defects that result in a dissatisfied customer and can hamper its ability to be defended in court or under inquiry by the State's Attorney General's Consumer Protection Department.

Below is the information that must be on each of the service documents that leads up to the final invoice. As each document is discussed, the corresponding Customer Service System stage is presented along with its Customer Interaction Step(s). Explained with each Customer Interaction Step is how the service consultant would communicate with the customer. Specifically, this textbook suggests what the service consultant might "do and say" as well as what "not to say" to achieve the desired result that assures the customer understands the repair process details.

The Estimate Document

The estimate is prepared by the service consultant for the customer. It has sections that need to be completed with the help of the customer. When preparing the estimate, there are two things a service consultant must accomplish. The first is to interact with the customer to obtain the information needed for the estimate. The second is to use the customer information to help carry the interaction forward. The sections of the estimate document are discussed first followed by details about what the service consultant will say. The estimate stage (first stage) contains the first four customer interactions of the nine customer interactions required to complete the Customer Service System's stages. Remember the idea in this stage is to collect information from the customer; therefore, the service consultant must ask a lot of questions.

Typically, the service consultant enters estimate information into a customer service computer program but at some service facilities the service consultant may use handwritten estimates. By definition, the estimate is the start of a contract that will become a repair order. It contains all needed information for a repair order with the related costs that are known at the time of its creation. This will be the basis for the final invoice charges.

An estimate typically contains the following sections, some of which are completed by the service consultant with customer input (see example in Figure 6-6):

- *Service Facility Information*: The service facility's name, phone number, address, the current date, and tracking number (invoice number), and name of the service consultant.
- *Customer Information*: Vehicle owner's name(s) and contact information (address, phone number, perhaps other electronic contact information), date/time promised (or waiting at the service facility) as well as time of the customer's arrival.
- *Vehicle Information*: The year, make, model, license number, vehicle identification number (VIN), current mileage.
- *Disclosure Information the Customer Must Sign*: The specific disclosure differs by state (if any) and includes:
    - *How repair charges are estimated* (The flat-rate method). Customers must know that the actual time to perform the repair may be less than the amount charged to them. This disclosure often follows state requirements and may not be on the estimate itself but rather presented on a sign posted in the customer service area for the customer to read prior to signing the estimate. Regardless, the service consultant should be sure to point this information out to the customer.
    - *State Attorney General's statement* (with or without additional insurance company clauses) that gives permission for employees

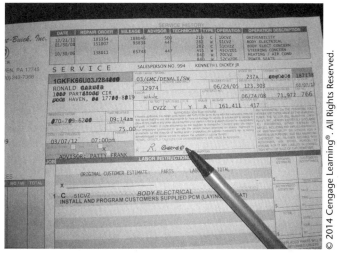

FIGURE 6-6 This estimate has the information required to be a legal contract in the state this service facility operates. Each state will have specific information that must be included on the estimate such as disclosure wording. The estimate layout will often vary by service facility and some will have additional information such as this dealer's computer program providing the customer service history at the top. The history tracks the date, past RO number, and operation description, among other information that may be helpful in serving the customer.

to operate and work on the vehicle as well as an agreement to pay the bill otherwise a lien (as per state law) can be placed on the vehicle. Again the service consultant should point this out to the customer.
- *Estimate options*: This varies by state regulations and typically permits the customer to choose between paying no more than an estimated dollar amount for a specific repair or allows the repairs to proceed without a written estimate up to a preset dollar limit. This is referred to as "waiver of the advance estimate." This is to be pointed out and discussed with a customer by the service consultant as necessary.
- *Permission to dispose of removed parts*: Often customers will not want their old oil filter filled with oil. However, a customer does need to know that some parts installed on their vehicle may have a core charge. Therefore, the customer agrees to pay a core cost if he or she wishes to retain the old part. This usually needs an explanation. In some cases, a customer will ask to "see an old part" so keep it on hand until payment is received.
- *Service Facility Warranty Information*: Disclosures must be provided on any limitations of a service facility repair guarantee based on mileage of use since the repair/replacement and/or on any time limits on

the use of the new, parts or after the labor was provided, or both. Sometimes a statement will include that in the event of a problem with a repair only the service facility can perform the warranty repairs and work performed by others will not be paid by the service facility. This often needs to be explained carefully to a customer in advance.

- *Warranty Payment*: Recognition must be noted on the customer's responsibility to pay any deductible associated with a warranty "paid" repair as well as repair charges that are denied payment under the warranty. For non-factory warranties, an agreement to pay for the entire repair charge when work is completed must be noted especially if the charge is greater than the warranty reimbursement the customer may receive. Waranty limitations are important for a customer to know and must be pointed out.

In addition to the information discussed earlier, the estimate must have the estimated repair costs (see example in Figure 6-7). The details of each are as follows:

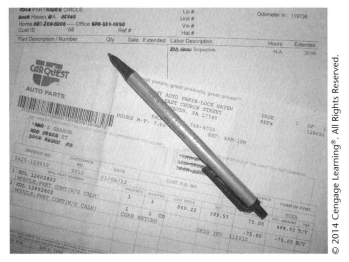

FIGURE 6-7 This estimate design has the labor estimate on the right and the parts estimate on the left. The labor estimate section has been completed and it is ready to have the parts estimate section complete. A parts delivery receipt is shown to illustrate that the parts estimate section will require the parts store to supply the part number, part description, quantity, and the sale price to the customer. The sale price of the part is the "list price" on the delivery receipt. However, management may require that the sale price be calculated by marking up the cost a specified amount (cost is "NET" on this delivery receipt). This delivery receipt shows a core charge for the part, so the customer must know that the part must be returned to the parts supplier unless the customer wishes to pay the core charge and keep the old part.

- Repair cost estimate (this is covered in further detail in chapter 7):
  - The part's sale information includes an itemized list of *all* parts and fluids needed for the job with the number of each item required; such as 5 quarts of oil. Often the part number and sometimes the vendor are included on the invoice along with the sale price of the parts to the customer. The parts information is usually listed in a separate section of the repair order from the labor information.
  - The labor information includes the customer complaint with a description of the labor operation to be performed, such as diagnose crank, no start. The labor sale price is calculated based on the flat-rate hours estimated for the job, such as replace water pump at 1.2 hours. The labor information with the labor sale price appears in a different section of the estimate than the parts information. Likely the service consultant will have to explain how the estimated number of labor hours was obtained.
- The sale price of subcontracted work, such as towing or machine shop operations, if required to perform the work, is listed on the estimate.
- Applicable taxes such as state and local sales taxes as well as other taxes, such as a state's tire tax.
- Any fees associated with the repair must be listed, such as any environmental or shop supply fee.
- A cushion, of perhaps 5%, to help cover anything unexpected may appear on the estimate or be "built into" the estimate at some service facilities. The method used to "build it into" the estimate will vary.

The Estimate Stage—Customer Interaction: The First Four Steps

The interaction between the service consultant and customer to obtain the information in the estimate stage involves four steps:

- The first is to greet the customer.
- The second is to obtain the customer and vehicle information.
- The third is to identify the customer concern.
- The fourth step is to confirm the understanding of the customer's request.

The first step to greet the customer was covered earlier. In the second step, which is to obtain the customer and vehicle information, repeat customers will likely have this information in the customer service computer program and it will merely need to be updated. For new customers, the customer information may have been collected prior to arrival at the service facility (over the phone) or will take place upon check-in. To obtain the information, the service consultant will merely ask for the information required for the estimate and type it into the customer service computer program or write it on the estimate form.

Often the vehicle information needed will require looking at the vehicle to obtain mileage and the VIN. Therefore, if the customer is at the counter, collect as much customer and vehicle information (make, model, and year) as possible. Then the service consultant with the customer will have to go to the vehicle and get the information but after completing step 5—which is obtaining customer authorization to conduct the work which is covered below. Do not leave the customer at the counter to get vehicle information and then return. This is both rude and inefficient.

At a service facility with a "drive-in lane," the vehicle information can be collected easily while the vehicle is "in the lane" and the customer is being served. Some facilities have the technician write down the information needed on the form when the vehicle is pulled into the shop. The service consultant then enters the information later. However, this is not desirable due to errors and lost technician production time. Ultimately, accurate vehicle information is required and how it is obtained varies by service facility but it is the service consultant's duty to get it.

In step 3 (Identify Customer Concerns) of the estimate stage, the service consultant will identify the customer concern. However, some customers may not have a concern that needs a diagnosis but rather already knows the repair or maintenance is desired. The differences of each will be discussed next.

Diagnosis Required

When a diagnosis is required, a technician needs to examine the automobile and then based on the repair suggested, the customer is contacted. A price for the repair is provided to the customer and the customer must give approval before the repair can be performed. To assist the technician in the diagnosis of the automobile, the service consultant must obtain accurate information about the customer concern that must be verified. This must be written on the estimate in a clear manner as discussed further in chapter 9.

Often, there is a charge for the diagnosis and the service consultants must disclose it to the customers so they can approve it. In some cases, a diagnostic charge may not be made if the customer has the repair made at the facility. The amount charged for the diagnosis depends on the customer's concern because a service consultant cannot estimate a cost to repair the problem without a diagnosis. The reason for the diagnosis charge is that it is to cover the time needed to look up the repair or diagnostic information, testing of the automobile's system(s), and any time needed to take apart (or remove) potentially defective components to test them. The diagnosis cost is difficult to estimate, so typically this charge starts with a basic fee required to cover the use of expensive equipment plus a minimum amount of technician time.

After a diagnosis, the diagnostic charges may be "added to" and included in the repair charges (common at service facilities with a variable labor rate system) or it may remain as a separate labor charge or it may be eliminated with approval of the repair. The nature of the work and the policies at the service facility will dictate the method used.

To obtain the information needed for the diagnosis to begin, after recording the customer's description of the problem, which begins with "The customer claims . . . ," there are some questions the service consultant must ask. These types of questions may be part of a "service script" or could be a separate series of questions management and technicians need to know.

Service Consultant Script for Diagnosis

The following is a script indicating what a service consultant might say to a customer when interacting with him or her on a diagnostic concern. The actual wording may vary depending on several factors; however, the purpose of the script is to obtain the information needed for the technician to fix the vehicle as well as disclose the Shop Production System's diagnostic process to the customer. The service consultant would start by stating:

> "To help our technician diagnose the problem, I would like to ask you for some additional information about your concern. Is the automobile currently having this problem?"

If *no*, then this will be much more difficult for the technician to diagnose because it is an intermittent problem. The service consultant questions then will have to be more focused on "when" relative to the environment, driving, vehicle condition. If *yes*, then the next question would be:

> "How long has your automobile had the problem?"

Then the next questions should be:

> "Can the technician duplicate the problem?"
> "When did this problem first occur and when did it occur most recently?"
> "What are the driving conditions when the problem occurs and what should we do to make it happen again?

Additional questions the service consultant might ask are:

- "Does the problem occur when the engine is hot or cold?"
- "What was the weather like the day of most recent occurrence?"
- "Does the vehicle perform differently when it occurs?"
- "As a driver, what do you do to try and make it stop when the problem occurs?"

These examples are merely general questions and more precise questions can be developed for more specific problems. In the entire process, the service consultant must pay attention, focus on the customer, and make sure a positive tone that conveys interest and a desire to help solve the problem is used.

After the questions have been asked and the service consultant has written (or typed) the answers to the customer concerns on the estimate document, they must be repeated back to the customer to make sure the customer agrees they are correct. This is the fourth customer interaction step in the estimate stage: confirm understanding of the customer's request. For example, after several questions about a check engine light flashing, the service consultant would repeat to the customer the main details, such as:

> "I want to make sure I understand this; your check engine light is flashing and started when you went up a hill at approximately 40 miles per hour. The vehicle lost power. Is that correct?"

On the estimate, the service consultant might write or type "Customer claims that under load at 40 miles per hour, the check engine light started flashing, vehicle lost power." Next the service consultant should follow a script that would disclose:

> "(Customer's name), in order to repair your automobile, we must begin with a diagnosis to identify the exact problem. Our minimum diagnostic fee is $_____ (assuming this is not covered by warranty). After the technician performs the necessary tests and analyzes the results, I will call you with the findings of our diagnosis as well as an estimate for the repair. Does that meet your approval?"

In some cases, the customer will want to know what tests will be performed. Therefore, the service consultant must either know the diagnostic procedures or a technician should be present to explain them. At the end of the diagnosis time, the customer must be called and told what the technician found. A problem may require more diagnostic time than originally agreed and the customer will need to authorize more time, such as an electrical wiring problem. The service consultant must then inform the customer that additional diagnostic time is needed and any additional charge to be made.

Repair Requests Interactions

Some repairs, such as brake lining replacement, may not require diagnosis. These repairs may not be "specials," but are often "package offers," meaning that the brand or quality of brakes to be installed on the car is predetermined. Service consultants save a lot of time when a job is a pre-priced package repair or special offer charge because they do not have to look up the cost of the parts, calculate the markup, and then calculate the total cost for an estimate. Service consultants, however, must have the information on the parts to be sold, such as the brand and product warranties to convey to the customer. Since this is done in the estimate stage, the service consultant will repeat the order back to the customer (step 4 of the customer interaction) to confirm the understanding of the customer request. It is as simple as saying "Let me make sure I understand your request, you would like for us to ..." (repeat diagnosis, repair, or maintenance request). Is this correct?" Make sure the customer agrees that his or her request is understood properly before moving on to step 5, approval. The final interaction step, which may occur after some additional discussion, is when the estimate document is accepted and signed by the customer to become a repair order document. Therefore, step 5—approval—is necessary to move from the estimate stage to the repair order stage.

Maintenance Request Interactions

Maintenance requests are often pre-priced maintenance jobs, such as an oil change with filter. The service consultant must be careful and describe the contents of all pre-priced sales. For example, a customer may wish to purchase the pre-priced special but then indicates that he or she prefers a specific brand of oil. Because the pre-priced special is probably based on the purchase of a large quantity of oil to reduce its cost, the request may not be possible at the special price.

A service consultant must be aware of any details associated with a special, such as the brand, weight, and quality of the oil. This is needed so the customer feels comfortable about buying it. The point is that the information on all products and their guarantees, especially those sold in pre-priced specials, must be available to the service consultant.

It is also appropriate to explain to the customer that the product, in this case the oil, was purchased at a reduced price and the savings are passed on to the customers. Similar to the repair request interactions covered prior to this section, the service consultant needs to repeat the order back to the customer (step 4) to confirm understanding before moving on to the next request or to step 5 (approval).

Preparation of the Estimate Before Step 5 Authorization

The service consultant must prepare an estimate for all maintenance, repair, and diagnostic work and review the charges with the customer before a repair order is prepared. Before the service consultant presents the customer with the estimate for a repair, a slight margin should be added to the price of parts and labor. This will ensure that costs will cover the charges and hopefully allow the invoice to be less than the estimate.

For example, a $150 estimate may be multiplied by 1.10 to 1.15 (such as 1.10×150.00 and $1.15 \times \$150.00$) to get a revised estimated amount ($165.00 to $172.50) for the customer. The additional amount ($15.00 to $22.50) will hopefully cover any unexpected expense (such as an increase in the cost of a part for a particular model of automobile), miscalculations (such as a slight mistake in the labor hours for the job), and complications encountered during the repair process (such as the time to remove rusted bolts or broken parts) and even the state's sales tax, if applicable. Of course, if problems do not occur, the invoice would be reduced from the $165 to $172.50 estimate to $150 plus the state's sales tax (if the sales tax is 6%, the total bill would be $150 × 1.06 = $159). When the bill is less than expected, the customer is usually pleased!

The service consultant should remember that the information on a repair order will be read by the technician and, in some cases, a warranty auditor (if the repair is to be paid by a warranty claim). Therefore, when a customer's claim has not been confirmed by a diagnosis, vehicle manufacturers suggest that *customer claims* be written before a description of the customer's problem. For example, "The customer claims the engine will crank but won't start until the car sits overnight."

An estimate for a diagnostic charge is difficult because customers want to know how much it will cost to fix, and not diagnose, a problem. In some cases, a repair labor guide may have a diagnostic charge for certain systems; for example, a check of the starting system is reported in some labor guides as a one-half hour job. This means the fee charged to the customer will be equal to half of the hourly labor rate (if the hourly rate is $80 per hour, the fee to check the starting system would be $40 or half of the hourly rate). Diagnostic charges for other concerns, such as drivability problems, emission failures, or electrical problems, do not have a set fee because the verification and diagnosis of the problem can take anywhere from a few minutes to several hours. Therefore, to ensure that the technician has enough time to diagnose most concerns, a minimum fee of one and a half (1.5) hours of labor may be a standard charge. In this case the charge would be $120 ($80 per hour × 1.5 hours = $120).

After accurate estimate calculation and four successful interactions during the estimate stage of:

1. Greet the customer,
2. Obtain customer information,
3. Identify customer concerns,
4. Confirm understanding,

the estimate as a document is completed. To become a repair order, the customer must authorize the work on the estimate (step 5). This means the customer is agreeing to the disclosures, information accuracy, and repair cost. Once the estimate has been signed, it will become a repair order (RO) for the technician to use. Starting the work but failing to obtain the vehicle owner's signature (assumed to be same as the customer) on the estimate can result in:

- Not getting paid for work performed (or being unable to enact a lien covered later),
- Unable to defend against any legal challenges in court or a government inquiry

- Not having written permission to work on or drive the vehicle can potentially cause insurance issues if there is a dispute on the cost of parts, part or labor guarantees, or the vehicle is damaged.
- A disagreement with a customer on repair details, such as services and final cost, that is not easily resolved.

Step 5—Approval and Entering the Repair Order Stage

Customer Interaction step 5—Ask for Approval: The customer must be given the estimate to review. After reviewing and answering any questions the customer may have, a simple statement by the service consultant is needed to start the repair order process, such as "I appreciate your business and to start the repair process, I need for you to authorize the repair order." The service consultant should then hand the customer a pen to sign the estimate. At the bottom of the estimate is a place for the customer's signature for approval. Upon receiving the signature, the estimate becomes a repair order (RO). At times, a deposit may be required on larger jobs. A service consultant statement such as "I appreciate your business, to start the repair process, I need a deposit of $___ as well as your authorization." Then the payment is to be recorded on the RO along with a signature. The payment in cash, check, or credit card follows the business operations procedures similar to the invoice stage discussed later.

If the customer does not wish to sign the estimate, then clearly he or she does not wish to have the repair performed at this time. The customer should be encouraged to take the estimate and consider it. Explain that when the customer is ready to have the work performed, they should call to schedule the job. Depending on company policy, the service consultant should inform the customer that the estimate is valid for ___ days and then write it on the customer's and facility's copy of the estimate.

In cases where the parts and labor for a repair cannot be determined, such as a custom car with unique components or a transmission overhaul with unknown internal parts failures, the estimate cannot be prepared nor can any information be placed on the repair order for labor/parts cost. If this is the case, service consultants should follow the directives of their state law. Most laws require the customer to sign a repair order waiver, meaning the customer agrees to have a repair done without a repair cost estimate. In some states, the law sets the exact wording to be used in the waiver but generally allows one or more of the following options:

- The customer can choose to allow repairs to proceed up to a certain dollar amount (the dollar amount is filled in prior to signing the waiver).
- Diagnosis and/or disassembly are to be performed for a certain dollar amount (or up to a certain dollar amount), but repairs cannot proceed until the customer is notified of the repair cost. Under

some state laws, if the customer does not want the repair performed, reassembly of the automobile to the condition it was in prior to disassembly is required (in some cases without additional charge to the customer).
- The customer agrees to pay all diagnostic and repair charges without any preset amount.

The Repair Order Stage

After the service consultant obtains authorization (step 5 approval), he or she will prepare the RO to enter the Shop ProductionSystem for assignment to a technician. Therefore, the service consultant would start by entering information into the Repair Order Tracking Sheet's first phase. This was covered in chapter 2 and 3 and will be covered further in chapter 11.

Then the "hard copy" version of the RO will be printed for the technician with a second "soft copy" for parts. At some service facilities, the RO will be a copy of the one signed by the customer. At other service facilities, it will be a handwritten "work order" with just the customer name, customer concern/request, and the basic vehicle information. In some cases, the RO could even be sent electronically to the tablet or shop computer of the technician who is assigned the job as well as to the parts department through the customer service computer program.

Once assigned to a technician, the RO enters the second phase of the Repair Order Tracking Sheet initial work. Upon RO assignment to the technician, he or she will "punch on" the RO using a manual time

FIGURE 6-8 These repair orders await assignment to the technicians.

card and time clock or the time will be electronically recorded if sent by the computer program to the technician's computer tablet or shop computer. As discussed in chapter 4, the technician will check over the vehicle (inspection) and perform any diagnosis requested by the customer. Then, before moving to any other service work, the technician will report his or her findings to the service consultant in writing on the back of the hard copy, the computer tablet, or tech's shop computer. When needed, the technician will obtain part prices and availability from the parts department, who will give the information to the technician or the service consultant depending on the Shop Production System.

The service consultant will meet with the technician as needed to review the repair order. In some cases, the service consultant may have to determine the cost of each repair before prioritizing them (covered further in chapter 9) for the customer to approve. If all or part of the suggested additional repairs is approved by the customer, the RO needs to be updated and then entered into the third phase of the repair order tracking sheet (additional work). Additional work as it relates to updating the repair order will be covered later. If not approved, any services from the initial work (second phase) will be completed by the technician before moving to check out (fourth phase).

It should be noted that once the technician has completed all of the work on the RO (this may include additional work), he or she will write up the service findings and repair details on the "back of the RO" or type them into the service computer program (see Figure 6-9).

FIGURE 6-9 This technician records his findings and suggestions on the back of the RO before punching off and giving it to the service consultant. He will then be assigned another job as discussed further in chapters 10 and 11.

The technician will use a time clock to "punch off the job" prior to returning the RO to the service consultant to create an invoice document (invoice stage).

An interesting note is that the fourth phase of the Repair Order Tracking Sheet (Shop Production System) overlaps with the Invoice Stage (Customer Service System) when the invoice document is created (see Figure 6-4). Also at this point in the process is that most customers react positively when a technician is present to review the proposed work with the service consultant. The technician can assist in explaining what was done in the diagnosis, the problem, and what will be done to repair the vehicle. This service is more likely to be possible in a smaller facility than a larger one.

Entering the Additional Work Stage

The additional work stage means the service consultant will work with the technician to update the repair order (Customer Interaction step 6) and obtain authorization, typically verbally from the customer. When the technician suggests the customer needs additional work, essentially the RO returns to the estimate stage (Customer Interaction steps 3 to 5). Communication with the customer is similar to what was covered in prior sections except the technician has identified the concern rather than the customer. Understand that if this is a warranty paid repair, the customer may identify only concerns, and a technician should be discouraged from finding additional concerns to be paid under warranty.

Just as in previous section discussions, the customer is presented with the repair suggestions, explanation about the suggestions as needed, and cost. Then the service consultant will allow the customer time to consider the suggestions before asking for approval. This is commonly done by telephone, unless the customer is waiting at the service facility and is done in person.

In cases where a telephone is the means of communication, a verbal authorization for the additional repair (or more diagnostic time) may be obtained from the customer over the phone. Most state laws have specific guidelines for the service consultant to follow. Because service facilities must comply with their state laws, they should refer to these regulations. For example, when taking a verbal authorization over the phone, many states require that service consultants make clear and detailed notes on the repair order as follows:

- Date
- Time
- Name of person authorizing the repair
- Work approved
- Total amount of the repair the owner agreed to pay
- Signature of the service consultant

Regardless of the law, a telephone approval for a repair is always a concern for the following reasons:

- Will the customer remember the amount of the repair?
- Is the person on the phone really the owner of the automobile?
- If a person other than the owner approves the repair, such as college student for a parent's car, will the owner (parent) honor the approval?
- What if the wife authorizes a repair for the husband and he is upset over the repair charge when he picks up the vehicle?
- What does the state law indicate about contracts and mechanics liens (covered later)?

When receiving approval for the work to be conducted (whether by phone or with the customer present), the service consultant should read all of the repair items on the repair order to the customers to remind them of the scope of the work. As each is presented (what is done and the total cost with parts) to the customers, the service consultant should make a checkmark. When the item is approved, the consultant should indicate it by placing an "OK" after the amount to be charged for the item. If an item is not approved, a line should be drawn through the item which will change the total amount of the bill.

ADDITIONAL DISCUSSION

The Irritated Service Consultant

A service consultant and technician must remember the process and that the details are important. A service consultant may become irritated when a technician, who completed a timing belt replacement asks, "The customer had an original water pump and maybe we should put one in? I'll wait until you find out." This is irritating because the customer has to be called again to approve more work. The time it takes to contact the owner adds down bay time to the job and, if approved, means the vehicle will not be done on time. If the owner declines to have the work done and two weeks later returns because the water pump is leaking, owners will likely not be pleased and may even be suspicious. If done properly, the technician and service consultant would have provided the customer with the recommendation during the additional work stage.

The Invoice Document

The invoice must contain everything that appears on the estimate with any additional work requested, as covered earlier. The service consultant should review the invoice with the customer to make sure it is accurate and is acceptable to the customer. In the labor sale area of the invoice, the customer's service request or complaint along with any additional work approved needs to appear. The technician's comments and suggestions on the RO need to be summarized along with any diagnostic findings (cause and cure of a complaint). The invoice comments must be brief and in terms the customer can understand. Any work that was not approved (additional work suggestions) as well as warnings (heater hoses are almost worn through against engine bracket), concerns (tires are feathered and will wear out sooner (should be rotated), or special instructions (new engine, change oil in 500 miles) are to be written on the invoice in order to help avoid potential conflict if a future problem occurs.

In the parts sale section of the RO must appear *all* of the parts used to service the vehicle. This list must be checked against the parts invoices to avoid errors. This means the names of the items, price, and part numbers must be checked. The number of the RO should be written on the parts delivery slips with a copy attached to facility's copy (not customer's copy) of the RO. Subcontracted work must appear in the proper section of the RO with a copy of the work conducted attached to the facility RO. The RO should also report any charges for shop fees and finally all applicable taxes (sales taxes and any special taxes).

The payment of the invoice will follow business operations procedures as discussed in the author's text *Managing Automotive Businesses: Strategic Planning, Personnel, and Finance* (for instance, invoice payments may require a cashier to receive payment) and must be marked with payment details (cash, credit card, check number). In addition, the customer must sign the final invoice again acknowledging the repair. While the original estimate was signed, the final invoice must be signed as well. In the event of a dispute, this final invoice will be needed in court and therefore should be signed.

Entering the Invoice Stage

The invoice stage contains customer interaction steps 7 to 9 and is the final stage of the Customer Service System. At the end of the invoice stage, the business operation will collect the customer payment (at some service facilities), deposit money, and record the transaction in the business operation's computer program. The customer communications during the invoice stage include:

7. Communicate the vehicle is finished
8. Present invoice and obtain payment (if not part of the business operations)
9. Present vehicle to the customer

The communication that the vehicle is finished can be done by text, email, phone call, or whichever method the customer prefers. The time of the contact should be recorded. For some repairs and some customers it is best to call them so that details about a repair can be explained in depth.

The presentation of the invoice depends on the service facility. At some facilities the service consultant will present the invoice (obtain the customer signature on the invoice) and obtain/process payment before sending the invoice and payment to business operations. Some service facilities will have a cashier handle step 8 as part of the business operations.

After the payment, the presentation of the vehicle to the customer may include little more than handing them the keys and directing them to the parking lot. At others, the service consultant will conduct a more extensive presentation to the customer. This textbook refers to this as "active delivery" (covered in chapters 8 and 10), especially for more expensive jobs and at luxury vehicle service facilities. The main idea in the invoice stage 9 is to convey an appreciation for the customer's business.

Paper Estimates, RO, and Invoices

Even though most shops use a customer service computer program, some service facilities use handwritten "paper" estimates and repair orders (see Figure 6-10). These forms typically have sections for the service consultant to "fill in" the necessary information. When the service consultant is ready to create a repair order, a numerically sequenced form with three or four "carbon copies" is used. The repair order form often follows the same format as the one used for the estimate. When the service

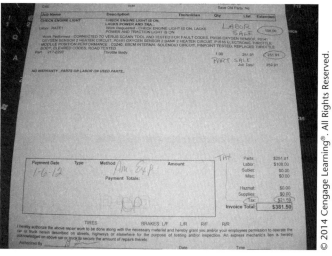

FIGURE 6-10 A final invoice is shown. Notice the comments about the repair by the technician are detailed and located below the original customer complaint. On this invoice design, the labor and its sale price are listed above the part sold to the customer and the part sale price. The tax appears above the invoice total (at the bottom) with the payment method appearing to the left of the invoice total. At the bottom of the invoice is the disclosure and customer signature.

consultant fills in the information on the top form, it is copied to the carbon forms underneath it. When the service consultant prepares a paper repair order, the customer's name, address, phone number, the repair request(s), parts, labor charges, and tax are written in the spaces provided. Once completed, it becomes a repair order when signed by the customer.

The form comes apart and the **technician's hardcopy** is used by the technician to punch on/off, and write all technical comments. The soft copy goes to parts to write down any parts that are sold to the customer. When the services are completed, the hard and soft copy come "back together" and become an invoice that customer pays. The invoice copies come apart again with the top copy given to the customer, the second copy sent to business operations, and the third copy is filed numerically by invoice number while the hard copy with the technician comments goes into a customer file. Thankfully computers make this process easier; however, this explanation makes it easy to "see" the process used by the computer program. Further, it is helpful to understand if a service facility still uses this process or aspects of it in their operations.

Nonpayment of a Repair

Repair orders are viewed as binding contracts between a customer and a service facility. Therefore, a customer must sign a repair order with an estimated amount for the service or a waiver if exact charges cannot be determined. Since a repair order is a contract, it signifies that a service facility has offered to make a repair for a stated amount of money. The customer's signature indicates the service facility has been given the legal authorization to make a repair using the stated parts and labor and the customer will pay the amount shown on the invoice.

If a customer refuses or cannot pay for a repair, many states (but not all) allow the service facility to hold the vehicle until payment is received. The action taken by a service facility is referred to as a **mechanic's lien**. The execution of such a lien depends on the state law and assumes that the service facility performed the work, corrected the problem (typically the lien can be only for repairs and storage—not diagnostic work alone), has an invoice that was signed by the customer, and followed all applicable consumer protection laws.

In addition, the offer and acceptance of a signed repair order means the guidelines of the state's consumer protection laws will be honored. For example, the services performed by the service facility must meet professional standards of workmanship. If standards are not met and/or the customer is dissatisfied, then consumer protection laws permit a customer to recover the money paid for a repair.

Legal Challenges to the Repair Order

The final invoice must be completely filled out to be considered a legal contract. (Contracts are covered further in chapter 14.) The final invoice is created differently than other legal contracts because it evolves in stages. Specifically it goes from the estimate stage to the repair order stage and finally the final invoice stage. Each stage must be completed in detail as discussed earlier; otherwise the final invoice can have defects that hamper its ability to be defended in court or in an inquiry by the State's Attorney General's Consumer Protection Department.

State Attorney General Inquiry

Occasionally a customer's invoice will be questioned by the State Attorney General's Office, typically for a consumer protection violation. A service consultant should defer all inquiries to the owners and managers. Generally what will happen is a staff attorney for the attorney general's office contacts the service facility typically by an official letter. Phone calls from anyone claiming to be a staff attorney should be treated respectfully but considered questionable because it is unknown whether the call is genuine. Therefore as a policy, the caller should be told that a repair or invoice dispute should be made in writing to the service facility owner(s).

The letter sent by the state attorney general will usually request information related to a specific customer, an invoice(s), and the complaint. Naturally, this will require a written response from the owners. Often, the inquiry is to determine whether state guidelines have been met in terms of customer disclosure, procedures followed, and completeness of the documents. If statutory violations are found, the service facility may need to pay money (fines and/or compensation to the consumer) or comply with certain mandates overlooked in the law, such as changing a customer service procedure among other remedies. If the concerns are answered completely and promptly, the matter may be dismissed if the service facility has followed state guidelines. Whether an attorney is required depends on the situation and how well the owner feels he or she can understand the complaint and express, typically in writing, his or her position.

Damage Claim and Insurance

When an inquiry about an invoice involves an alleged injury or damages as a result of the service facility's actions, the service facility's insurance agent should be notified so that a claim can be filed. The service facility must supply all documents upon request to the insurance adjuster. The service facility should record any details, such as the sequence of events or other personal observations in a diary for review later, if necessary. The facts will dictate whether the advice of an attorney is required. Again, with prompt and detailed records of the situation, the claim adjuster can hopefully remedy the situation and close the case quickly.

Court Complaint

At times, there will be a dispute between a customer and the service facility over nonpayment (customer's check did not clear and

FIGURE 6-11 Court is where a complaint about non payment of services may need to be resolved. As the plaintiff, you sit at the table on the right. The customer is the defendant and sits at the table on the left. The judge is behind the bench at the front. If the customer files a complaint about your service facility, the roles described are reversed.

bounced—Figure 6-11) or a customer's expectations was not met (vehicle is still not working properly after a repair). When it does not involve physical damages (insurance matter), then it is best if the matter is resolved between the customer and the service facility owner. If that is not possible, then the court may become involved. Usually the dollar amount of the dispute must be under a certain limit to be heard by a district justice (DJ), magistrate, or justice of the peace. A service facility owner or customer can file a complaint with the court and pay the court fee for a hearing date to be set by the court clerk. On the day of the hearing, both parties will present their complaint to the DJ or "judge."

If either party does not show up on the day of the hearing, then the ruling will be against the party that did not attend. The court will review documents, hear testimony, allow cross-examination of witnesses before applying the applicable law to make a ruling in one party's favor or the other. Depending on the outcome, the service facility may need to pay money (including court costs) to the customer or they may be able to collect money owed from the customer (including court costs). In either case, the matter should be resolved.

When preparing for court, a service consultant can help a facility owner by obtaining all related documents (invoice, estimate, RO's technician notes, inspection sheets, parts invoices, among other information). If the service facility loses the case, the amount of money owed to the customer, assuming the customer paid the invoice prior to the court hearing, is the entire invoice amount plus the court costs and possibly attorney fees. During the hearing, a possible argument to reduce the amount owed to the customer is to request that the cost of the repair parts be removed from the judgment because they were necessary to complete the repair and can't be returned once installed on the customer's vehicle. Therefore, the repair parts are the property of the service facility after it pays the final judgment, unless the amount owed is reduced by the repair parts cost. The judge can take several days to decide a case and the ruling can be appealed to a higher court process. This process may include an arbitration board but depends on the state judicial system.

Each state's or even court's legal process may differ but there is a court to handle legal challenges. Collection of the judgment, however, can be a different matter that is beyond the purpose of this textbook. However, in general, the court ruling can be turned over to the constable or sheriff to retrieve property to pay a debt after certain legal criteria are met.

Some cases, such as a dollar amount in dispute, are more than the lower court's limits so an attorney will be involved and the matter will rise to a higher court. The primary point to understand is that the invoice and related documents must be complete and correct to withstand challenges brought forth by the attorney general's office, insurance company claim, and court hearings. Therefore, the preparation of the invoice and the information required by your state's law must be complete.

Review Questions

Multiple Choice

1. A customer enters the service area while the service consultant is on the telephone with another customer. Which of these should the service consultant do?
   A. Finish the conversation with the telephone customer first.
   B. Place the telephone customer on hold to take care of the walk-in customer.
   C. Acknowledge the walk-in customer with a wave and finish with the telephone customer.
   D. Ask the telephone customer if he or she will call back.
2. A customer recites a list of symptoms to the service consultant. What should the service consultant do next?
   A. Try to write down everything the customer says exactly the way the customer says it.
   B. Try to give an estimate of the repair cost over the phone.
   C. Try to offer suggestions about what the problem might be (over-the-phone diagnosis).
   D. Ask "open-ended" questions to try to narrow down the customer's problems and establish a priority of what he or she needs so the service facility can be of greater assistance.
3. Which of the following is customer information?
   A. The complete VIN
   B. The customer's name
   C. The customer concern
   D. The vehicle's license number

4. What are some concerns about an approval for a repair taken over the telephone?
   A. Will the customer remember the amount of the repair so that when it comes time to pay he or she remembers the cost?
   B. Is the person on the phone really the owner of the automobile?
   C. If a person other than the owner approved the repair, will the owner honor the approval?
   D. There are no concerns with over-the-phone approvals.

5. If a repair order cannot be signed by the customer and the approval has to be taken over the phone, which of the following does not need to appear as detailed notes on the repair order?
   A. Date and time
   B. Name of the person authorizing the repair
   C. Work the person approved and the amount
   D. How the customer plans to pay the bill

6. A customer has just given approval for a repair of his or her vehicle. Service consultant A says that the technician should be provided with a copy of the approved repair order. Service consultant B says documentation of the customer's approval should be on the repair order. Who is right?
   A. A only
   B. B only
   C. Both A and B
   D. Neither A nor B

7. A service consultant has just completed compiling and writing up a customer's concerns. Which of these should she do next?
   A. Dispatch the work order to the technician.
   B. Arrange for a ride home for the customer.
   C. Offer an estimate for the repairs needed.
   D. Confirm the accuracy of the information with the customer.

8. Service consultant A says that when greeting customers, service consultants should offer their name and a handshake. Service consultant B says that when greeting customers, service consultants should make eye contact and smile when welcoming them. Who is right?
   A. A only
   B. B only
   C. Both A and B
   D. Neither A nor B

9. A customer calls with a shopping list of problems with his vehicle. How does the service consultant put this information in a format that will help the technician find the customer's problem?
   A. Write down everything the customer says in the order he or she says it.
   B. Ask open-ended questions regarding each item to determine the problem.
   C. Ask the customer to boil the problem down to a specific system on the car.
   D. Verify that each item on the repair order is a symptom or a maintenance request.

10. Service consultant A says that a customer service computer program is needed for the service facility to save time. Service consultant B says the use of pre-priced jobs can save time. Who is correct?
    A. A only
    B. B only
    C. Both A and B
    D. Neither A nor B

11. A potential customer calls very concerned about an estimate received from another shop. Which of these should the service consultant do?
    A. Suggest that the other shop is probably too high and to make an appointment.
    B. Look the job up and offer an estimate.
    C. Offer a discount if the customer brings the vehicle into your shop.
    D. Show empathy for the customer and offer an appointment for a second opinion.

Short Answer Questions

1. What is a greeting and how is it done?
2. How should a service consultant obtain and document customer information?
3. What type of vehicle information is needed and how is it obtained?
4. How can customer concerns be categorized?
5. Why is repair authorization important?
6. List different telephone skills and explain each.

Class Activity

Class Activity 1: To know what to say a new service consultant should practice shaking hands with another person by extending the arm and at the same time saying,

"Welcome to (the name of the automotive repair business)." "I am the service consultant and my name is __(Paul)___ and your name is _____?" (Wait for an answer then *remember* it.) Next, the service consultant should repeat the name and say: "Mr./Mrs./Miss/Ms./Dr. (customer's name), how may we help you?"

In class, ask two students to demonstrate the preceding script in front of the class. Critique and adjust the interaction, then ask the students to pair up and have one student play the customer and the other the service consultant. Have the students change roles, then change partners and repeat this activity.

Class Activity 2: In pairs ask students to write a service script for steps 1 to 5 of the customer interaction. Ask a pair of students to demonstrate their script, then critique it after the demonstration. After the critique of the demonstration, ask the student pairs to adjust their script based on the feedback provided. Instruct each pair to practice it and take turns in the role of customer and service consultant. Finally, ask each pair to demonstrate their script in front of the class allowing the students who are watching to provide feedback as to the script content and its execution. Try to perfect the script and each student's ability to execute the script as a service consultant until flawless. Also work on non-verbal cues as well, such as shaking hands, eye contact, body language, and enthusiasm as well as voice clarity, tone, and volume.

Activity

Activity 1: Work with your librarian and search your state's website as well as the FCC.gov website for laws concerning appropriate telephone solicitations. Also search Google for telephone solicitation laws as well as email and text for federal and your state's regulations. Determine the rules that you must follow.

Activity 2: Research the guidelines of your state for a legal repair order. Ideas to obtain research include information consulting with your librarian, Google search, study of the state attorney general website, interviewing local service facility owners, examining invoice templates found on a customer service software program or sold by reputable printing companies (found online) for use by service facilities in your state. You can also ask other professionals such as an attorney or the justice of the peace if necessary. Write up your findings in a memo to the owner of what an estimate/invoice must contain, including any customer disclosure wording.

Activity 3: Research your state's law (statutory and/or case law) on a mechanic's lien if there is one for your state. If not, there is likely one for contractors or construction that "applies" to automotive repair. Ideas to conduct research include consulting with your librarian, Google search, study of the state attorney general website, interviewing local service facility owners, You can also ask other professionals such as an attorney or the justice of the peace if necessary. Write up your findings in a memo to the owner of the details about your state's law.

Activity 4: Obtain two to four invoices from past repairs to your vehicles (or another person's). Examine invoices for completeness and ability to defend from a challenge if necessary. Summarize your finding as to the completeness of the documents and any areas that should be improved.

Activity 5: Go to your local small claims court and obtain a court complaint form that would be filed to apply for a court hearing. Ask the court clerk the amount of money paid to file the complaint and any special procedures you must follow if any. Examine the court complaint form and attempt to fill it out using your name as the plaintiff and a fictitious defendant. Provide to your instructor a completed court complaint form.

Activity 6: Many disputes do not require court proceedings but rather a letter informing a customer that the invoice has not been paid and that your service facility is requesting payment. Create a business letter addressed to you as the customer. Mail this letter to yourself both by regular mail (with proof using a USPS "certificate of mailing") and a second letter by certified delivery to understand the USPS delivery process. Provide these letters (you received) to your instructor for review.

Activity 7: Obtain a blank "hand-written estimate (RO)" that meets the legal guidelines of your state (a photo copy of an old one provided by your instructor is fine). Fill it out completely with your name for the customer and fictitious repair information that requires parts, such as a water pump replacement and antifreeze flush. Make sure you include all parts such as fresh antifreeze and obtain your part prices from an online parts store. Mark up "your" cost of parts 2X. Look up labor times from a labor time guide or computer program supplied by your school and use a labor rate of $50 per hour. Make sure the written repair order meets the legal guidelines of your state, such as including your signature as the customer. After completing the hand written repair order, if available at your school, create a computer generated repair order. Hand in both copies to your instructor for review.

CHAPTER 7

WORKING OUT SERVICE DETAILS WITH CUSTOMERS AND INDUSTRY BUSINESS PRACTICES

OBJECTIVES

Upon reading this chapter, you should be able to:

- *Explain why and how alternative transportation is provided (Task A.1.7).*
- *Communicate completion performances to the customer (Task A.1.11).*
  - *Identify labor operations (Task D.3).*
  - *Provide and explain estimates (Task C.1).*
- *Identify and recommend service and maintenance needs (Task A.1.10).*
  - *Describe the elements of a maintenance procedure.*
  - *Identify related maintenance procedure items.*
  - *Locate and interpret maintenance schedule information.*
  - *Communicate the value of performing related and additional services.*
  - *(Task C.4).*

CAREER FOCUS

Your career advancement depends on how well you give advice that can help the customer and generate sales. To do this well earns respect, increased wages, and advancement. However, knowing what to sell and how to go about selling it is the key. As you read this chapter, think about the sale of services as an opportunity to help customers solve their problems by keeping their vehicles in top condition. When sales are done properly, your efforts will be noticed and result in an appreciation for your work that can elevate customer and management opinions about your concern for others and the company plus your knowledge and abilities.

Introduction

When working with customers, service consultants must take care of numerous details from the time of the initial communication and greeting to the presentation of the final invoice. This chapter is devoted to these details. Specifically, the service consultant must serve as a host for the customer, have answers about parts availability and time to perform the repair, cost information, and knowledge about the maintenance requirements or specials that will benefit the customer's vehicle. In addition, the service consultant must be able to keep the customer informed and be prepared to take care of unexpected problems and complications when automobiles are serviced. Further, service consultants must be ready to answer questions and offer advice about repairs and maintenance.

The purpose of this chapter is to discuss the details on how service consultants must be ready to take care of customers when their automobile is serviced. To explain these responsibilities, the chapter is divided into three sections. The first concerns the need to serve as a host to customers while their automobile is at the service facility. The second regards the cost and time required to make repairs and respond to customers whose estimates are not accurate. Finally, the third examines regular and preventive maintenance.

Serving as a Host to Customers

The service consultant is the "host" when customers are waiting at the service facility. Customers are inconvenienced when their automobile is being serviced. Therefore, after a service transaction, the customers' comfort and, possibly, the need for alternative transportation must be addressed.

To determine the customers' transportation needs, service consultants must ask whether the customers plan to wait at the automotive service facility, have a ride to their next destination, or prefer to have alternative transportation. Alternative transportation means that the service facility has a courtesy shuttle to take customers to their next destination, or that loaner cars or rental cars are available. Loaner cars are typically given to customers for a short period of time and are free of charge while their automobile is being serviced. Rental cars, in contrast, cost the customer money, but in some cases, a warranty company absorbs the cost.

If customers plan to stay at the service facility while service is performed, service consultants should make sure they are comfortable. Specifically, service consultants should make sure their customers have newspapers and magazines to read, know how to operate the television, and have fresh coffee to drink. In addition, service facilities should have books, toys, and activities to occupy the customers' children. Helping customers pass the time comfortably while waiting for their automobile is important but service consultants must not let this interfere with the performance of their job. For example, at a service facility in New Jersey, customers can use a set of its golf clubs to play virtual golf at no charge while waiting for their automobile.

Towing Arrangements

Unfortunately, not all customers will drive their vehicle to the service facility. Sometimes they are broken down. Some customers may have towing coverage through their insurance or an organization, such as the American Automobile Association, most commonly known as the AAA, which will take care of the details. Other customers will call the service facility and the service consultant must arrange for towing. When towing is to be arranged for the customer by the service consultant, he or she must make sure to obtain all of the details from the customer to give to the towing company and/or tow truck driver. Problems occur when the service consultant does not get the exact location of the vehicle and its keys as well as the make, model, color, and license number. Also customer contact information, such as cell phone number, must be up-to-date and accurate.

Detailed information collection is important because different vehicles require different types of tow trucks. All-wheel drive vehicles and full-time four wheel drive vehicle require a rollback or car carrier tow truck (Figure 7-1). A wheel lift tow truck that lifts only one set of wheels can be used provided it has a dolly to put under the second set of wheels before towing (see Figure 7-2); if not, it cannot be used. Failure to get the correct information can result in vehicle damages, delays in retrieving the vehicle, and even mistakes such as towing the wrong vehicle. The idea is to get the customer off the road and back to the service facility as quickly as possible. If the tow truck is delayed and the customer is waiting with the vehicle, the shop should consider sending the courtesy shuttle to pick up the customer so he or she is safe at the shop unit until the vehicle arrives.

FIGURE 7-1 Rollback or car carrier–style tow truck picking up a broken down motorist.

FIGURE 7-2 Wheel lift–style tow truck. This tow truck can be used to tow this all-wheel drive vehicle provided the dolly shown behind the cab of the truck is used under the rear set of tires.

Repair Cost and Time Promised

Customers generally have two major concerns when they enter a service facility to have their automobile repaired. The first is the cost and the second is when their automobile will be ready. Accurate estimates are important because customers do not like to be "surprised" when an invoice shows costs that are more than expected.

As noted in chapter 6, the cost estimate should be slightly higher than anticipated. This is because an estimate, which includes labor, parts, and taxes, may be compromised because of errors in calculation, unanticipated problems when making the repair, and the cost of parts.

Service consultants must try to inform customers when their automobile will be finished. In most cases, the service consultants can estimate the amount of time a repair will take by looking it up in a labor time book or a computer database. To locate this information, they must be sure to have the correct year, make, and model of the automobile as well as other features, such as engine size and transmission type. Unfortunately, the availability of parts and repair complications (a part is rusted and cannot be removed quickly) can extend the time it takes to make a repair. When this occurs, the customers must be notified immediately because they do not like to be "surprised" when told that their automobile is not ready when promised.

Labor Times: Where Do Labor Times Come from?

Labor times are divided into two categories. The first are times studied by the factory and used for warranty claims at new-car dealers. The second are aftermarket labor times published in books sold to service facilities that are derived from the factory times. These times are usually a little higher than factory times because it is felt that aging parts require more time to remove, such as rusty bolts. Commonly, aftermarket labor times are used by nondealership repair facilities and by many dealers for "customer paid" repairs. Labor times are studied by the manufacturer and the actual repair is conducted by technicians who work for the manufacturer. They conduct the repairs in a shop environment located at one of the manufacturer's test facilities. The repair is performed multiple times. Extra time is added to account for obtaining parts, normal work breaks, and if required, setting up the vehicle on the lift. When applicable, the time usually does not include the "test drive" because the service consultant is assumed to perform this function. Diagnostic time is often included in the repair time, so is the removal of the related parts to perform the repair. The manufacturer's technician uses the same special tools for the job as the dealer is required to purchase for the service department. It is assumed that special tools are organized and readily accessible to the dealership technician. Furthermore, the repair process uses hand tools and not power tools during disassembly and reassembly process (see Figure 7-3).

The entire repair process is monitored so that the technician follows the same procedure as found in the repair information and the study can even result in improvements to the service information for dealership technicians. A specific repair is repeated more than once so that the labor time established for warranty claims reflects what an experienced technician can do when assigned the same job multiple times in his/her career. For example, the first time to remove an engine during a time study might be 12.5 hours and the second is 11 hours; the second labor time will be

FIGURE 7-3 At a dealership, a technician performs an engine repair. The time for the repair was established by the process described.

used in the labor time manual. The actual labor time research method may vary by manufacturer but the end result is a repair time that is considered adequate for performing the same repair in a similar shop environment. When dealers feel the repair time is not adequate, there is a formal process to petition the manufacturer to review the labor time for an increase. Also it is possible for a labor time to be decreased if it is found that fewer parts than necessary needed to be removed to perform the repair. When the labor time concern affects previously submitted claims, the manufacturer can charge back dealership service departments if the labor times were too high or give credits if found they were too low.

The Labor Times and Labor Sale Price

The labor sale price is based on the labor time (flat-rate time) to complete a specific type of repair. The labor time is obtained from a factory flat rate or aftermarket computer program database (perhaps a paper manual). For a specific year, make, model, and options, the service consultant will obtain the labor time to do the job. The labor time will then be multiplied by the service facility labor rate to obtain the labor sale price for the customer estimate. To illustrate how this works, next is an example that came from an aftermarket database for the vehicle listed.

- *Vehicle:* 2003, GMC Yukon, 6.0 liter engine, with active suspension
- *Repair Request:* Replace both front shocks
- *Labor Time:* 1.4 hours (add 0.2 hour for active suspension)
- *Labor Rate:* $90 per hour

Service Consultant Calculation of Labor Sale Price for the estimate:

> 1.6 hours (1.4 + 0.2 for active suspension) × $90 (Labor Rate) = $144 labor sales price

The service consultant will put on the estimate under the labor sale area in a statement such as:

> "Replace Both Front Active Suspension Shocks (1.6 hours)" $144

Provided there are not any additional comments or suggestions from the technician, this statement will follow through to the invoice.

Parts Cost and the Parts Sale Price

An estimate presents the prices for parts and labor to a customer. The estimate assumes that the information received on the cost of the parts to a facility is accurate. To arrive at a price to charge a customer (parts sale price), the service consultant begins with the cost the service facility must pay for the part. The part is then resold to the customer at a price that is **marked up** a specified amount or percentage (mass retailer method) over the cost. **Markups** are needed to cover the expenses that a facility incurs to obtain or pick up part as well as earn a profit from the sale. In some cases, the markup covers any parts that must be prepaid before delivery (such as engines) and cost to store them in the inventory, the time the parts specialist takes to look up the parts and order them from a supplier, and the time managers or owners spend to make arrangements with parts suppliers. In addition, the markups must cover the work associated with core returns or the proper disposal of old parts in accordance with local, state, and federal laws.

The part number and price to be charged to the customer are shown on the estimate as well as on the repair order once the customer authorizes the work. Therefore, the amount of the markup must be carefully calculated. To relieve the service consultants from calculating parts markups, many customer service computer programs calculate them when a part or its cost is entered. The price charged to the customer is then added to the estimate.

Information about the amount a service facility pays for a part is obtained from the parts supplier, an Internet connection to a parts supplier's computer database, or a computer link to a parts supplier inventory system. Of course, the critical task is for service consultants to be sure that the vehicle information needed to order parts is correct. This may seem to be an easy task but a part for the same year and make of automobile may be different for the various models, engine size, and so on. If an incorrect part is identified, a difference in the cost of the part to the service facility is very likely.

Parts Availability

The next piece of information that is critical to the service consultant is whether the part is "in stock" (called parts availability) or if it must be ordered. Customers must be told when the part is not immediately available because they may need to return the automobile at a later date for the repair.

ADDITIONAL DISCUSSION

The Parts Markup Calculation

The author's second textbook *Managing Automotive Businesses: Strategic Planning, Personnel, and Finance* goes into accounting and markup further. However, a service consultant should know how parts markup is calculated. The terms markup and gross profit margin are specific terms, but may be calculated differently by managers, bookkeepers, and accountants. Specifically, some calculations use equations that require "multiplication factors," some use "division factors," and some use a "grossing up" method (where the cost of parts is held at 100% and the sale price percentage is a percentage over 100%). Therefore when the term "markup" is used, one must determine the method of markup and the definition the person is using before proceeding.

Most parts stores, dealers, and service facilities use the mass merchandiser "retail" method. This means markup is the difference between the sales price of the part (sale price percentage held at 100%) and the cost of the part with the cost being less than 100%. To illustrate this concept:

Sale Price	$25	= 100%
Part Cost	$10	= 40%
Gross Profit	$15	= 60%

The "retail markup" is 60% (the difference between the sale price and part cost) and the markup in this example is the same as the gross profit margin. It is more important for the parts specialists and service consultants to be able to calculate the proper markup for each part to be sold. For example, if the service facility bought a part for $10 and the owners directed management to ensure a markup of 60% was obtained, then the following equation would be used:

Sale price = Part cost/(1 − Markup)

Therefore, if a parts specialist has a part cost of $10 that is to be marked up 60% or 0.6 (which equals the gross profit margin ratio), then that part must have a sale price of $25. This is calculated as follows:

Sale Price = $10/(1 − 0.6) = $10/0.4 = $25

In the next example, assume the owners desire a 40% gross profit margin for a part that cost $35. The markup would also be 40% (*note:* a parts specialist would use 0.4 when the percentage is converted into a decimal so that it can be used in the equation). For the $35 part, it must be sold for:

Sale Price = $35/(1 − 0.4) = $58.33

A mathematical "proof" for this example would be as follows:

Sale Price = $58.33 = 100%
Part Cost = $35.00 = 60% ($58.33 × 0.6 or 60%)
Gross Profit = $23.33 = 40% ($58.33 × 0.4 or 40%)

Therefore, the availability of the parts must be confirmed before customers are told when they can pick up their automobile. The availability of a part is also needed to ensure it arrives prior to the start of the repair. For example, if a technician has removed an old part from an automobile and does not have the new part, a service bay is tied up and time is lost on the repair. This costs the service facility and even the technician money, and the customer will not have the automobile when expected. This means the owner, manager, technician, and customer will be unhappy! A service consultant who is on top of his or her job knows that to avoid parts availability problems, he or she will examine

> **CAREER PROFILE** **The Parts Specialist**
>
> The parts specialist's job can be challenging yet appealing for many graduates of automotive technology programs. There are two basic types of parts operations: the dealership parts department and auto parts stores. The customer service skills, knowledge of automotive systems, and the ability to accurately look up and locate parts are similar. However, they differ in the scope of the job.
>
> Auto parts stores will typically deal in aftermarket and some factory or original equipment manufacturer (OEM) parts that customers need for a variety of makes and models. This means there are often options for the customer to choose, such as the quality of the parts and brands, which have different prices. Therefore, the scope of knowledge to work in an auto parts store is broad. Especially since the parts specialist serves retail customers who tackle small automotive repairs and also service facilities that need parts to finish their customers' vehicles' repairs. Often there is a wider variety of inventory at an auto parts store to serve a wide range of customers' needs. Most employees of parts store enjoy it very much because of the frequent customer interaction and advancement in large companies.
>
> In terms of parts, it is impossible to have all of the parts for all vehicles given the variety of vehicles and number of parts. Therefore, a desired part may need to be ordered because it might not be in the store's inventory. Typically, this is not a problem because of parts proliferation with multimillion dollar inventories at regional warehouses to back up local dealerships and parts store inventory; parts for service facilities are usually available the next day.

the jobs scheduled for the next day (or the coming week) and assure parts are at the service facility before the vehicle arrives.

The dealership parts department is similar to the auto parts store but typically the only parts sold are OEM parts or parts made by the manufacturer of the vehicle. All of the parts for current and many late model vehicles are available from the dealer. However, as a model's design changes, parts are discontinued and not available any longer. Like a parts store, only certain parts are kept in stock and some parts need to be ordered. The complexity of looking up parts (see Figure 7-4) is not part of this textbook, but training and experience is needed to become proficient. Proficiency, whether at a dealership parts department or auto parts store, means that the parts specialist knows:

- What vehicle information must be obtained to properly look up the part,
- How the catalog system works,
- How parts are numbered and organized within the catalog system,
- What parts interchange to other parts or new numbers,
- How to locate parts that are in stock,
- How to locate and order parts that are not stocked on the shelves.

When working with customers at either the dealership parts department or auto parts store, the operation has two types of customers: the retail customer and the wholesale customer (see Figure 7-5). The retail customer may not know very much about cars. There are many

FIGURE 7-4 A dealership parts specialist and service consultant work together to check on a customer's parts order. They have the same goal to get the correct parts in a timely manner to assure the customer's expectations are met.

FIGURE 7-5 This parts specialist works at a parts store that specializes in high-performance parts. Retail customers ask his opinion about parts applications relative to their special vehicle needs. This is a very specialized parts specialist position that requires a unique understanding of the high-performance parts industry.

questioning techniques used to help retail customers get the correct parts needed that are beyond the scope of this book.

Wholesale customers typically know more about vehicles because they work for service facilities that buy parts for technicians to install. The person who calls the parts store or parts department may be a parts specialist at the service facility who already has the part number and vehicle information. However, the person who calls may also be a technician, service consultant, or other employee of the service facility. They may or may not readily have the information needed to look up the part. However, with more online resources such as parts illustrations, pricing, inventory information (availability), and online ordering, it is more common for them to obtain the needed information before calling the parts store. This is becoming more common even for the retail customers to use online resources to know what they require in some cases before they come to the parts store.

However, in other cases the wholesale and retail customers may not know what they need because parts have become more complicated and sometimes whole assemblies or entire components, such as CV axles and brake calipers must be ordered rather than obtaining a part to repair an assembly. At times, a parts specialist may find a kit to rebuild the components or assemblies is available as well as units that have been already reconditioned (returned to working order), rebuilt (parts that typically wear out are replaced), or remanufactured (all parts are replaced except for major components that are inspected and reconditioned). There is a lot of detail to this job and it is more than a clerical position where the part specialist waits for an order (see Figure 7-6). It is challenging and requires technical knowledge, training, and experience to do it effectively.

FIGURE 7-6 The parts specialist first checks the computer inventory. Seeing that the part is in stock, he goes to the correct shelf and pulls the part by matching the part number he looked up with the part number on the box. At times the computer's inventory list has incorrect information and the part is not in stock or the part was placed in the incorrect location. For this reason the inventory is counted periodically and the part numbers on the shelves are checked against the computer's inventory list.

Working Out the Details with the Customer

After an estimate of the repair cost and the time it will take to complete the repair have been calculated, the service consultant must take time to inform the customer of the approximate time of completion. The service consultant should begin by explaining the charges to the customer by stating, for example:

> "According to our database, this repair will take _X_ hours and the parts, labor, and tax will be less than $_ (estimated amount). Assuming the parts are available and the work in our shop stays on schedule, your automobile should be available by ____. Is this okay with you?"

The estimate would then move to the RO stage as discussed in chapter 6. However, if the time frame is not acceptable or the amount of money is too much, then the service consultant must review options with the customer. In terms of promise time, the service consultant can suggest a rental car or other transportation options to the customer. If those are not acceptable, then perhaps the job should be rescheduled. If the estimate is too expensive, then the service consultant might suggest the credit cards the service facility accepts. If cost continues to be a problem, then an examination of what services might be removed and performed at a future date. If nothing can be cut from the estimate (lower cost parts, delay in performing certain services, among other options), then the job should be rescheduled for when the customer can afford the repair. Again, the service consultant must remember that the estimate should have a cushion as explained in chapter 6 so the estimate is slightly inflated in order for the final invoice to be less than expected. Also, service consultants must not "overpromise" the time when the automobile will be available. The automobile repair should be completed and ready to be driven off the service facility's parking lot when promised. Therefore, a customer should never have to wait until it is taken out of the technician's bay, test driven, and cleaned up.

When the Estimated Cushions Fail

Unfortunately, despite the service consultant's best efforts, a final invoice may exceed the estimate or the automobile will not be ready when promised. The reasons for this are often due to various factors, such as ever-more-complex automobile systems. The key to handling these unexpected problems and complications is to keep the customer informed.

For example, a single defective part may have several different replacement choices, each with a different price and availability. In addition, a complex system may cause technicians to overlook related components that are defective or are on the brink of being defective when the diagnosis is conducted. These situations result in errors, which, in spite of the best efforts of the service consultant and technicians, will exceed the estimate and the time needed to make a repair will be longer than expected.

How these differences are handled depends on service facility policy. When a technician runs into a problem, the service consultant must be made aware of the situation immediately. Then various options should be considered to rectify the problem. For example, the technician and the service consultant need to work together and redo the estimate. Hopefully, they will be able to come up with options that can be presented to the customer. Then the customer must be called to explain the situation and the possible options available.

Selling Automobile Maintenance

The AAA estimates that the majority of automobile owners ignore the maintenance recommendations made by manufacturers. The United States federal government knows this all too well, so it empowered the EPA under the Clean Air Act to oversee the implementation of emission inspections and maintenance (I/M) programs. These programs require the emission inspection of automobiles in highly and densely populated areas and require repair if the automobiles do not pass inspection. The inspection regulations vary from state to state and even from area to area within some states.

Owners neglect the maintenance of their automobile for different reasons. For example, they may not want to spend the money, or they may be unaware of the maintenance needed to keep their automobile in sound operating condition, or they may not believe maintenance is really necessary. Service consultants must advise their customers of the maintenance services recommended for their automobiles and explain how each benefits them in terms of avoiding future repairs and improving vehicle reliability.

Pre-Priced Maintenance Menu Sales

When advising customers, service consultants must explain that the maintenance and repairs needed for older and higher mileage automobiles are different from newer automobiles. The sale of maintenance packages at certain mileage intervals, for example, 15,000 miles, works best for relatively new automobiles such as those that are less than four years old and under 70,000 miles. These packages are often based on the manufacturer's recommended maintenance schedule, which is designed to keep the customer's automobile in top mechanical condition. Naturally, the sale of a maintenance package assumes the customer with a newer vehicle wants to keep it in top condition and wants the manufacturer maintenance services done at the suggested mileage intervals.

To obtain the maintenance schedule for the customer's automobile, the service consultant can look in the owner's manual, an aftermarket labor time publication, or a computer database that lists the manufacturer's recommended maintenances by mileage. Once the labor time is known for each of the maintenance items, the service consultant can calculate the cost of the maintenances. The service consultant should not forget to add the cost of parts, fluids, and supplies.

To speed up the process, many service facilities use **pre-priced maintenance menus**. As shown in the example in Figure 7-7, such menus may list the different maintenances found in the manufacturer's manual down on one side with the mileage intervals across the top. The center of the chart shows the total price for each service with the total cost of the combined service for every mileage interval. Figure 7-7 is a miniature example of a service menu. When a customer decides to order one of the services, the service consultant will identify the service and enter it on the estimate.

To suggest these services to the customer, a service consultant may consider the following approach after addressing the customer's automobile repair needs:

> "Now that we have the repair cost estimate finished, how many miles does your vehicle currently have?"

This will help to establish the approximate miles so that maintenance needs can be matched to the vehicle. Using a menu such as Figure 7-7, the service consultant will relate the current mileage reported by the customer (if not already obtained in the estimate stage) to the maintenance requirements of the vehicle. Often the menus are specific to an automobile design, such as Chevy Trucks. A service consultant would say:

> "The factory recommends the following services at _X___ miles."

		MILEAGE		
SERVICE	3,500	7,000	10,500	15,000
Oil change	$22	$22	$22	$22
Lubrication		$5		$5
Tire Rotation		$16		$16
Alignment				$80
	$22	$43	$22	$123

FIGURE 7-7 A vehicle maintenance schedule.

Next the service consultant should show the customer the maintenance menu, the computer database printout of the services, or the page from the manufacturer's and owner's manual. Then, the service consultant might offer the following to the customer:

> If our technician could have the recommended services completed before you pick up your automobile, would you like for us to complete them for you?

If the customer agrees to have the maintenance done, it should be entered on the repair order.

The method described earlier works well for newer automobiles whose owners want to ensure that every maintenance procedure is done at the interval recommended by the manufacturer. However, service consultants may find that customers with older or high mileage automobiles are not interested in the same menu. This is mainly because these customers are more concerned with "keeping their automobile going" with the repair and maintenance of systems that prohibit function of what they consider "critical components." Therefore, these customers may be more receptive to maintenance specials packages discussed next.

Maintenance Specials Sales

Typically, maintenance specials are for the more common maintenance services, such as an oil change. In states that require an annual state safety inspection, the inspection service can be a time when customers also receive seasonal maintenance services. A seasonal maintenance service includes such services as antifreeze flush in fall or spring. When a seasonal maintenance is combined with other services, such as an inspection, a discount is often offered and is referred to as a maintenance special. This is often popular and has a broad customer appeal.

To suggest maintenance specials to customers, service consultants should remember that part of their job is to provide advice. The suggestion about maintenance is intended to help customers keep their automobile in top condition. To help a customer aware of maintenance specials, the service consultant might ask:

> Now that we have addressed your concerns, (Note: this question would likely be asked in the estimate stage 1 after the service consultant has discussed the cost of repair with the customer), we have a special that gives you. . . (provide an appropriate service recommendation such as) . . . an oil change and antifreeze flush for $__. If our technician could have this service finished before you are ready to pick up your automobile, would you like to have that done?

> If the customer agrees to have the service done, record the sale on the estimate.

Service consultants should not be discouraged when customers refuse to have a recommended maintenance performed. Some customers may want to think about it until the diagnosis or inspection is completed and they know more about the cost of repairs. This is not a cause of worry because there will be another opportunity to promote the suggested maintenance again when the service consultant calls the customer with the results of the technician's findings (additional work: stage 3).

Finally, a method used to promote maintenance specials is to keep regular customers informed about the services their automobile needs when they schedule an appointment or arrive at the service facility. The information needed to provide this type of customized assistance comes from the customer service computer program database. Most databases contain a wealth of information, such as past estimates given after a previous repair was completed, past maintenance services performed, and naturally, the vehicle's age as well as past mileages. This information can be used to focus attention on critical repairs or maintenances that the customers' automobile needs. For example, if the database shows that there has been 10,000 miles driven since the customer's last tire rotation, the service consultant should suggest a tire inspection and rotation while the automobile is in the shop.

Preventive Maintenance Sales

Preventive maintenance usually involves the inspection and replacement of parts before they break. This is an important service not often considered by automobile owners but is common among fleets because they have data to support failures that regularly occur over a certain mileage or vehicle age. One of the reasons these services are not considered is that many owners do not know what needs to be changed on their automobile and this is where a service consultant can help.

For example, assume an automobile engine has 75,000 miles and the automobile manufacturer recommends a timing belt replacement at 60,000 miles. The service consultant should examine the customer's file in the database, and if the belt has not been replaced in the past, it should be brought to the customer's attention. The service consultant should advise the customer that the belt should be changed and explain why.

Other items to check in a customer's file in the database include the most recent brake pad thickness readings at the last service relative to the current mileage of the automobile. If the lining was thin and the automobile was driven a lot of miles since the last check, it should be inspected again. In addition, the facility should keep track of when each customer's state safety and emission inspection is due as well as when the engine had its last oil change, the spark plugs replaced, the radiator hoses and fan belt changed, and the fuel filter replaced. Because computer programs can identify customer automobiles based on past estimates and services, some facilities send out reminder notices. These notices should remind the customer of any maintenance services needed and the specials that are being offered.

Related Services and Parts Sales

When customers request repairs, service consultants should always consider any maintenance services that should be performed at the same time. For example, when a water pump that is driven by the timing belt is replaced, it may make sense to replace the timing belt. This will help avoid a breakdown after the old belt is re-tensioned to the new water pump. The opposite may also hold true because old water pumps with loose bearings may seize soon after a new timing belt is installed and re-tensioned. Needless to say, old water pump bearings that seize will ruin a new belt. Therefore, the service consultant must help the customer understand the relationship between the system parts and should ask the technician to pay special attention to the condition of related parts.

Likewise, when a repair is performed, such as the replacement of a head gasket, it makes sense to change the oil, drain and replace all of the antifreeze, and maybe even replace the spark plugs at the same time. These services will reduce the likelihood of a costly comeback from oil contaminated with small particles of dirt and small amounts of antifreeze that have drained into the oil. These services also benefit the customer because he or she saves money if the work is done when the head gasket is replaced. This is because the antifreeze is drained from the engine and the spark plugs are readily accessible when the head gasket is removed. Therefore, the labor cost is minimal because most of the charge is for replacement parts (spark plugs and oil filter) and fluids (antifreeze and oil).

Tracking Parts

Often a service consultant will need to track parts when the service facility does not have a parts specialist. The concept of tracking parts is based on the need to know the parts supplier where the part was ordered, the part number that was ordered, an approximate delivery time (ordered or in stock), as well as the cost information for comparison against the part delivery receipt. This is needed to orchestrate the parts arrival with the technician's schedule and the customer's promise time. An example of the parts specialist tracking sheet is shown in Figure 7-8.

PARTS SPECIALIST	PS = Parts Specialist	SC = Service Consultant		
Parts to be Ordered	RO = Repair Order	Tech = Technician		
TRACKING SHEET		TL = Team Leader		
Customer LastName	Anderson	Anderson	Peterson	Thomas	Guzman
RO#	1234	1234	1235	1233	1239
Vehicle year	2012	2012	2013	2015	2011
Vehicle Make	Ford	Ford	Toyota	Chevy	Chrysler
Vehicle Model	F250	F250	4 Runner	Impala	Caravan
TECH Name	Tom	Tom	Frank	Bill	Frank
TECHNICIAN PRICE REQUEST					
For additional work					
Technician Repair Operation	brake pad	Calipers	Head light	VC gasket	Lug nuts
Number required	1	2	1	Pair	7
Vendor	Carquest	Carquest	NAPA	Dealer	Dealer
Part Number	XX-1234	998/997	FB-89	XXX	YYY
Core charge	No	Yes	No	No	No
Estimated delivery time of ordered parts	3pm	3pm	11am	Tues PM	2pm
Cost of Part (per unit)	27.66	30.02	5.6	14.98	1.05
Sale Price of Part (per unit)	69.15	60.04	14	37.45	2.62
Total Sale Price of all units	69.15	120.08	14	37.45	18.34
(When SC duty) labor time to install part	1.0 hr	2.1 hr	0.6 hr	1.6 hr	0.3 hr
Authorization					
Time SC placed part order with PS	12:15	12:15	11:30	12:05	1:10
Time part arrived to PS from Vendor	2:50	2:50	11:37		Pick up
TIME PS texts (SC+TECH) pick up parts	3:12	3:12	11:46		
Check Out					
TIME PS supplies parts receipt copies to SC	3:33	3:33	11:55		
Core Returned to Vendor	NA	Yes	NA		
Receipt number of core return	NA	2224	NA		
Returns properly credited to account	NA	Yes	NA		

FIGURE 7-8 Shown is a parts tracking sheet that a parts specialist would use. It can be adapted for a service consultant to use as well. Study of the sheet shows various parts orders. The parts specialist keeps track of each part ordered so the shop can get the parts needed to keep the jobs in the workflow process moving forward.

Review Questions

Multiple Choice

1. Service consultant A says that telling the customer when the vehicle will be ready at the time he or she drops it off creates expectations. Service consultant B says that accurate completion time can only be determined after vehicle inspection and diagnosis. Who is correct?
   A. A only
   B. B only
   C. both A and B
   D. neither A nor B
2. When a vehicle is found to need maintenance work that was not requested by the customer, a service consultant should recommend it because:
   A. It provides additional income for the service facility.
   B. It is the responsibility of the service facility to advise the customer of his or her vehicle needs.
   C. The customer must have this work done to maintain the vehicle's warranty.
   D. It keeps the vehicle in good working order.
3. A service consultant needs to find alternative transportation for a customer who is under 21 years old. Policy dictates that rental or loaner cars cannot be given to anyone under age 21. What do you do?
   A. Offer a reduced rate rental car.
   B. Lend the boss's vehicle.
   C. Offer directions to the nearest bus stop.
   D. Offer to provide a ride.
4. Service consultant A says that providing an estimate is required by law in most states. Service consultant B says that explaining the details of an estimate helps to add value to the services the customer is buying from the shop. Who is correct?
   A. A only
   B. B only
   C. both A and B
   D. neither A nor B
5. Service consultant A suggests that offering a customer a ride home or to work represents alternative transportation. Service consultant B says that driving the customer to the bus stop is providing alternative transportation. Who is correct?
   A. A only
   B. B only
   C. both A and B
   D. neither A nor B

Short Answer Questions

1. Explain why and how alternative transportation is provided.
2. How are completion performances communicated to the customer?
3. How are labor operations identified?
4. How should estimates be presented and explained?
5. How can a service consultant identify recommended service and maintenance needs?
6. What are the elements of a maintenance procedure?
7. How can maintenance schedule information be located?
8. How is the value of performing related and additional services communicated to the customer?
9. Examine the tow truck picture provided and explain which tow truck can be used to pick up an all-wheel drive vehicle?

Activity

Activity 1: Use the web to find an auto parts store that will let you look up parts online. Then choose your or a friend's car to look up the following parts. Start with some common parts such as headlight bulb, air and fuel filter, and wiper blades. Then try to find some mechanical parts, such as ball joints, radiator, and alternator. Finally, look for some difficult-to-obtain parts from an auto parts store, such as the driver seat belt assembly.

- Report for each item whether it is available or not.
- If the part is available, is there a choice in brands?
- Next, report the part number, price, and whether it is in stock; if there are choices in brands, choose one to report.

Activity 2: Some of the parts in activity 1 may not be available. Contact a dealer to determine if the part can be obtained and report your findings. Some dealers have an Internet site that you can use; try one out. If the part cannot be obtained from the dealer, report how it possibly might be obtained.

Activity 3: Use a labor time guide (database) supplied by your school and the vehicle from activity 1 to look up the labor times to install the parts listed in activity 1. List each labor operation and create a labor estimate for a customer using a labor rate of $90 per hour.

Activity 4: Calculate the part sale price to a service facility customer for each part in activity 1. Use a markup of 60% and show your calculations as shown in the chapter. Then add the labor cost from activity 4 to the part sale price to determine the total piece of each labor and part operation. Finally, add up the sale price of all of the labor and parts operations and multiply by 1.15 to establish a cushion and total estimate price that includes tax (assume the tax is 6%). Examine your estimate and determine if there are any additional parts, fluids, or related services (such as alignment) that may be required to complete the vehicle repair estimate.

Activity 5: Use a vehicle owner's manual to determine the required maintenance for a vehicle. Report what is recommended by the manufacturer at 30,000 miles.

CHAPTER 8

CLOSING A SALE AND SUGGESTING ADDITIONAL WORK

OBJECTIVES

Upon reading this chapter, you should be able to:

- *Explain how to promote the procedures, benefits, and capabilities of the service facility (Task A.1.8).*
   - *Close the sale (Task C.7).*
   - *Explain product and service features and benefits (Task C.5).*
   - *Overcome objections (Task C.6).*
- *Identify and prioritize vehicle needs (Task C.2).*
- *Explain how to present customers with the work to be performed and related charges, and review the methods of payment (Task A.1.16).*

CAREER FOCUS

The more you sell, the more you make as a professional service consultant. However, recognition for your sales ability goes beyond your income. It is prestigious and you will be looked up to by your technicians, management, and owners as you close sale after sale. Even your customers will recognize your talents as you present them with recommendations that are well organized and demonstrate your ability to explain highly technical problems in terms they understand. As your customers approve repairs you recommend, you are not a salesperson or clerk but rather a valued consultant who is on the same level as other professionals who provide high-quality advice to clients. This chapter presents the concepts that can help you enhance your talent to become a skilled professional.

Introduction

One of the responsibilities of service consultants is to promote service facility sales. To accomplish this, service consultants must prescribe appropriate services that help customers solve their problems. More commonly, service consultants suggest services that help keep customers' automobiles in top condition, such as the maintenance specials discussed in chapter 7.

When promoting sales, however, service consultants must avoid **overselling** to the customers. In other words, they should not try to sell customers any services their automobile does not need. When suspected of doing this, they can lose the confidence of their customers and may even break the law.

To avoid the temptation to oversell, service consultants must put their customers first and be committed to offering recommendations that will enable their automobiles to run longer and better. To assist them, service consultants should refer to the recommendations made by the engineers who manufactured the automobile, the expert advice of the service facility technicians, and management's seasonal specials aimed at assisting customers at discounted prices.

The purpose of this chapter is to describe some of the procedures and techniques service consultants should follow when selling services to customers. First, to make a sale, service consultants must understand the products and services sold by their facilities and how they benefit customers. Next, service consultants must be able to explain clearly the benefits of the service to customers in relation to the needs of their automobile and how the work will be performed. Finally, in order for a facility to be a "full-service facility," service consultants must be able to offer and explain the different methods customers can use to pay their bill.

Closing a Sale

The previous chapters discussed the tasks and duties of the service consultant from the greeting of customers to the presentation of the invoice plus the reasons why complications occur during this process. At any time during this process, however, customers may walk away and the service consultant will lose the sale. The service consultant must be mindful that one of the duties is to get customers to accept his or her recommendations and approve an estimate.

To get approval for a service (signature on the estimate so it becomes an RO) means the sale was closed, also called **closing the sale.** To do this successfully, the service consultant must be genuine in his or her desire to help the customer. If not, he or she may be seen as an insincere "snake oil" salesman, meaning a person who sells something that is worthless. One way to communicate sincerity is to show a willingness to help customers improve their automobiles' performance so it will last longer and function better.

Therefore, a service consultant must think of customers as clients or friends and recognize that their automobiles are important personal possessions, often viewed with pride. For example, for one reason or another, such as body style or gasoline efficiency, customers typically handpick their automobiles when they purchase them. A service consultant must never put down the customer's selection of an automobile because to do so implies the customer had made a wrong decision.

In addition, when treating customers as clients or friends, a service consultant must explain the work to be performed on their automobile in a patient, calm, and logically prioritized manner. Then the service consultant must explain how the service facility is equipped to perform the work and why the technicians are the best at performing the services needed on the customer's automobile. This, of course, is easier said than done when a service facility is busy, but the techniques must still be practiced.

Customer Communication Methods

To close a sale with customers who "call" (this is meant in the broadest sense, it may be telephone, text, or email among other means) for information, the caller must be enticed to come in for service. For example, as mentioned previously, many customers seek assistance over the phone, by email, Internet, or text before coming to the service facility. In some cases, the customers may try to shop for the best prices for a repair. This may occur because they don't know what other questions to ask the service consultant. In these cases, service consultants must try to convince the person on the other end of the phone or computer to bring his or her automobile into the service facility for a proper inspection and diagnosis. This is often difficult because the difference between obtaining the customer's business and losing it depends on how well the service consultant is able to project an image of caring about the customer's problem. Thus, this step is important and difficult to do over the phone or other electronic means. For a successful communication, the service consultant must use a tone of voice and the words to convince customers that the service facility can help them solve their problems and relieve them of their concerns. The words and phrases selected must match the tone of the message or "voice" on the telephone. For instance, a service consultant should take a positive position when telling customers how the service facility can solve their problem or improve their automobile performances (assuming that is true) if they bring their automobile into the facility. For example, statements that imply doubt, such as "I think we can take care of your problem," or any negative thoughts, such as "Well, if we cannot fix your car, you can trade it in for another one," should never be used.

When customers agree to bring their automobile into the shop, they should be provided with directions to the service facility. If the vehicle is broken down, the service consultant may arrange for a tow truck to pick up the automobile. As explained in chapter 6, when customers arrive at the facility, the service consultant must be ready to greet them properly, prepare the estimate, and then close the sale with an authorization for the repair.

Selling Repairs: Prioritizing Automobile Problems

One recommended approach to working with customers to sell the repairs needed on their automobile is to use a method that prioritizes problems from the most dangerous to the least problematic. This method explains the work to be done, and allows the service consultant to promote the ability of the service facility to do the work. In addition, this method presents the opportunity for the service consultant to overcome any customer objections when each finding is discussed and before the customer is asked for approval.

Suggested Repairs from Technicians

A common method to obtain the repair or service information needed to present to a customer in the additional work stage (3) is for a technician to examine the automobile (see Figure 8-1) and then provide a list for the service consultant to review. The technician and the service consultant should then take the suggestions and prioritize them from the most important (#1 is the highest priority) to the least important. For example, a safety problem, such as an axle that has too much play compared to manufacturer's specification and may fall off (Figure 8-2), will have a higher priority number (say a number 1) than a piece of trim that is loose.

FIGURE 8-1 Technicians making a repair.

FIGURE 8-2 Levels of repairs.

In between the most and least important repairs are other service suggestions of varying importance, such as tire wear at the edges and small oil leaks at the valve cover. An example of a list provided by a technician might be as follows:

1. Tie rod is very loose and the ball is ready to pop out of the socket
2. A tire is worn to less than 1/32 and the cords are about to show at the outer edge
3. The battery fails its load test and will not crank the engine quickly enough

4. The alternator serpentine belt is showing wear and cracking due to its age
5. The air conditioning is low on refrigerant charge and cycling too fast
6. Oil is leaking at a valve cover
7. A cracked engine cover for a component for decoration

Ordering Suggested Repairs by MUST, SHOULD, COULD Categories

Another approach to suggesting repairs is for the service consultant to use **repair categories**. The categories for the repairs would be:

MUST perform
SHOULD perform
COULD perform

The MUST, SHOULD, and COULD categories effectively help the customer understand the seriousness of the different problems. The explanation for each category is presented next. These criteria also help a technician and service consultant work together by placing each suggestion in a proper category. The presentation then is easy to follow so the customer can make decisions faster about whether to approve a suggestion or not.

The MUST Category Includes

1. *Imminent Danger*—Safety repairs that will cause harm if not taken care of immediately. An example would be steering components that are very loose and ready to break.
2. *Hazardous Danger*—Safety repairs that could cause harm if not repaired soon. An example would be a tire with cords showing.
3. *Imminent Malfunction*—Nonsafety concerns that will soon cause a breakdown if not repaired. An example would be a battery that fails its load test.

The SHOULD Category Includes

4. *Potential Malfunction*—Nonsafety concerns that may lead to a breakdown if not repaired. An example would be an alternator serpentine belt that is worn and cracked due to its age.
5. *Nonessential malfunction*—Nonsafety concerns that will cause a system to not operate but will not lead to a breakdown. An example would be an air-conditioning system that has a leak.

The COULD Category Includes

6. *Nonessential Concern*—Nonsafety concerns that may cause a system not to operate as intended but will probably not lead to a malfunction. An example would be a slight oil leak at an engine valve cover.

7. *Noteworthy items*—Nonsafety concerns that a technician notices but really are not important to the function of any system. An example would be an engine cover that is cracked and covers a component mainly for decorative purposes only.

A presentation to a customer might appear as follows:

MUST suggestions:

- Left tie rod is very loose and the ball is ready to pop out of the socket, causing steering problems and tire wear;
    - needs tie rod end and alignment
- Left front tire is worn to less than 1/32 and the cords to show at the outer edge;
    - needs a new P205, 75 R 15 tire, the other three appear fine
- The battery fails its load test and will not crank the engine quickly enough;
    - needs a battery

SHOULD suggestions:

- The alternator serpentine belt is showing wear and cracking due to its age, may fail in the future;
    - needs a belt
- The air conditioning is low on refrigerant charge and cycling too fast to cool efficiently;
    - suggest air-conditioning service to check for leaks and top off the system

COULD suggestions:

- Oil leak at a valve cover is slight or minor, starting to make a mess on the engine;
    - needs gasket and inspect for sludge in return passages under the cover
- A cracked engine cover; covers engine for decoration purposes;
    - cannot be fixed, must order new cover if customer wants it replaced

The MUST, SHOULD, COULD presentation lists the suggestions in the proper category along with what is wrong (concern), why it is a problem (cause), and what needs to be done to fix it (cure). The presentation would help to assure a customer that the service consultant is not exaggerating or understating the seriousness of a problem, and makes sure the customer understands the relative importance of the repair. When done properly, value is added to the customer's service visit because the inspection of the automobile follows "seven criteria levels" of possible problems to produce the MUST, SHOULD, and COULD format.

Technician Inspection Limitations

Of course, some technicians do not have either the ability or the willingness to examine a vehicle for potential problems. For example, a technician may not think the problem is important to point out because it is not something that he or she would fix on his or her own car. Another common reason may be that the technician does not want to do the repair because it is not the type of work he or she likes to perform. When a technician has this attitude or lacks the ability to identify potential problems, he or she should be given work that does not require the inspection of automobiles. Rather, after a competent and conscientious technician checks an automobile, the work may be passed on to a less capable technician.

Maintenance Sales: The Feature-Benefit Method

Convincing customers to purchase maintenance services is often difficult. As a result, the sale of maintenance services depends on the service consultant's ability to convince customers that these services will help them. One technique is for the service consultant to focus on what the recommended maintenance will provide to the customer (feature) and how the maintenance will be an advantage to the operation of the customer's automobile (benefit). This technique is referred to as **feature-benefit selling**.

The feature-benefit selling technique relies on the service consultant's knowledge of the maintenance schedules recommended by automobile manufacturers and the seasonal maintenance specials offered by the facility. Then, when discussing any services needed by a customer, the service consultant attempts to sell these features, meaning what the manufacturer recommends, the seasonal specials offered by the service facility, plus any recommendations made by the technicians conducting a diagnosis or inspection. To accomplish this, the service consultant must carefully describe to the customer how a maintenance service will benefit him or her. Common benefits may be extending the life of the automobile, safety, improved operation of the vehicle, and reduced operating costs.

An example of feature-benefit selling is a recommendation for a battery service because of corrosion. The feature, of course, is that the battery terminals and cables will be cleaned. To make the service relevant to the customer, the service consultant also explains the benefits of the recommended service. In this case, the customer should be told how the battery service helps extend the life of the battery and reduces the possibility that the battery may fail to start the engine. As a result, the battery service (feature) can prevent an inconvenient and possibly costly breakdown (benefit).

Feature-benefit selling is a common technique because it does not enter into the technical details about how an automobile works but states simply and briefly the work to be done (feature). For instance, too often, a service consultant tries to impress a customer about how technically difficult or complicated a maintenance or repair job is and the customer loses interest. The service consultant then may lose the sale because the customer becomes overwhelmed by the details. Therefore, this selling approach helps simplify what the customer wants to know; specifically, how the service can help the customer and why it should be purchased (benefit).

If the benefit is understood and appreciated, customers will most likely purchase the recommended maintenance. These sales are the backbone of a service facility because they help to maintain stable sales income. In other words, a service facility can hire the best and fastest technicians, but the service consultant must make maintenance sales to keep the technicians working.

Of course, in order for service consultants to use the feature-benefit selling technique effectively, they must have something to sell. This requires a list of manufacturer recommendations for automobiles, seasonal special offers by the owners or managers, and recommendations from the technician about the automobile's condition. Therefore, when examining automobiles for repairs, technicians must also inspect them for possible repairs and maintenance work. This does not mean that technicians should remove any parts to identify needed services but should simply examine each customer's automobile and point out items that need repair. The inspection forms that can be used for these reports are shown in chapter 4. From the preceding example, the following services were suggested and the features are evident but the benefit is what sells the service. To illustrate what this means:

MUST suggestions:

- Tie rod is very loose and the ball is ready to pop out of the socket:
    - Feature: The tie rod must be replaced and the vehicle aligned
    - Benefit: To assure you, the driver, can maintain steering control and handling at all times
- A tire is worn to less than 1/32 and the cords to show at the outer edge:
    - Feature: The tire must be replaced after the tie rod has been replaced
    - Benefit: Avoid an inconvenient road side repair or call for assistance because it blows out and needs to be changed
- The battery fails its load test and will not crank the engine quickly enough:
    - Feature: Replace the battery
    - Benefit: To avoid not being able to start the engine and causing an inconvenient breakdown

SHOULD suggestions:

- The alternator serpentine belt is showing wear and cracking due to its age:
  - Feature: Replace the serpentine belt
  - Benefit: May perform adequately for an unknown period of time but may break unexpectedly causing an inconvenient needed repair
- The air conditioning is low on refrigerant charge and cycling too fast:
  - Feature: Test for leaks and determine any needed repair before recharge
  - Benefit: It is more pleasant to drive the vehicle when it is hot outside and the air conditioning performs well.

COULD suggestion:

- Oil leak at a valve cover is slight or minor:
  - Feature: Replace the valve cover gasket
  - Benefit: To keep oil from leaking onto other engine parts that may cause an unpleasant odor and possibly avoid oil on your (the customer's) driveway if the leak becomes more serious over time
- A cracked engine cover that mainly covers a component for decoration:
  - Feature: Repair of the crack or replacement of the cover
  - Benefit: While the cover does not need replaced for proper vehicle function, some customers are bothered by broken parts that may fall off, rattle, or not keep the vehicle in pristine condition. This is truly a personal decision about whether this inspection observation is of value to address at this time

Whether the customer will approve all, none, or some of the suggestions depends on many factors.

Some customers wish for their vehicle to be perfect and therefore will want the entire list done. Service consultants should be mindful that their own personal preferences should not enter into the suggestion process by talking "customers" out of services they may desire because the service consultants "would not do them on their own car." Likewise, they should not "oversell" and present information in a more serious light than appropriate, such as "everything is a MUST do" suggestion. Everyone is different and customers, perhaps with limited funds, will focus only on the MUST suggestions. Present the suggested feature and benefit accurately and then let the customer think and decide. If they wish more information, such as the opportunity to talk to the technician or see the problem, arrange a meeting and/or show them the problem so they can understand your suggestions.

To help customers understand the inspection process, it sometimes helps for the service facility to have a waiting room that overlooks the service bays. This allows customers to watch how technicians are going about their examination. More recently, service departments are placing cameras in the bay with a connection to the Internet so customers can log in and watch their vehicle being serviced. When appropriate, customers who are waiting can be shown the defects found.

Four Opportunities to Sell

Closing a sale often takes persistence. However, knowing when there is an opportunity to make a sale comes before persistence. There are four opportunities for a service consultant to make a sale to a customer.

First Opportunity

The first opportunity to sell occurs when a person calls or emails the service facility and communicates with the service consultant. The customer's name should be entered into the customer service computer program to create a database and an appointment should be scheduled. During the conversation, the service consultant can suggest maintenance specials, but the important objective is to get the person and the automobile into the facility.

Second Opportunity

The second opportunity to sell is at the end of the estimate stage (1) just before the repair order is created in stage (2). This opportunity allows the service consultant to make customers aware of specials and to suggest maintenance services specific to their automobile. A benefit presented to the customers should include the opportunity to "keep your vehicle in top mechanical condition."

Another service to be offered during the initial write-up is to inspect the various vehicle systems. The benefit is to "provide you with information about your automobile." For example, assume Mr. Hoffman is having his vehicle diagnosed for a serious engine problem. He may want to have information about other vehicle systems, such as the brakes or air conditioning. At the very least, this information can help him make a financial decision about the overall condition of his automobile to determine if the engine is worth repairing. In addition, if he decides to have the engine repaired and the technician finds that the air conditioning does not work, Mr. Hoffman may want it fixed at the same time when the engine is being repaired.

Third Opportunity

The third opportunity to sell is when the service consultant contacts the customer in the additional work stage (3) for approval to perform a repair that a technician discovered. The purpose of the additional work stage (3) is to give the service consultant the opportunity to point out any other items the technician may have found in need of repair or replacement while working on the automobile, such as replacing a windshield wiper and replacing a bad tire. At this time, the service consultant may ask the customer to reconsider previously suggested services or repairs that were declined during the first and second sales opportunities. If the job does not require any additional work, then this third opportunity to sell would come before the invoice stage (4) when the service consultant would inform the customer of "when a repair or service will be completed."

Fourth Opportunity—The Active Delivery Process

The fourth opportunity to sell is when customers pick up their automobiles. This is the invoice stage (4) or following the Repair Order Tracking Sheet, the fourth phase—check out. In an ideal situation, the service consultant should present the keys of the automobile to the customers and then walk them to it, preferably near the service entrance. This is called an **active delivery**. During active delivery, the service consultant has a final opportunity to "connect" with the customer and discuss any issues of importance.

In the first part of active delivery, the service consultant should convey to the customers positive points about their automobile. This is suggested because during the sales process, customers have been told of all the things that are wrong with their automobile. This can have an unpleasant effect on customers about the condition of their automobile and the wisdom of the payment they have just made for repairs. In some cases, they may even feel like they are "driving junk." No one who has just paid a service facility for repairs should think he or she is leaving with junk. So the service consultant should mention some of the positive features of the customers' automobile. For example, the service consultant may point out the automobile's interior, gas economy, paint job, stereo system, the attractiveness of the model of automobile, and so on. In addition, the service consultant should note how much better the automobile will perform now that the repair has been made and how the customer will not have to worry about another repair for quite a while. They may even discuss the vehicle systems that were checked and seem to be in "good condition."

In the second part of active delivery, the service consultant might try to sell to the customer, one last time, any needed services that were previously declined. This can be accomplished by asking customers if they would like to schedule a future appointment to perform a previously suggested repair or to schedule the next regular maintenance, such as an oil change. In some cases, showing customers any worn or bad parts that need to be replaced in the future can make a future sale that was declined earlier. To make the sales opportunity work, customers must be shown (not told) about their vehicle's problem(s). Even if customers decline the repair a second time after "seeing the part," they should be impressed (again) that the service consultant is interested in helping them with their automobile problems.

While many service consultants are not opposed to active delivery, they often feel they barely have enough time to take care of their other customers who need help. Therefore, at some service facilities customers see a cashier to pick up their automobiles. If this is the policy of the automotive service facility, then service consultants should at least call their customers after the automobile's repair is finished. Then they should engage customers in a discussion described in the active delivery conversation, such as pointing out the positive points of their automobile, and then try to schedule additional repairs or maintenance. This is not to say that time is still not at a premium for service consultants, but it can cost the service facility sales and reduced customer satisfaction if the active delivery conversation is not attempted.

Impact of the Environment on Sales

Service facilities operate in **open business environments.** This means there are forces outside of their control that affect their sales potential. Service consultants, managers, and owners should always attempt to take advantage of the positive environmental features and counteract their negative influences. Two of the business environments important to sales are the tactical and the operational environments.

The Tactical Environment

The **tactical environment** refers to the resources and the support arrangements needed to conduct business. For example, an automotive service facility needs to have auto parts to conduct business. The availability of one or more auto parts vendors, or suppliers, can influence the service facility's ability to make a sale. A sale is more likely if parts are immediately available than if they will not be delivered for a week. A negative influence from the tactical environment must be counteracted by internal strategies, for example, how the facility can get the part sooner.

At the same time, a facility with a supplier next door can use that to its advantage by telling customers how soon it can get the parts needed to make repairs. Also, a service consultant at a facility with several parts suppliers available to it can tell customers how he or she will "shop around" three parts suppliers to get the best price.

The Operational Environment

The **operational environment** concerns factors related to conducting day-to-day business activities and relies on the flow of customers. These factors may be the local demographics, economics, and sociocultural features of a community. Obviously, a business must appeal to the people in its community and recognize their interests, preferences, economic status, and so on.

For example, people who live in a community with a very high average income may own or lease new automobiles that have comprehensive warranty contracts. Sales for repairs may be limited in such a community. On the other hand, used cars that are not under a new car warranty may be found in a town or city with residents with lower incomes. In this case, repair work may be in greater demand. Likewise, if a town factory closes and a large number of people lose their jobs, the economic environment will change and lower sales can be expected.

Service facilities should always consider ways to take advantage of or have an influence on their operating environment. One example is to open a new service facility in or near a populated area with easy access to and from a heavily traveled road. Another is to ensure that the facility's building is large enough to handle the number of automobiles needed to generate the sales needed for a profit. This would include room enough for vehicles in the process of having long-term work done on them as well as those needing short-term maintenance work. At the same time, the property (not just the building) must be clean, attractive, and landscaped (see Figure 8-3) and the interior should be clean and appealing to customers (not to the owner or workers). Old parts, parts that are being returned to a vendor, and parts that are delivered to a facility should never be lying around where customers can see them. Customers must always feel comfortable and never uneasy when they are at a service facility.

Of course, some features, such as location, may not be able to be changed but some, such as cleanliness, can easily be improved. Another option that influences sales is the facility's operating image. This could be improved, for instance, by increasing advertising, offering shuttle services, promoting maintenance specials and discounts, or adding new services (such as a car wash or quick lube station).

FIGURE 8-3 An attractive service building.

Selling and Methods of Payment

Closing a sale often depends on the method of payment that a customer can use. Service facilities must provide customers with as many alternative methods as possible to pay their bills. Then service consultants must understand the financial options that customers can use to pay for their service. When customers need some other means to pay for a repair that will cost a lot of money, a sale may be closed because the service consultant can make the arrangements for them to finance the charges.

Therefore, when service consultants are trying to close a sale, they should be able to offer several methods of payment that a customer can use. These methods may range from short-term credit provided through the facility (such as credit provided to fleet vehicles) to long-term loan arrangements through a bank. At Renrag Auto Repair, many of the larger repair jobs, such as an engine replacement or overhaul, were sold because customers were able to use one of these methods to finance the repair (see Figure 8-4).

One of the most popular methods used to pay for automobile services is by credit card. Although this option costs the service facility money when it processes a charge, it is necessary for a business to be successful. Service consultants who act as the cashier must be able to use the credit card processing equipment and programs.

For customers who do not use credit cards, service facilities must be able to accept personal checks. However, because a customer's check can "bounce," meaning the person does not have enough money in the checking account to cover the payment of the check, a facility may use a check cashing service. For a fee, this service guarantees the facility will receive payment when a check bounces. Service consultants must be sure that all checks are properly written for endorsement and deposit.

FIGURE 8-4 A service consultant explains payment methods to a customer.

In addition, service facilities should have financial arrangements available for customers making a large purchase (such as a set of tires) or needing an expensive repair. In these cases, through arrangements with a finance institution, the service consultant could offer customers a 30-day same-as-cash option or low monthly payments. These approvals typically take half an hour or less so that customers can have the work started on their automobiles right away.

Large financial loans, for example, for the replacement of an engine and transmission costing more than $3,000, can be offered by a service facility through a bank. The loan application is usually made at the service facility and then faxed or taken to the bank for approval. The bank checks the customer's credit and, in most cases, the approval is obtained before the end of the working day. When the loan is approved, the bank will send a check to the service facility to be endorsed by the customer and the service facility manager or owner. The technicians can then begin the long-term repair and parts can be ordered. To make this type of financial arrangement appeal to customers, it must be convenient and must have reasonable finance charges. Needless to say, the service consultant must carefully explain the process, loan agreement, and benefits to the customer.

Review Questions
Multiple Choice

1. After inspecting a vehicle the technician recommends the following: replacement of a damaged driver-side seat belt, cooling system flush, replacement of brake pads that only have 1/32 inch remaining, and an oil change that is 1,500 miles overdue. Which of these represents the best way to prioritize this list to the customer?
   A. Brake pads, oil change, seat belt, cooling system service
   B. Seat belt, oil change, brake pads, cooling system service
   C. Seat belt, brake pads, oil change, cooling system service
   D. Oil change, cooling system service, brake pads, seat belt

2. If a customer objects to the cost of a necessary repair, which of these is the best response that the service consultant can give?
   A. Offer the customer a discount on the repair.
   B. Reschedule for a later time.
   C. Explain the reasons for the cost and benefits of the repair.
   D. Remove it from the repair order immediately.

3. Service consultant A says that providing a ballpark estimate is a useful tool to close a sale. Service consultant B says that being friendly and asking for an appointment will close a sale. Who is correct?
   A. A only
   B. B only
   C. both A and B
   D. neither A nor B

4. A 30,000-mile maintenance procedure is being performed. Which of these is a benefit of performing the service?
   A. The cooling system gets flushed.
   B. The transmission fluid gets changed.
   C. The maintenance will help the vehicle to continue to deliver dependable service.
   D. The completed checklist is given to the customer.

5. Service consultant A says that an example of a feature of an oil change is the weight and brand of oil used. Service consultant B says that an example of a benefit of an oil change is longer engine life. Who is correct?
   A. A only
   B. B only
   C. both A and B
   D. neither A nor B
6. An upset customer comes in when the service department is very busy. Which of these is the best way to work with this customer?
   A. Listen to the upset customer before offering an explanation or solutions.
   B. Ask the customer to come back later.
   C. Tell the customer he or she is wrong.
   D. Offer the customer a discount.
7. A customer has come to pick up his or her vehicle when the service department is very busy. Which of these is the best way to handle the situation?
   A. Direct the customer to the cashier to cash him or her out.
   B. Advise the customer that you are very busy.
   C. Review the work performed and the invoice with the customer.
   D. Ask the customer to come back when it is quieter.
8. When customers pick up their vehicle, Service consultant A says that it is important to take the time to explain the work performed in as much detail as the customer requires. Service consultant B says that if customers ask questions, it indicates they do not trust the shop or dealership. Who is correct?
   A. A only
   B. B only
   C. both A and B
   D. neither A nor B

Short Answer Questions

1. List the ways in which a service consultant can promote the procedures, benefits, and capabilities of the service facility and its employees.
2. What does "close the sale" mean?
3. Provide examples of service features and benefits.
4. Provide examples of the features and benefits that might be used to up sell a premium part.
5. List ways to overcome customer objections.
6. Explain how to identify and prioritize a customer's vehicle needs.
7. Explain how to present customers with the work to be performed and related charges.
8. Explain how different methods of payment can help close the sale.

Activity

Obtain inspection results for a vehicle that you or someone has conducted or "made up." Prioritize them for a customer using the MUST, SHOULD, COULD methods. In a report describe each service suggestion's feature and explain how the benefit will be presented to the customer.

CHAPTER 9

WRITING FOR THE TECHNICIAN: COMMUNICATING TECHNICAL DETAILS

OBJECTIVES

Upon reading this chapter, you should be able to:

- *Demonstrate how to effectively communicate customer service concerns and requests to the technician (Task A.2.1).*

- *Demonstrate how to assist service facility personnel with vehicle identification information.*
    - *Locate and utilize the vehicle identification number (VIN) (Task B.7.1).*
    - *Locate the production date (Task B.7.2).*
    - *Locate and utilize component identification data (Task B.7.3).*
    - *Identify body styles (Task B.7.4).*
    - *Locate paint and trim codes (Task B.7.5).*

Introduction

After a customer approves a service by signing the repair order, a copy of the repair order is given to the technician. This copy may be called the repair order hard copy, or hard copy. The popular term used for this copy, however, is *technician hard copy*. This is because before computerized repair orders were popular, the last copy in a set of preprinted repair forms was made of light cardboard and was for the technician.

Today, some computer forms print the technician's hard copy on regular paper. Other customer service system computer programs will electronically post the RO to the technician's computer. Still many shops choose to use the heartier cardboard that is either printed from the computer program or handwritten. This is because it holds up in a shop environment, and the heartier material keeps it from getting bent and smudged from being around liquids and lubricants in the shop. The term hard copy is used in this chapter to represent the technician's copy of the work order. When paper or a hard copy is used to further protect the technician's copy, it should be placed on a clipboard. In many facilities, the clipboard is placed on a wall peg for the technician.

The front of the hard copy is typically an exact duplicate of the repair order and must have the entire automobile and customer information needed by the technician. The purpose of this chapter is to cover the information that the service consultant must place on the technician's hard copy. This includes the information that the service consultant wrote down about the customer's concern in terms that the technician can understand to service the automobile. It also is to include information on the automobile.

Ensuring that all of the required information is entered on the repair order in the correct space is important to the efficiency of the shop. Technicians cannot use their time to hunt for the information on the form, to interpret what is written, or to look for information on the automobile. This is because the technician is paid by the hour and there is an hourly cost associated with every repair bay. Lost bay time costs the service facility money. In addition, if the technician is paid on a flat-rate basis (explained in chapter 3), this lost time costs the technician money, and the service facility also loses money. Service consultants must have their technicians in a position where they can immediately begin to work on a customer's automobile as soon as they receive the hard copy.

Recording Customer Information: What and When

When a customer service computer program is used, the computer program provides the service consultant with space for a limited number of characters, such as 100. Therefore, what a customer has to say may have to be boiled down to 100 alphabetical letters, periods, and spaces. Not all systems operate like this but the point is that the service consultant must be brief yet accurate in reporting the customer's complaint, service needs, and any other necessary information. Therefore, this limitation is actually preferred because it is inefficient for technicians to take the time to read a long report. To keep the description of the problem to the point, service consultants should focus on *what* the problem is and *when* it occurs. The service consultant's choice of words to state *what* and *when* is important because of the limited space provided.

For example, assume that Mr. Berger has a problem that needs to be diagnosed. The concern has not been verified and, therefore, the first words to record are: "Customer claims," which contains fifteen characters. The next part of the sentence needs to tell the technician *what* happens to Mr. Berger's automobile. In this case, assume that Mr. Berger states that his automobile will not properly shift from second to third gear.

The service consultant must ask Mr. Berger more questions to determine *when* the problem occurs. As the service consultant asks more questions, Mr. Berger explains that the automobile's engine speeds up before it goes into the higher gear. To put Mr. Berger's entire explanation of the problem into writing may exceed the maximum number of characters the system will permit, plus the service consultant must select terminology used by technicians to describe the sensations felt by the customer. Specifically, the service consultant may use the word "slips" to describe *what* happens when the automobile's transmission shifts gears. To describe approximately *when* the problem occurs, the service consultant writes down when the problem occurs such as, "when the car shifts from 2nd gear." As a result, the description that is written on the repair order for the technician states: "Customer claims the transmission slips when the car shifts from 2nd gear." The statement is under 100 characters and conveys to the technician *what* the problem is and when it occurs.

The beginning of the sentence states "Customer claims" to indicate that the problem must be verified. As explained in chapter 5, the verification of a problem is very important because under many warranty contracts, as well as state lemon law statutes, a problem does not exist until a technician verifies it.

Communicating Other Information and Warnings

In some cases other information must be given to the technician, such as the wheel lock "key" to remove special lug nuts is "in the glove box" (i.e., location of the wheel lock key). This information may be used to help the technician save time and avoid additional communications with the service consultant or may be a warning that the technician needs to know. Remember, the vehicles that are brought to the service department are broken and that means they will not operate as they were designed.

To illustrate this point, a customer installed a starter motor in his truck. After a week, the starter engaged and cranked the engine with the key off. The truck owner disconnected the battery to make it stop cranking. Unable to fix the problem, the truck was towed to a service facility with the battery unhooked. The tow truck driver left the manual transmission truck in gear so it would not roll backward but did not engage the emergency brake. After a lengthy discussion with the customer, the service consultant wrote up the repair as merely, "No crank" and gave it to the technician without any additional information. When the technician went to where the truck was parked, he noticed the battery was unhooked. He installed the battery cable on the battery post and the starter motor engaged. Since the truck was in gear after being dropped off by the tow truck driver, it lurched forward crushing the technician between the truck and the vehicle parked in front of it. The technician was seriously injured.

This situation would involve a worker's compensation insurance claim, explained more in chapter 14. The importance of this example is the service consultant has a duty to assure the information from the customer is provided to the technician. The repair should have been written as perhaps:

"Warning, starter engages when the battery is hooked up. Diagnose problem." This description has a better chance of a technician understanding the significance of the problem than just "No Crank." Hopefully, with additional detail in the problem's description, he would take additional precautions when examining the vehicle. Realizing the seriousness of the problem, the service consultant may go further and decide that a "written warning" is not enough and the danger must be communicated verbally. Therefore, a service consultant may deliver the RO in person with a verbal warning or may write on the RO simply, "Diagnosis—MUST see service consultant for more information." This allows the service consultant to relay what he or she knows directly to the technician and even the team leader or shop foreman who may want to be informed.

Duplicating the Problem

After receiving the hard copy, a technician makes some preliminary checks on Mr. Berger's automobile, such as checking the transmission fluid level and condition. Next, he tries to duplicate the problem by taking the automobile for a drive but the problem does not reappear. Mr. Berger's automobile does not seem to have a problem when shifting from second to third gear. As a result, the technician must write down on the hard copy that the problem was checked and could not be duplicated. The hard copy is then returned to the service consultant, who presents the information to the customer.

When a problem cannot be confirmed, a number of actions may be taken. For example, the service consultant and the technician may take the automobile for a drive to see if the problem reoccurs. An option may be to ask Mr. Berger to drive the automobile with the technician riding as a passenger or vice versa. Another alternative is for the service consultant or technician to take the automobile for a drive after the lunch break or to take it home overnight. The objective in all cases is to duplicate the problem and obtain an accurate diagnosis.

Selecting Appropriate Terminology

Some service consultants may lack the technical knowledge or have trouble choosing the right words to describe a relatively detailed customer concern. In these cases, the service consultant should have access to a publication such as Delmar's *Automotive Dictionary* by South and Dwiggins (Thomson Delmar Learning, Clifton Park, NY, 1997). For service consultants at new automobile dealership service departments, the automobile manufacturer typically recommends terminology to use to describe certain concerns in the narrative, such as the transmission fails to engage upon acceleration, or service operation codes may be used.

Customer-Provided Vehicle Information

Two types of customer-supplied information include the registration card issued by the state and the insurance card issued by the customer's car insurance company. The registration may be a "card" or in some states a "printed Internet registration." It should list the owner, which may be a leasing company, the address of the owner, and vehicle information: year, make, model, Vehicle Information Number (VIN) (discussed next). The card will usually have an expiration date among other information, such as county. If the customer is not the person listed on the registration, he or she may not be the owner of the vehicle and this could have ramifications for repair approval. In other words, if a car needs thousands of dollars of work and the customer's name does not match the name on the registration, the customer may not be the owner of the vehicle and may not wish to pay for the repair. Care must be taken because a lien to get paid for the repair cannot be asserted against a vehicle where the owner has not approved the repair.

The insurance card contains similar vehicle information as the registration form as well as the insurance company's name and contact information. Depending on the type of insurance the garage carries (covered further in chapter 14), the insurance card may be required in the event the customer's vehicle is damaged. For some repairs, such as state safety inspection, a valid insurance card and registration may need to be presented before an inspection is conducted. In some cases, only the registration is required, such as state-operated but federally mandated vehicle inspection and maintenance programs (I/M), also known as an emission inspection, SMOG test and tail pipe testing. Some states such as California have CARB or the California Air Resource Board, and its standards exceed federal requirements.

Vehicle Identification Information

Service consultants must become very familiar with the location and formats of the VINs. The information contained in these numbers and letters is critical to the service consultant and technicians. For this reason, these numbers are to be included on all estimates, repair orders, and invoices. The VIN is important to submit warranty and insurance claims, obtain correct parts, identify service procedures and specifications, and obtain correct information about recalls, campaigns, and Technical Service Bulletin (TSB) information.

Vehicle Information and Creating an Estimate

Additional information that should be recorded in the customer service computer program is any information on the automobile's optional equipment shown in Figure 9-1, such as air conditioning or power steering. This is important because when completing an estimate for the replacement of an engine, the connecting and disconnecting of the air-conditioning equipment adds time and cost to a repair. If the service consultant does not include this information in the estimate, the service facility, and possibly the technician, will lose time and money on a repair.

Vehicle Identification Number (VIN)

The best method to obtain vehicle information is to use the VIN. This works best when the service consultant has access to a factory database

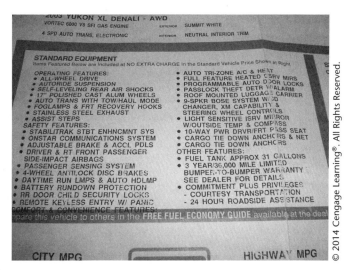

FIGURE 9-1 A vehicle can have options that require special parts and will cost more money or will take more time to replace. This manufacturer's sticker provides an illustration of some of the special options on this truck. Options can be identified in some cases by "looking" at the vehicle (such as air conditioning); in other cases the VIN will be needed to pull up manufacturer "build" data, and a few manufacturers have an options sticker on the vehicle.

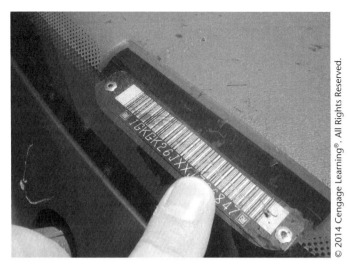

FIGURE 9-2 This service consultant writes down the VIN found under the driver-side windshield. The 8th digit provides the vehicle's engine size (J is a 7.4 litre) and the 10th digit indicates the year (X is 1999).

and can access all information related to the vehicle that includes accessory options, warranty information, as well as factory bulletins.

The VIN is a 17-digit number, and a number is assigned to each automobile. The service consultant can locate the VIN at a number of different places on the vehicle (door jam or windshield shown in Figure 9-2 among a few possible locations) and it is always found on the state-issued registration card. Once it is located, it should be carefully recorded in the customer's file in the computer database. Some warranty procedures require that the VIN be obtained directly from the vehicle. When copying the VIN, the service consultant must be sure to write down each letter and number accurately. Failure to record all 17 digits accurately will cause several problems, such as ordering wrong parts, rejection of warranty claims, inaccurate repair times, and even state law violations regarding emission or state safety inspection. One suggestion is that after the VIN is copied on a piece of paper, the service consultant should count the numbers to be sure there are 17.

What the VIN Digits Represent

An example of an automobile's VIN (1G2FS32P8RE100000) is shown in Figure 9-3. Note that this is a number used by General Motors. Manufacturers may use different letters and numbers in some categories. For example, the digits shown in Figure 9-3 create the following 11 categories: origin (where the car was manufactured), manufacturer, make, vehicle line, body, restraint, engine, check digit, year, plant code, and sequence number.

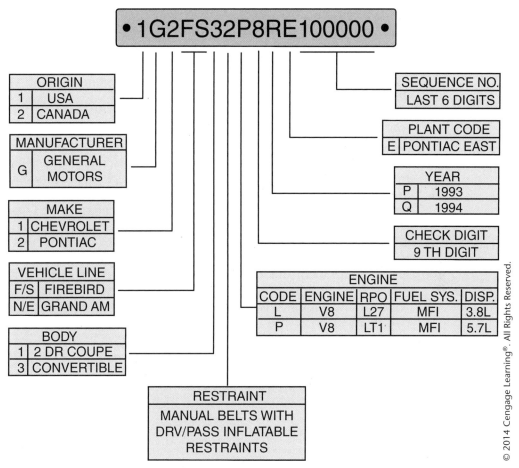

FIGURE 9-3 This VIN shows that the automobile was made in the USA (1) by General Motors (G). The vehicle make is a Pontiac (2) and the line or model is a Firebird (FS). The body style of the Firebird is a convertible (3) that has manual belts with air bags on the driver and passenger sides (2). The engine in this Firebird is a V8, 5.7 liter (P). The check digit for diagnosis is the 9th digit (8). The vehicle was manufactured in 1994 (R) at the Pontiac plant on the East Coast (E).

Again, caution is required when interpreting what a digit represents. For instance, the number for the body style in Figure 9-3 is a 3, which indicates a convertible. This number may not represent a convertible for other manufacturers because the numbers used to describe a particular body style may be different for different manufacturers. The numbers may be different even within a manufacturer. Also, the number used to describe a particular body style may change from year to year and may be different in a manufacturer's car lines and truck lines. Consequently, the meaning of each letter or number must be found in a shop manual or database.

Date of Production and Door Sticker Information

The production date of a vehicle is most commonly needed to order certain parts. For example, there may be two brake pad part numbers for the same automobile. The production date of the automobile is needed to decide which part number to use. The month and year an automobile was built can be found on the jamb of the driver's door along with the VIN, Gross Vehicle Weight Rating (GVWR), and vehicle tire sizes with inflation requirements (see Figure 9-4).

Other information from the "driver door jamb" sticker is needed for various reasons. If the service facility rents trailers, the towing capacity of the vehicle must not be less than the trailer capacity (specialty databases are available for this information as well). Furthermore, the GVWR will be needed to determine if a vehicle is subject to certain state inspection requirements, such as undergoing emission testing. In addition, the GVWR must be explained to customers who want to know how much they can "load up" their vehicle with materials before their automobiles reach the limits of the vehicle design. For example, a plumber bought a long bed mini pickup truck, put a cap on the bed and then loaded it with tools like it was a one-ton full-size van. The springs sagged and he had many steering and suspension complaints. It was the wrong truck for the application and his anger should not have been directed at the

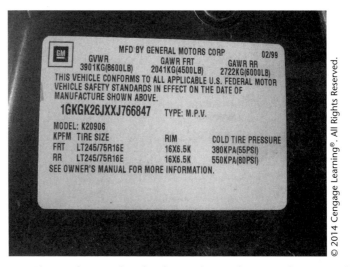

FIGURE 9-4 The production date for this truck is in the top right corner along with gross vehicle weight rating and tire size with inflation requirements in addition to the VIN. The GVWR for this truck is 8,600 pounds, which is the maximum weight of the vehicle after it is fully loaded with people and cargo. This truck current weighs approximately 6,900 pounds unloaded and about 1,700 pounds can be loaded into the vehicle including passengers.

service consultant but perhaps toward the salesperson who sold it to him. However, in the automotive industry, this isn't how it works and the service staff may need to educate the customer about how much he or she can haul to avoid damaging the vehicle, or worse, causing an accident.

Vehicle Model Year

Exercise caution when using the production date to determine model year because at one time vehicles made after September were assigned to the next model year. In other words, if a vehicle was made in November of 2012, it was considered a 2013 model. This is not the case anymore because the model year for an automobile changes when the manufacturer determines the "time is right." The timing for the change of model year is dictated by a number of factors that are associated with the manufacturing of the automobile. As a result, a manufacturer may decide that a new model year may begin at any point in the year, such as June.

To obtain the correct model year for an automobile (see Figure 9-5), service consultants may look it up on the automobile owner's registration card, the 10th digit of the VIN, or under the hood of the automobile on the emission decal (see Figure 9-6). When the emission decal is read, it will state that the automobile "meets the Environmental Protection Agency emission standards" for "model year," unless the hood has been replaced or the sticker is missing.

FIGURE 9-5 At a dealership service department, the production date along with a multitude of other important information, TSB, customer warranty information, among other data can be obtained by entering the entire VIN into the manufacturer's database.

FIGURE 9-6 The emission decal (VECI—Vehicle Emission Control Information) lists the model year as 2003. The emission parts are listed under the word "CATALYST" (SFI—sequential fuel injection, HO2S—oxygen sensors, TWC—three-way catalysts). Assumed to always be on the vehicle but often not listed on the emission decal are fuel inlet restrictor, evaporative canister, fuel cap, and PCV—positive crankcase ventilation.

FIGURE 9-7 These industry publications offer information about state emission programs, technical and emission components information by model year of a vehicle, and procedures to reset OBD II readiness monitors.

Model year is important when it comes to state-operated emission programs (see Figure 9-7). The model year will change the inspection test requirements. The vehicle emission testing programs vary by state, so it is impossible to cover them in this textbook. However, research on the web can be conducted to find the program requirements for a particular state.

Parts Identification

Often service consultants and parts specialists need more specific component identification information to order a part. The reason is because even with the model year of the automobile, production date, and complete VIN, there may still be too many different parts from which to make a selection. Again, the dilemma the service consultant or parts specialist faces is that a mistake in a part order will cause delays, and customer expectations will not be met. In addition, the price for an incorrect part may cause a problem in the estimated repair cost.

Admittedly, in some cases, the quickest way to obtain the part information is from the old part itself. If a number cannot be found, an option is to remove the part and take it to the parts supplier for comparison to a new part or a picture in a parts book. If these methods fail, another option is to obtain the part number from a source that is specific to a manufacturer. General Motors, for example, uses production option codes to describe the specific characteristics of each automobile produced. The list of codes on the sticker can be looked up in the manufacturer database or production code manual, such as B72, which describes a type of molding used and is one of approximately 50 to 100 production codes for every General Motors automobile. The codes describe every characteristic of the automobile, from its wheelbase to mechanical parts to the trim and paint. The codes are so detailed and the list is so complete on each vehicle that it is difficult to find two General Motors vehicles that have all of the same codes. Location of the GM Production Option tag varies from model to model. The most common areas to find the General Motors tags are the glove box or spare tire cover as shown in Figure 9-8.

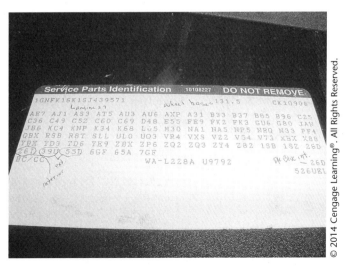

FIGURE 9-8 A General Motors option code sticker.

CAREER PROFILE: Field Service Engineer

In some cases dealers need help beyond the manufacturer's technical hotline. In these cases a manufacturer representative, called a field service engineer (or similar name), will be dispatched to provide assistance. In some cases the factory representative (various titles) will also be the field service engineer, if the representative has the proper credentials. Typically, the field service engineer credentials include an AAS degree in automotive or a B.S. degree in automotive with a business and technical emphasis. They also have several years of experience with the manufacturer's technical hotline (or other technical position) along with extensive manufacturers' technical training among other criteria.

The field service engineer is responsible for several dealers in an area. They will visit each dealer regularly and also when there is a technical problem that requires their attention. The technical problems they assist with can include advice to technicians who have problems with diagnosis test results or following procedures. They are often also involved in more complicated concerns that include collecting information for engineers on an accident blamed on a vehicle's design or on lemon law issues. Under lemon law there could be pressure added to the repair process because the customer's new vehicle has had too many repair attempts or is at risk of being out of service too many days.

The field service engineer is also available to help dealers with special tools or equipment needs as well as warranty issues that can include angry or confused customers. In a few cases, a field service engineer may teach training courses for technicians or have technical hotline duties, especially if they work for vehicle manufacturers and importers with a smaller market share.

Review Questions

Multiple Choice

1. A customer calls with a shopping list of problems with his or her vehicle. How does the service consultant put this information in a format that will help the technician find the customer's problem?
   A. Write down everything the customer says in the order he or she says it.
   B. Ask open-ended questions regarding each item to determine the problem.
   C. Ask the customer to boil the problem down to a specific system on the car.
   D. Verify that each item on the repair order is a symptom or a maintenance request.

2. Which of these is a common location for the production date?
   A. On the valve cover
   B. Inside the driver door pillar
   C. On a sticker on the radiator support
   D. Inside the gas door

3. Service consultant A says that before computerized repair orders were popular, the last copy in a set of pre-printed repair forms was made of light cardboard and was for the technician. Service consultant B says the word "hard copy" often is used to describe the technician's copy of the work order. Who is correct?
   A. A only
   B. B only
   C. both A and B
   D. neither A nor B

4. Service consultant A says that when replacing an engine, the air-conditioning equipment will *never* add more time or cost to a repair. Consultant B says when

information such as air conditioning is not considered in an estimate, the service facility, and possibly the technician, will lose time and money on a repair. Who is correct?

A. A only
B. B only
C. both A and B
D. neither A nor B

Short Answer Questions

1. Explain how a service consultant can effectively communicate customer service concerns and requests to the technician.
2. List ways a service consultant can assist service facility personnel with vehicle identification information.
3. What does VECI stand for and what publications, databases, or websites can you find to obtain information about VECI information such as the abbreviations?

Activity

Activity 1: Using an automobile, find the following vehicle information decals. Record the information found on the decal. Then use industry publications and databases to determine what the information (as much as you can find) means?
A. The VIN under the windshield
B. The "driver's door" decal
C. The VECI

Activity 2: Research the state safety and/or emission program in your state (if any) and determine the documents that the customer must take to the inspection center. If your state does not have any programs, follow your instructor's instructions, which may include researching another state such as Pennsylvania.

Pennsylvania has a state safety and separate emission programs with program information and inspection manuals readily available on the Internet to serve as an example of guidelines that other state might use.

Pennsylvania Safety inspection links are found at: http://www.dmv.state.pa.us/inspections/safety_stationmech_faq.shtml

The Pennsylvania Enhanced Safety Inspection manual for reconstructed and modified vehicles is found at:

http://www.dmv.state.pa.us/pdotforms/inspections/enhanced_si_guide.pdf

Pennsylvania Emission inspection links are found at:

http://www.dmv.state.pa.us/inspections/emission_stationmech_faq.shtml

CHAPTER 10

WORKFLOW: PRODUCTION CAPACITY AND SCHEDULING

OBJECTIVES

Upon reading this chapter, you should be able to:

- *Explain the relationship between company policy, operational procedures, and operation manuals.*
- *Define workflow.*
- *Present a plan to manage customer appointments (D.5).*
- *Describe techniques used to schedule the appropriate amount of workflow (D.1.)*
- *Explain why scheduling enough workflow is important (A.2.5).*
- *Describe how to estimate workload for today and tomorrow.*

CAREER FOCUS

You know that a satisfied customer is priceless. You also know that everyone at the service facility makes money when the technicians are busy with the right kind of work. You have to be on the top of your game because it gets complicated when your customers are at different stages in the customer service system and the technicians are in different phases of the shop production systems. You must keep everything under control because in the seamless service system the service consultant is on the front lines scheduling the work and pushing through the workflow. When there is not enough work, the company loses money because the technicians aren't busy. When there is too much work, the technicians may get the wrong kind of work for their skill level and the customers' expectations may not be met, especially if their vehicle is not ready for delivery when promised.

To be the best in the business, you must keep the schedule tight so there is enough work for all technicians until the end of the day. Each day that you accomplish this task the reward is . . . you make money, the technician makes money, and the business makes money . . . "We all make money!" The money is the customer's appreciation for a job well done by the service team! When you buy food at the end of the week, imagine your boss and the customers standing at the end of the checkout line. As you check out, they step up and say, "we got that" as they hand the money to the cashier to buy your food. Your money comes from satisfied customers and a shop with busy hardworking technicians.

Introduction

Workflow is the processing of work from the initial contact with the customers to the return of their automobile. The ultimate objective for an automobile service facility is for all work to flow smoothly through the service system. The service system is an orderly arrangement of procedures that are linked together to form a process followed by employees and used by management to control customer service, shop production, and business operations. To accomplish this, service consultants must control the volume of work taken into the facility each day and monitor the flow of work through the customer service, shop production, and business operations systems. The objective is for the customer's invoice and serviced automobile to be ready for pickup as soon as work is completed.

For service consultants to control and facilitate the flow of work, they must have a firm understanding of the facility's mission, which is what the type of work the business performs (this is covered in more detail in chapter 16). Also they must understand how the service facility's process and procedures form the service system. To discuss these processes and procedures, the chapter begins by describing the creation of **policy** and how it is used to write **operational procedures**. Next, the chapter explains how and why operational procedures become the guide for the creation of **operation manuals** that are to be used by employees to process work through a facility. In other words, these components are and should be connected to form a service system. They do not stand alone. This chapter builds upon the service consultant's duties presented in the earlier chapters and the workflow process, such as the procedures service consultants follow to make appointments and prepare repair orders. In order to meet the objective to have all work flow smoothly through the facility, the service consultants' duties must be linked together into a seamless service system. This seamless system is the heart of the operations manual, which is like an automobile service manual.

The Service Consultants' Responsibilities

The workflow responsibilities of the service consultant fit into each of the three systems (customer service, shop production, and business operations) with the objective of keeping the service team working. In other words, the technicians and the parts specialists cannot simply stand around while the service consultant prepares the paperwork needed to make up repair orders. They have to get the repair orders for the business to earn money. To make matters more complicated, while the service consultant may have technicians and parts specialists waiting for repair orders, additional work authorizations from customers may be needed to move through the shop production system. This means the service consultant has a critical role as he or she constantly prepares and processes repair orders. Keeping the repair orders organized as they enter and exit various stages and phases in the larger service system is challenging. Some repair orders will be ready to be given to a technician. Other repair orders will be waiting to have the additional work authorized by the customer so the technician can get back to work on the job. In addition, there may be repair orders that have to be completed later because parts have not arrived or there may be repair orders that were just completed by the technician and waiting to move into the business operations system.

Needless to point out, the service consultant in a busy shop does not have time to stand around and chat. If a service consultant does find spare time, he or she needs to be thinking ahead about what will be coming next. Service consultants' days usually fly by and before they know it, the day has ended. In this chapter, scheduling and the control of work flow is the focus while in chapter 11 monitoring workflow with the Repair Order Tracking Sheet is the means to keep the work flowing. This is why some franchise service facilities have carefully developed operation manuals and require each independently owned franchise to use them. The benefit is the assurance that their systems are seamless, that workers are not standing around waiting for paperwork to be completed, and that customers are provided with exactly the same services or products each time they come into one of the facilities, regardless of the location. Because of the success of some franchises, their operational procedures have set the standard for their industry. For example, Ray Kroc founded McDonald's and worked hard to make sure customers knew and got what they expected at each McDonald's franchise around the world. Their procedures made McDonald's a leader in the fast-food industry. As a result, the concepts of exact procedures that make up a system were copied by other industries and businesses. The procedures that make up a formatted system originate with the company's policies.

Company Policy

The owner, owners, or a corporate board of directors set **company policies with input from management**. These policies indicate how the owners want to conduct business. In addition, the basis for some polices comes from a law or legal regulations. In other words, when a service consultant works with customers, some polices will be based on the state's consumer protection laws, lemon law, lien statutes (perhaps case law), as well as other regulations that govern shop operations, such as California's Bureau of Automotive Repair (BAR) regulations. The shop's policies, for technicians to follow, may also be based on other regulations such as those for state safety and emission inspections, factory repair standards, and warranty or insurance requirements. Employees must realize that a "policy statement" may explain why the owners want the policy to be followed, but this is not necessary, because owners and boards are not required to justify their directives to employees.

Policies should also describe the authority and responsibilities of the managers, assistant service managers (ASM), service consultants, lead technicians, and technicians, which is discussed further in chapter 13. The descriptions should be in reference to the positions of the managers relative to each of the employees, such as the service consultants, in the organizational structure, which was discussed in chapter 1. As a consequence, when service consultants begin a new job, they should carefully review their responsibilities, which are usually shown in a job description. In fact, the service consultants should review the job descriptions of all employees with whom they work. These descriptions are typically approved by the owner(s) or board and are critical to the workflow process.

As shown in Figure 10-1, company policies lead to the operational procedures of the service facility. The operational procedures, which are usually written by the owner(s) or managers in charge of operations, are the basis for the operation manuals (the rule books) to be followed by all employees. The operations manuals direct how the work is to be processed by each work area. Therefore, to ensure that the owner(s) or board expectations as well as the laws will be met there is to be a direct link between the operations manual and company policies. As shown in Figure 10-1, there may be one manual for the entire facility with sections for each work area or a separate manual for each work area in the facility, such as the parts department. This chapter will assume that there is a manual for each work area.

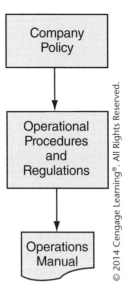

FIGURE 10-1 The connection between policy and operations manual.

An example of a policy that goes into an operation manual is as follows:

1. A company policy set by the owner states that all parts put on a customer's automobile must be purchased through the facility (perhaps for a liability concern) at a profit (probably a financial requirement to stay in business).
2. Operational procedures and regulations set by the owner, management team, or the service director (senior management) direct that to manage the business effectively, all parts needed for customer automobiles must be ordered by the parts specialist and marked up for resale.
3. The operation manual directive written by the management or the area manager for the employees to follow directs that:
    i. All parts will be ordered by the parts specialist within one half hour of the order being placed by the service consultant or the technician.
    ii. The parts sale price to the customer must follow management guidelines and customers may not supply or bring their own parts to the service facility for installation.

Employees must understand the full sense of the policy in the operations manual that contains specific guidelines for them to follow. Some directives in an operations manual for a work area will overlap with other work areas. In the example above, the operations manual for the service consultant, the technicians, and the parts department specialist would all state: "Customers may not bring parts they have purchased to the service facility for installation."

Beginning the Workflow Process: Scheduling Work

Scheduling of work can be one of the most difficult jobs for a service consultant. When customers call, the first question a service consultant must answer is whether the work is appropriate for the shop. Specifically, is the work within the mission of the business and does it follow the operations manual directives that are based on company policies and procedures. Details considered are whether there are technicians, tools, equipment, information systems, and procedures available to handle the job. For example, a specialty of the author's was engine performance, electrical, and computer system diagnosis. The ability to do the work was not an issue but rather the expensive computer interface equipment to do the work was not economical to buy for the small number of certain vehicle makes that were owned by customers. As a result, the mission was to do this type of work, but certain makes or models were not scheduled for certain types of electronic repairs when the equipment to do the repair was not available.

The second question a service consultant needs to answer is whether the shop has the capacity to do the work. Specifically, are there enough technicians hours available and do the technicians available have the necessary skills to do the work? In some cases, the limiting factor is the equipment availability. For example, an alignment takes about one hour to complete. If the shop has only one rack with a qualified technician to do the work, the maximum of eight alignments could be scheduled each day. A fortunate service consultant might have enough qualified technicians and equipment to schedule most types of work without regard to limitations. Regardless of the details, the service consultant must know that if too much work is scheduled, some customers will be disappointed when their expectations are not met. Too little work and the technicians will run out of work causing the shop to lose profit. It is up to the owners and managers to set capacity and tell the service consultants when to "turn away" work. If an owner catches a service consultant turning away work that should be sold to a customer, he or she can expect to be terminated.

How Work Is Scheduled

How work is scheduled depends on how the repair facility receives customers. Some repair facilities prefer and may even insist on customers waiting for their vehicle to be completed. Therefore customers are on a first-come, first-serve basis. If the line is too long, they will leave and hopefully return at some point in the future. Commonly maintenance operations, such as fast-lube garages operate in this manner. Occasionally, tire, brake, muffler, steering, and suspension service facilities operate similarly. In some cases, the customer may not "wait" at the service facility but "stay in the area," such as a mall, until contacted by the service consultant by cell phone, text, or beeper (sometimes owned by the service facility), such as Sears Auto Repair. For the service consultant, a busy day will be hectic as they try to process all of the customers. Stress will run high as they try to appease waiting customers and deal with unexpected problems. Some service facilities will have multiple service consultants who will attempt to help out during busy times, such as weekends. Other service facilities prefer to have the customer make an appointment to work on the vehicle. Appointments are usually scheduled for a couple days to weeks in advance.

Appointments

The benefit for making appointments is to control the workflow. When there are too many customers, some may not have their work completed by the end of the day. Furthermore, the rush to complete too many jobs may cause the technicians to have quality problems and/or not check vehicles thoroughly, missing services to suggest to the customer. If there are too few appointments, some technicians may not earn their expected wages for the day and service facility sales will be lower. To further complicate the process, if a service facility has some technicians who perform maintenance work and some who perform repairs, the service consultant must actually schedule work for two operations.

Making Appointments

To illustrate how this works, first, when scheduling an appointment, the service consultant might say, "I can schedule you to come in at either 8:15 a.m. or 9:45 a.m. on Friday. Will either of those times work for you?" Experienced service consultants know that, if possible, a customer should be given two times to choose from because it encourages them to make a decision and it gives them a choice.

Service consultants also know that to get the "work moving" in the morning, they need to schedule customer appointments in 8 to 15 minute segments within the first few hours of opening. This means they must process 4 to 8 "check ins per hour" as discussed in chapter 6 (the Customer Service System—Stages 1 and 2) and each customer must be tracked on the Repair Order Tracking Sheet (each customer starts on "first phase—check ins"). Longer time segments may be needed when more in-depth problems must be discussed with the customer, such as larger jobs with many operations. If appointments are not used, customers will be impatiently lined up at the service consultant's workstation. This means a service consultants should never tell customers to "just stop in when the shop opens at 8:00 a.m."

To schedule customers for appointments, a computer or an appointment book is used to keep an appointment schedule. The advantage of a computerized appointment schedule is the ability to look at the customer's history in the database. As a result, when an appointment is being scheduled, suggestions for any additional services can be made prior to arrival at the service facility. Also a computerized appointment schedule can be updated from one day to the next and appointments can easily be changed.

When an appointment book is used, such as the one shown in Figure 10-2, each sheet in the appointment book represents a new day,

APPOINTMENT SCHEDULE

Monday	DEC	12	2014			
Time	Customer Name	Phone Number	Year & Make of Vehicle	Services	Time Promised	Est. Work Hours
8:00	Zell	555-1111	2010 Honda	ft. brakes	wait	2.2
8:15						
8:45	Plessinger	555-2222	2012 Ford	30K service	3pm	3.2
9:00	Piccari	555-3333	2013 Infinity	State Inspection	noon	.7
9:15						
9:30	Pyle	555-4444	2012 Toyota	Warranty install ordered parts	5pm	0.9
9:45	Kranz	555-6666	2011 Jaguar	Wiring diagnosis	1pm	1.2
10:00	Haas	555-5555	2015 Chevy	Engine gaskets	5pm	3.5
10:15						
10:30	Shaeffer	555-7777	2013 Lexu	Trans rebuild	End week	14.4
10:45	Louderback	555-8888	2014 Subaru	4 tires/align	3pm	2.4
11:00	Runowicz	555-9999	2013 Dodge	Coolant leak	wait	0.8
11:15	Bounds	555-1010	2014 Ford	Steering parts	5pm	2.7
11:30	Miller	555-0000	2009 Hyundai	Exhaust R&R	2pm	1.6
11:45	Bolio	555-2020	2014 BMW	Fix lights	3pm	0.7
NOON	Renn	555-3030	2015 Mercedes	15K service	4pm	2.2

FIGURE 10-2 The appointment schedule sheet shows the customers who will "check in" and what time they will arrive to meet with the service consultant.

and there should be enough sheets in the book to cover the current as well as the next month. Also, since an appointment book requires the service consultant to write down the information, a pencil should be used so that changes can be erased. Pencils must be sharp and the handwriting must be neat so that others can read the entries. If a consultants' handwriting is illegible, then all entries must be printed.

When the service consultant does not have control over when customers come to the shop, they are prone to arrive at the same time and must wait in line. At some shops this is the way business is conducted. At other shops, this can cause the customer to not receive the attention expected or for phone calls to be missed. For these service consultants the focus is on control of work so each customer is served to a certain standard. While there still may be "high-traffic" periods such as early morning, the idea is to schedule each customer to arrive at a certain time (e.g., in 8-minute increments) and to know in advance how much work is coming into the shop for the day.

While the service facility that prefers appointments will serve "walk-in" customers for perhaps emergency repairs, maintenance, or other minor service issue, the customers must wait until a technician is available. A few garages do not take "walk-in" customers because they are so backed up with work that the customer must wait several weeks for an appointment. These service facilities could be specialty operations such as a restoration business where demand is greater than the labor hours the service facility has available.

Managing Today's Work Schedule

A lucky service consultant will have a full schedule of work before the workday starts. However, the reality is there is often unsold technician hours (called undercapacity) and this means there is not enough work to keep the technicians busy all day. Therefore, the service consultant will have to rely on "walk-in" business or emergency repairs, such as a "tow in" to keep the shop busy. But how much work is needed? To determine how much work is needed requires some information. First, the target flat-rate hours of the technicians working today needs to be determined. This can be obtained by adding up each of the technicians' targeted hours for the day. For this example, three technicians are working and they have targeted hours that total 26 hours. Next, the service consultant needs to determine how many jobs are leftover from yesterday and how many hours are needed to finish them. Often a quick meeting at the end of the previous day with the team leader or the technicians can determine this detail. For some jobs, a technician may be waiting for a part and the job will be finished in a few minutes. For other jobs, the technician has encountered a problem and the job will be delayed. A thorough but quick assessment either the night before or that morning can help determine how many hours of leftover work will reduce the day's target. For example, assume the leftover work was 7 hours and the capacity for the coming workday (today) is 19 hours. As an equation the capacity for today would be:

26 Target hours (hours available today)
−7 hours (hours to finish "left over" work)
=19 hours (capacity for today)

As the day progresses and jobs arrive, the service consultants will adjust the 19 hours. To do this, the Repair Order Tracking Sheet (see Figure 10-3) has rows that can be totaled and will tell service consultants

	A	B	C	D	E	F
1	SERVICE CONSULTANT	PS = Parts Specialist		SC = Service Consultant		
2	REPAIR ORDER	RO = Repair Order		Tech = Technician		
3	TRACKING SHEET			TL = Team Leader		
4	**FIRST PHASE - check in**	TODAY IS:	Monday			
5	Customer Last Name	Mr. A	Mr. B	Mrs. C	Mrs. D	**TOTAL**
6	RO #	123	124	125	126	
7	Vehicle year					
8	Vehicle Make					
9	Vehicle Model					
10	Time Promised or Waiting					
11	Initial hours sold	0.3	1	1.1	0.8	(3.2)
12	Time SC gives RO to TL and PS					
13						
14	**SECOND PHASE - initial work**					
15	Text message from Tech: TIME started job					
16	TECH Name	Frank	Omar	Omar	Ted	
17	TIME Tech reports in person to SC with RO					
18	Additional Work (Hours) recommended?	3.1	2.5	2.2	1.5	X
19						
20	**THIRD PHASE - additional work**					
21	TIME SC got Customer Approval					
22	Hours approved (change time promised)	2.5	2	2	1.5	(8)
23	Time SC gave updated RO to Tech (or TL)					
24	Time SC placed part order with PS					
25	Estimated delivery time of ordered parts					
26	TIME PS texts (SC+TECH) pick up parts					
27	Text from Tech to SC additional work started					
28						

FIGURE 10-3 Repair Order Tracking Sheet used to help determine capacity to schedule more work.

the hours that have been sold. Referring to the Repair Order Tracking Sheet, row 11 (first phase—initial work—initial hours sold) and row 22 (third phase—additional work—hours approved) are each totaled (row 11 = 3.2 hours; Row 22 = 8 hours) and added together (total = 11.2 hours) to determine the current sales.

Once the current sales are determined and subtracted from the capacity for the day, the remaining capacity can be determined. In this case, there were 19 hours of capacity for today and currently 11.2 hours were sold. After adjusting the hours there are 7.8 hours of remaining capacity. As an equation the remaining capacity would be:

Capacity for today (19 hours) − Hours sold for today (8 hours) = Remaining capacity for today (11 hours)

The service consultant has 7.8 hours that remain and depending on the time of the day, this may be a big problem. If it is early in the morning, hopefully work will come into the shop whether from "walk-ins," phone calls, or other electronic communication means to make up this shortage (undercapacity). If it is late in the morning or early afternoon, the service consultant may need to worry. The reason for concern is the service consultant has to remember that technician time is like milk at the store; it expires and unsold technician capacity can never be regained once time passes. If the technician's time is not used, the inventory of technician time is lost and cost is incurred. Therefore, depending on the amount of the shortage and time of the day, the service consultant may start to contact commercial accounts, other managers (at the dealership), or even past customers to get more work. Perhaps, a fleet customer may want to bring in a job scheduled for tomorrow so it can be finished sooner. Depending on the shortage and time of the day, the service consultant may need to alert management of a shortage and perhaps get permission to authorize the technician(s) to start "online" training, shop equipment, company vehicle maintenance, facility cleaning, checking inventory, or even organizing the toolroom. The most desired action, is to have more repair work but if not possible, then other options must be explored before sending a technician home.

Immediately sending technicians home when work slows down should be avoided. While it is true that this will reduce payroll costs (if technicians are paid hourly), the business still has other overhead costs that need to be paid. The only way to pay the overhead costs is to wait for work to come into the shop. If technicians are sent home, there is not any way to get new work done. The reality is, the automotive repair business does not typically have a constant flow of work; it is cyclical. When work "gets slow," the service consultant must be patient and wait to see if any new jobs come into the shop unexpectedly. If the amount of time to get the work done exceeds the remaining capacity of the day or requires a qualified technician who is not available, do not turn the job away or promise that it will be done today. Instead try to convince the customer to leave the vehicle overnight and it will be part of tomorrow's workflow.

When it comes to technicians' qualification, the repair order tracking sheet in row 16 lists the technicians' name. The service consultants must know each technician's qualifications to perform different jobs. For instance, Omar has extensive experience on BMW engine work but not on Frank or Ted. If a BMW engine repair job comes into the shop, unfortunately, Omar is busy with 6.1 hours of work. This job may need to be scheduled for tomorrow or perhaps Omar can start working on it later today and promise it to the customer for tomorrow. Other options may be available, such as another technician starting the job for Omar, but this concept is not without concerns and potential problems that are not limited to qualifications, technician relations, and responsibility for errors among some of the issues.

Managing Tomorrow's Work Schedule

The capacity for tomorrow is relatively simple in theory. The basic equation has two parts, the first part is:

> Target Hours available for tomorrow
> <less> "Left over" hours for tomorrow
> = Capacity for tomorrow

For example, today is Monday. The service consultant knows that none of the technicians are off tomorrow and their flat-rate target is 45 hours. Therefore, this is the number of shop hours available for tomorrow (Tuesday). By talking to the team leader and the technicians, it has been determined that the amount of leftover work to be finished tomorrow is 8 hours. Note that work often remains unfinished until tomorrow because parts must be ordered, work was started late in the day, or other problems have pushed back the promised delivery time until the next day. In this example, the result is that approximately 37 hours will be leftover for work (hours) to be finished tomorrow (Tuesday). Throughout Monday, the service consultant can schedule appointments for tomorrow (Tuesday) and other days in weeks that the customers desire. However, the question is, how many jobs should be scheduled? If too many or few are scheduled for tomorrow (Tuesday), there will be production problems.

Scheduling Work Hours per Vehicle to Determine Capacity

Theoretically, the service consultant can schedule up to 37 hours for tomorrow (Tuesday). Commonly, however, there will be already some appointments for tomorrow from perhaps last week. To assure customers remember their appointment, the service consultant should call Tuesday's customers on Monday to remind them of their appointment. Knowing the type of work to be done helps the service consultant estimate how much of the 37 hours will be sold. In some cases, there is diagnostic work as well as basic maintenance work. Likely additional work will be sold as a result of inspections, diagnosis, and maintenance specials.

To make it easier to determine how much of the 37 hours will be used, the average flat-rate hour per vehicle can help determine the number of hours. This calculation is discussed in chapter 15, but for these examples assume that the average hours to be sold for each vehicle is 1.6 hours. Therefore, if there are 20 appointments for tomorrow, then 32 hours will likely be used (20 cars×1.6 hours per car=32 hours). In reality, the service consultant will review the work and, based on experience, will determine whether 32 hours is realistic and adjust as needed. For example, if one job is a large engine repair that will take 7 hours, then that vehicle may be removed from the calculation. Therefore, 19 cars × 1.6 hours = 30.4 hours + 7 hours for the engine job = 37.4 hours. As an equation, the calculation would be:

37 hours of capacity for tomorrow
−37.4 hours sold for tomorrow-estimate
= −.4 hours of overcapacity

From the equation, the remaining capacity has been met and the work scheduled is overcapacity by –0.4 hours (negative numbers are overcapacity and positive numbers are undercapacity). However, there are a number of assumptions in this method. Assumed is that all 20 cars are either dropped off (see Figure 10-4) before Tuesday morning

FIGURE 10-4 This drop off box outside the service facility entrance allows a customer to drop off for his or her vehicle for service when the business is closed. The customer will open the drop box door and inside is an envelope. The customer will fill out the information on the envelope cover that requests the customer's name, contact information, and the work to be performed. There is a disclosure statement and a place for a signature to authorize the work (make sure there are pens in the box). Finally, the customer will put the vehicle keys in the envelope before sliding it through a slot into the building for safe keeping. In the morning, the service consultant will retrieve the envelope and keys. As a courtesy and often common practice, the service consultant will call the customer to confirm the information and service request before starting the repair order.

or show up for their appointments on time. Further assumed is the average work required of the 19 cars will equal the "historic" average of the 1.6 hours except for the engine job that will take 7 hours. Assumed also is that the skills of the technicians match the jobs. Overskilled technicians are not difficult to keep busy with lower technical work but lower skilled technicians may not be able to do some of the more complex work, such as the engine job. Therefore, careful examination of the work relative to the technician skills is needed. Also, it is presumed that each technician will come to work on time and stay until the end of the day.

When scheduling work, it is good on one hand for all of the capacity to be used to its maximum but in practice, it is a rough guess and may not work out as desired. Whether a cushion should be added or subtracted from the capacity for tomorrow depends on management. Commonly, management will direct whether to schedule more or less work. Managers and owners who have run the business for a long period of time intuitively know whether there is enough work or not. Therefore, as a tool to understand the principles of the process, the equations provided earlier are useful. However, in practice there are a lot of considerations and service consultants, who are new to a service facility, need to learn the factors that affect scheduling at a particular business.

Scheduling Work, Over- and Undercapacity, and Other Factors to Consider

Whether the service manager or owner mandates over- or undercapacity depends on many factors. Overcapacity means that all the capacity has been used and there is a shortage of technician hours to do all of the work. Often the decision to operate at overcapacity is because the technicians will get the work done faster than expected. In addition, management may want "leftover work" for the next day with some customer promised times set for the next day or further out in the future. Commonly, management will track how much "leftover" work is remaining at the end of the day and examine how much work is on the schedule for the days ahead.

If business has been slow, management may direct service consultants to "pack the shop" and not worry about leftover work or overcapacity because the shop needs to reach certain performance targets. Therefore, more work is scheduled to "force" increase production which may help to overcome a performance problem, such as income losses. Overcapacity may also be prescribed by managers because the shop has a history of customers "not showing for their appointments" or the expected number of "walk-in" customers is low. In addition, the number of hours sold per vehicle may not be reliable or a recent economic downturn has reduced the ability to close "additional sales." Therefore to overcome this, management will direct the service consultant to schedule more work.

Undercapacity means there is unused capacity and there is an excess of technician hours to do all of the work. This can be attributed to several factors. Management may decide that the shop has quality issues and wants technicians to take more time on each job. Seasonal trends, an upcoming holiday, or impending poor weather conditions may cause a significant rise in "walk-in" or "tow in" business and extra capacity is required to take advantage of business opportunities. In some states with decentralized state-mandated safety or emission inspections (covered in chapter 1), as the month comes to an end some customers forget to have their vehicle inspected and the shop may need extra capacity. Also, at some luxury dealers, there is desire to have undercapacity to assure that customers who have unexpected vehicle problems are promptly served.

There are also other reasons for undercapacity, such as technicians who have been ill and not able to work at capacity or the shop has equipment problems. Furthermore, new technicians will require extra time to get settled into the job or a high number of complex jobs are being worked on simultaneously causing management to order undercapacity. However, it should be noted that undercapacity results in lost income. Unfortunately, undercapacity may be desired by some employees who do not want to work very hard for personal reasons, such as an upcoming holiday party. In these cases, service consultants who wish to keep their job need to listen to owners and managers when it comes to reducing capacity; otherwise they may be looking for a new job if they give in to employee pressure.

The Operations Manual

When a customer's automobile is ready to receive a service, typically the service consultant must start with the estimate and from that document the repair order is created. This is part of the Customer Service System and it was discussed in chapter 6. For a service consultant, the specific procedures followed in the Customer Service System will differ depending on the service facility. The procedures for a service facility should be written down in a manual called an operations manual. The operations manual will give the service consultant specific instructions about how to interact with the customer, what paperwork is to be filled out, and how it is to be used. The exact way a service facility's Customer Service System functions will be as unique as the service facility itself. Therefore, a new service consultant should read the operations manual. If there is no operations manual, then the manager or owner must be asked how to do the job so that job performance expectations are met. An operations manual may be created for each position at a service facility or there may be one for each work area. There may be one for customer service that the service consultants, assistant service manager(s), and even the receptionist or cashier will follow. Then there may be another for the production staff that the technicians, team leader, and the parts specialists will follow. For technicians, the operations manual may include specific procedures about how to perform certain repetitive jobs, such as diagnostic procedures for no-start, no-crank problems or maintenance services such as oil changes; it must also state how work is to be processed after it is received from the service consultant. An example of how work is processed can be followed in a flowchart and can be a useful visual aid for employees to refer to when learning or doing their job. A flow chart is especially helpful when several employees must interact as a team within a "service system," such as the Shop Production System that uses the repair order tracking sheet. Figure 10-5 shows how a repair order is to be processed by the different employees through a four-phase process of the Repair Order Tracking Sheet. The Repair Order Tracking Sheet used in the Shop Production (Operations) System will be examined in more detail in chapter 11.

In the flowchart shown in Figure 10-5, there are several important features to be noted since they would not be the same at all service facilities.

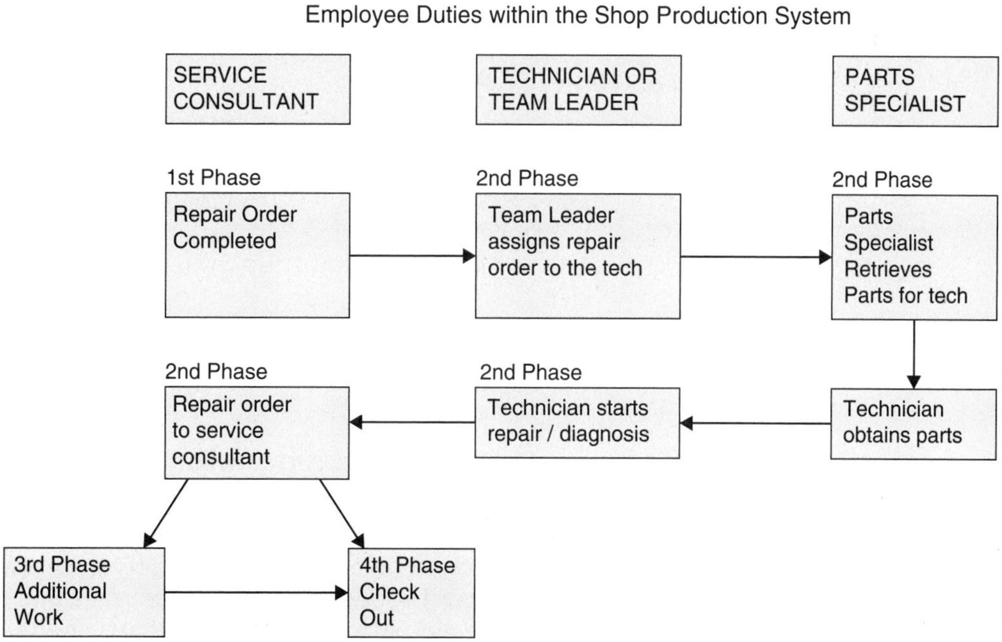

Employee Duties within the Shop Production System

SERVICE CONSULTANT	TECHNICIAN OR TEAM LEADER	PARTS SPECIALIST
1st Phase Repair Order Completed	2nd Phase Team Leader assigns repair order to the tech	2nd Phase Parts Specialist Retrieves Parts for tech
2nd Phase Repair order to service consultant	2nd Phase Technician starts repair / diagnosis	Technician obtains parts
3rd Phase Additional Work	4th Phase Check Out	

FIGURE 10-5 Employee duties within the Shop Production System

1. There is a team leader who receives the repair order from the service consultant and this concludes the first phase.
2. The team leader in this textbook's examples assigns the work to the technicians and this starts the second phase (other service systems will vary).
3. The Service Consultant in the first phase repair order tracking sheet (discussed in chapter 11) places the initial order for parts with the parts specialist.
4. The technician obtains parts prices for any additional repairs from the parts specialist.
5. The technician returns the repair order to the service consultant with any additional work recommendations, parts prices, and part availability.
6. If the customer approves the additional work, it will go into the third phase—additional work and then to the fourth phase—check out. If the customer does not approve any additional work suggested, it will go to the fourth phase—check out (skipping over the third phase).

By following a flowchart shown in Figure 10-5, a written set of procedures for an operations manual could be prepared. For example, the operations manual would start with the service consultant generating a repair order, which goes to the team leader, who assigns the work to technicians. The operations manual should also describe how technicians obtain the vehicles for repair, check them over, write down their comments about the repairs needed, return repair orders to the shop leader, and pull automobiles out of the shop; where to park them; and how to get another repair order.

Active Delivery

As discussed in chapter 6, the end of the Customer Service System process occurs when customers are presented with an invoice and the keys to their car. This delivery is an important function that is often overlooked by service facilities. The service consultant must remember that for most customers, the automobile is the second most expensive item they own (a home is the most expensive item) and they depend on it to take care of themselves and their family.

Pre- and Post-Inspection of Customer Automobiles

Of course, before a service consultant makes an active delivery, the customer's automobile must be checked. This requires service consultants to have the pre-inspection form (shown in chapter 4) on which the technician notes the general condition of the automobile, including damages and missing or broken parts. Service consultants should then inspect the automobiles and compare them to the pre-inspection report before delivery to the customer. Next, consultants should ensure that all repairs and services were performed, the vehicle was cleaned, no tools or old parts were left in it, and it was detailed (assuming this is a customer policy).

Therefore, the care and cleanliness of the customer's automobile should be monitored throughout the service process. This requires the use of fender covers, seat covers, floor mats, and steering wheel covers. Needless to say, upon delivery of the automobile to the customer, the automobile should be at least as clean as when it arrived.

Shop rags and old parts should not be left in the automobile (or under the hood) and all protective covers should be removed for the customer's convenience. A customer is not impressed when the paper floor mats are left in the car because there is no place to put them. Customers are even less impressed when there are oil stains on the carpet and not on the paper floor mat. In addition, the automobile should not have any grease stains on anything that a technician touched, such as white wall tires or door handles.

The Delivery

The primary goal behind an active delivery is not so much an action (such as washing of the automobile) but an attitude taken by the service consultant. The service consultant should take the time to add a personal touch to the delivery process.

Active delivery can include retrieving the automobile for customers before escorting them to their automobiles, which should be in front of the facility's entrance. The service consultant should express appreciation for their business. Then provide the customer with the opportunity to check over their automobiles, and allow them to state concerns, ask questions, or make requests they may have thought of after the repair was completed. This then permits the service consultant to correct any misconceptions or problems the customer may have as well as to point out the positive points about the customers' automobile. Then the service consultant should ask if the customer would like to schedule any additional work that perhaps was declined previously or set up an appointment for a future service such as an oil change. In a few cases, the service consultant can show the customer a problem that was found, such as a worn tire. If allowed by company policy, the service consultant might use a smart phone to show the customer a picture of the concern found on a vehicle. Always remembering that the idea in active delivery is to add extra value for an expensive repair that often cannot be "seen" by the customer. Therefore, remembering to be informative, positive about the vehicle, and appreciative for the business is key for a successful customer send off.

Review Questions

Multiple Choice

1. A customer has just given approval for repair of a vehicle. Consultant A says the technician should be provided with the approved work order. Consultant B says documentation of the customer's approval should be on the work order. Who is correct?
   A. A only
   B. B only
   C. both A and B
   D. neither A nor B

2. During quiet times when customer and phone demands are low, a service consultant should:
   A. Examine the progress of customers' car repairs and try to anticipate any problems that might occur.
   B. Get a head start by examining what aspect of their job can be done before the next wave of customers arrives.
   C. Take a break and either sit in the manager's office or visit the technicians.
   D. Either A or B is correct.

3. A customer approved the diagnosis of a problem but has denied approval for repair of his or her vehicle. Service consultant A says the technician should not perform the repair. Service consultant B says documentation of the customer's denial to have the repair performed should be written down on the work order. Who is right?
   A. A only
   B. B only
   C. both A and B
   D. neither A nor B

4. The workflow process begins with the scheduling of customers for work. Service consultant A says if there are too many appointments, some customers may not have their work completed by the end of the day. Service consultant B says if there are too few appointments, some technicians may not earn their expected wages for the day. Who is correct?
   A. A only
   B. B only
   C. both A and B
   D. neither A nor B

5. When booking an oil change appointment for a good customer, the consultant finds that his or her highly paid drivability technician is the only one with openings on the day the customer wants. Which of these is the best solution to this problem?
   A. Book the job for the drivability technician.
   B. Add it to the lube tech's schedule and tell the customer you will work them in.
   C. Offer the closest time that does not conflict.
   D. Move the appointment of a new customer.

Short Answer Questions

1. Explain the relationship between company policy, operational procedures, and operations manuals.
2. What is workflow and why is it important to the service consultant?
3. How might a plan to manage customer appointments be different at a fast lube service facility versus a dealership?
4. What is the difference between scheduling workflow for today as compared to tomorrow (future date)?
5. Why does a consultant need to monitor workflow?
6. What is an appointment schedule and how is it used?
7. What are the benefits of active delivery?
8. How can a repair order tracking sheet be used to determine how many work hours are available for customers who might "walk in" for service?

Repair Order Tracking Sheet

	A	B	C	D	E	F
1	SERVICE CONSULTANT	PS = Parts Specialist		SC = Service Consultant		
2	REPAIR ORDER	RO = Repair Order		Tech = Technician		
3	TRACKING SHEET			TL = Team Leader		
4	**FIRST PHASE - check in**	TODAY IS:	Monday			
5	Customer Last Name	Mr. A	Mr. B	Mrs. C	Mrs. D	**TOTAL**
6	RO #	123	124	125	126	
7	Vehicle year					
8	Vehicle Make					
9	Vehicle Model					
10	Time Promised or Waiting					
11	Initial hours sold	1.0	2.1	7.3	.6	(11)
12	Time SC gives RO to TL and PS					
13						
14	**SECOND PHASE - initial work**					
15	Text message from Tech: TIME started job					
16	TECH Name	Frank	Frank	Omar	Ted	
17	TIME Tech reports in person to SC with RO					
18	Additional Work (Hours) recommended?	X	X	X	X	X
19						
20	**THIRD PHASE - additional work**					
21	TIME SC got Customer Approval					
22	Hours approved (change time promised)	3	1.5	0	2.5	(7)
23	Time SC gave updated RO to Tech (or TL)					
24	Time SC placed part order with PS					
25	Estimated delivery time of ordered parts					
26	TIME PS texts (SC+TECH) pick up parts					
27	Text from Tech to SC additional work started					
28						

FIGURE 10-6 The Repair Order Tracking Sheet used to answer the Activity questions on page 222.

According to the Repair Order Tracking Sheet shown in Figure 10-6, 18 hours have been sold. The three technicians have a capacity of 21 hours (7 hours each). From the information provided answer the following questions:

Question 1: If a three-hour job comes when only one technician has the time to do it, who is that technician? What must a service consultant consider before accepting the job; think in terms of equipment and technicians' qualification among other factors?

Question 2: It is possible that either Mr. A's or Mr. B's cars will not be done because of the number of hours required is greater than the technician's (Frank) capacity limitations. If, after discussion with Frank, it is likely Mr. A's or Mr B's vehicle won't be done. What options might the service consultant consider in choosing which vehicle should be done first? Reconsider your answer if Ted has similar qualifications as Frank and the equipment required for the job is not a factor.

Question 3: Is Omar going to get Mrs. C's car done today? What might the service consultant do if Omar can't get it done today?

Question 4: If Omar and Frank cannot get their jobs done (each completes 7 hours of work today), how many hours will theoretically go into tomorrow?

Question 5: If Ted cannot help either Omar or Frank, and no other "customer pay" work is available, what might the service consultant do to keep him busy?

CHAPTER 11

WORKFLOW: MONITORING REPAIR PROGRESS

OBJECTIVES

Upon reading this chapter, you should be able to:

- *Explain how workflow is monitored using a repair order tracking sheet.*
- *List what occurs during each of the four Repair Order Tracking Phases.*
- *Describe how the tracking sheet times are used to find delays in the repair process.*
- *Describe techniques to manage workflow problems identified by the tracking sheet.*
- *Explain why monitoring workflow is important to meeting customer expectations.*

CAREER FOCUS

You know that the service system is an orderly arrangement of procedures that are linked together to form a process. You know how to follow it and how get other employees to follow it as well. Furthermore, you can effectively work with management to control the three Service Systems (shop production, customer service, and business operations) so that the desired outcomes are reached: profit is earned from a satisfied customer who will return for more work. You are important as the service consultant because you are the pivot in this process and direct the information needed for the three systems to function.

Specifically you are the master in the Customer Service System because you know how to work with the customer in each stage of the process to create a perfect repair order for the Shop Production System. You are the expert when working with your service team in the Shop Production System. Not only can you track the progress phases of the repair order using the Repair Order Tracking Sheet but you also know how critical it is to keep the customer informed of any additional work needed, delays, or problems in the repair process. Finally, you know how to create a flawless invoice for the Business Operations System. Above all, you can repeat the processes automatically and know how to run the systems even when all of the repairs are in different stages and phases. At times you are the busiest person in the business and as hard as it is to accomplish, you know you must keep an eye on production; otherwise, the technicians may run out of jobs to work on. The service team counts on you and whether the service system works or not is up to your skill and "know-how" to keep the work coming even when you are too busy to notice!

Introduction

As the chapter is read, take special note of the activities associated with the each of the four phases. They will help you master these concepts. Specifically, when instructed in the reading, stop and perform the activity that is found at the end of this chapter.

The service consultant has two basic jobs to perform for customers. The first is to help them to understand the services their vehicles need as covered up to this point in the textbook (the Customer Service System). The second is to keep track of the customers' repairs in the repair process so that the business outcomes are achieved (the Shop Production System). At a service facility, the desired business outcomes are to meet the customers' delivery expectations and the owners' productivity requirements. When a problem arises in production, the service consultant must have timely information to be able to take appropriate actions before it is too late. There are a variety of service facility "handwritten" worksheets that focus on information, such as worksheets for scheduled appointments, work scheduled forms, accounting sheets, repair order logs, daily production control sheets, technician work distribution documents, shop work capacity forms, and the list goes on (see Figure 11-1).

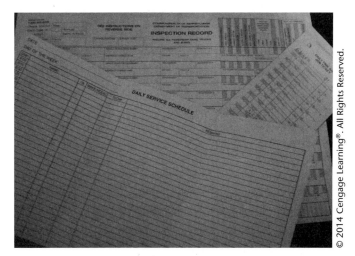

FIGURE 11-1 Examples of some blank handwritten service documents that might be found at a service facility to schedule, track, and record details of work performed.

A service consultant, dispatcher, and manager could spend a significant amount of their work day filling these out. The main purpose of any worksheet is to support a service system. Worksheet information provides management with details about the service system's performance so efficiency can be tracked (covered in chapter 15) and problems can be uncovered before they affect the customers' expectations or the owner's production requirement. Therefore, when used properly, worksheets help service consultants do their job better and can assist with management decisions so the customers and business owner are satisfied.

Whether the service system and resulting production is monitored with software (automated or not), handwritten worksheets, or self-created documents, the most important point is to get the job done on time so the customers' expectations and owner's productivity requirements are met. If a shop can "get the job done" with a piece of notebook paper on a clipboard, then the only limitation would be how many customers might be served before it fails to work as needed. When a higher volume of customers are serviced, perhaps to meet the owner's gross sales requirement, a new method may be needed. The point is that the service system must fit the volume, size, and type of service facility. The various worksheets and software programs are tools that make the service system work and are not an end in themselves as the authors have seen from experience. More specifically, filling out sheets and collecting information is not a service system and consequently will not fix a failed service system. A service system is a series of actions (procedures) and interactions that occur in a logical order (process). When the complexities of multiple jobs in different phases of production are executed with minimal errors so that the desired management outcomes are obtained, this is a system. For a service facility, the desired management outcome is to make sure the customers' expectations and owner productivity requirements are met.

The Repair Order Tracking Sheet Phases

The latest software systems can cut down on the time to enter, access, and share information with others. However, for students who study these tasks, understanding the basic repair process is important. To do this, the first edition of this textbook presented a Repair Order Tracking Sheet. This sheet had many applications; such as allowing two or more shifts of service consultants to work the same service desk. While similar to some industry documents, the sheet was developed by the authors to allow students to learn the repair process and, when on paid internship, the students could "share" the job of a service consultant. Specifically, the students had to be able to leave work to attend class and then return without missing details that happened with their customers while away from the shop. As a teaching tool, the sheet has taken various forms over the years and shown below in Figure 11-2 is the latest version.

To illustrate the basic function of repair order processing in the Shop Production System (see Figure 11-3), there are four phases that occur in the service of vehicles as needed by the service team (Service Consultant, Team "technician" Leader, Technician, and Parts Specialist).

1. The first phase is the Check-In phase. This phase records important information from the service consultant's work with the customer during the Customer Service System's First and Second Stages. All of the collected repair information, and initial work sold, will be on the Repair Order (RO) and transferred to the team leader and parts specialist at the start of the Shop Production System.
2. The second phase is the Initial Work Phase where there is communication between the service consultant and the technician who was assigned the job. During this phase additional inspections by the technician may reveal more work is needed. This is brought to the service consultant's attention to be suggested to the customer when the service consultant returns to the Customer Service System's Second Stage. If the vehicle does not require any additional work or the service consultant is unable to obtain approval for additional work, the repair will skip the third phase and go to the fourth phase; check out.
3. When more work is approved by the customer, such as a new starter is requested after diagnosis, the third phase of additional work will be entered. The third phase will monitor the hours sold and interaction between the service consultant, parts specialist, and the technician. Times are monitored closely and compared to the time promised to the customer to assure the customer's delivery expectations are met.
4. The fourth and final phase is Check Out. This phase monitors the interaction between the service consultant, technician, and parts specialist as they work to complete the final invoice and finalize the delivery of the vehicle to the customer on time.

	A	B	C	D	E	F
1	SERVICE CONSULTANT	PS = Parts Specialist		SC = Service Consultant		
2	REPAIR ORDER	RO = Repair Order		Tech = Technician		
3	TRACKING SHEET			TL = Team Leader		
4	**FIRST PHASE - check in**	TODAY IS:				
5	Customer Last Name					
6	RO #					
7	Vehicle year					
8	Vehicle Make					
9	Vehicle Model					
10	Time Promised or Waiting					
11	Initial hours sold					
12	Time SC gives RO to TL and PS					
13						
14	**SECOND PHASE - initial work**					
15	Text message from Tech: TIME started job					
16	TECH Name					
17	TIME Tech reports in person to SC with RO					
18	Additional Work (Hours) recommended?					
19						
20	**THIRD PHASE - additional work**					
21	TIME SC got Customer Approval					
22	Hours approved (change time promised)					
23	Time SC gave updated RO to Tech (or TL)					
24	Time SC placed part order with PS					
25	Estimated delivery time of ordered parts					
26	TIME PS texts (SC+TECH) pick up parts					
27	Text from Tech to SC additional work started					
28						
29	**FOURTH PHASE - Check Out**					
30	Time Tech reports in person to SC with RO					
31	TIME PS supplies parts receipts to SC					
32	TIME final INVOICE has been completed					
33	TIME customer notified that vehicle done					
34	TIME vehicle Picked up					
35	SC was able to provide Active Delivery?					
36	Thank-you note/survey sent to customer?					

FIGURE 11-2 This is a blank Repair Order Tracking Sheet that we will examine in more detail in future chapters.

228 THE SERVICE CONSULTANT

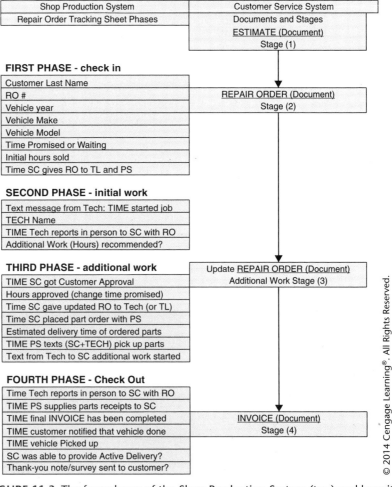

FIGURE 11-3 The four phases of the Shop Production System (top) and how it coordinates with the Customer Service System (bottom).

The First Phase of the Repair Order Tracking Sheet: Check In

The first phase (see Figure 11-4) of the Repair Order Tracking Sheet is check in. The service consultant performs the normal meeting, greeting, and information collection process in the Customer Service System (see chapter 6). Once the interaction with the customer has been completed, the estimate accepted, and the repair order created, the service consultant will fill out the Repair Order Tracking Sheet information in the first phase.

The items completed will be Rows 5 to 12. Information for Rows 5 to 11 will be information from the repair order for reference by the service consultant later. Summary of the information typed into the cells includes:

- Row 5: The customer's last name (needed when a customer calls into the shop for information).
- Row 6: The RO number (needed when working with parts and the cashier).
- Rows 7, 8, 9: The year, make, model of the vehicle (needed when working with the technicians and for reference when dealing with other departments such as parts).
- Row 10: The time promised to the customer when his or her vehicle will be done or if the customer is waiting at the service facility.

	A	B	C	D	E	F
1	SERVICE CONSULTANT	PS = Parts Specialist		SC = Service Consultant		
2	REPAIR ORDER	RO = Repair Order		Tech = Technician		
3	TRACKING SHEET			TL = Team Leader		
4	**FIRST PHASE - check in**	TODAY IS:	Friday			
5	Customer Last Name	Fosko	Pyle	Dame	Byrne	Stewart
6	RO #	5882	5901	5913	5921	5922
7	Vehicle year	2015	2013	2011	2012	2013
8	Vehicle Make	Honda	Toyota	Ford	Cadillac	GMC
9	Vehicle Model	Accord	Land Crus.	Taurus	SRX	Denali
10	Time Promised or Waiting	Waiting	Waiting	3pm	5pm	Mon 9am
11	Initial hours sold	0.9	1.2	0.5	0.3	1.5
12	Time SC gives RO to TL and PS	9am	9:15	9:30	9:45	10am
13						

FIGURE 11-4 This Repair Order Tracking Sheet shows the first phase information filled out by the service consultant.

This information in Row 10 is important because it establishes the goal for the service team. All of the other event times will be compared to the Row 10 time. To illustrate the relationship, the service consultant, when using the tracking sheet, is like a coach on the side lines of a football game. They are looking at the team on the field executing plays that moves the football (repair order) down the field toward the goal

	A	B	C
1	SERVICE CONSULTANT	PS = Parts Specialist	
2	REPAIR ORDER	RO = Repair Order	
3	TRACKING SHEET		
4	**FIRST PHASE - check in**	TODAY IS:	Wed
5	Customer Last Name	Hiney	
6	RO #	5890	
7	Vehicle year	2014	
8	Vehicle Make	Ford	
9	Vehicle Model	Focus	
10	Time Promised or Waiting	2:30pm	
11	Initial hours sold	1.3	
12	Time SC gives RO to TL and PS	8:45am	
13			
14	**SECOND PHASE - initial work**		
15	Text message from Tech: TIME started job	9:05	
16	TECH Name	Brian	
17	TIME Tech reports in person to SC with RO	9:30	
18	Additional Work (Hours) recommended?	2.5	
19			
20	**THIRD PHASE - additional work**		
21	TIME SC got Customer Approval	10:15am	
22	Hours approved (change time promised)	2.1	
23	Time SC gave updated RO to Tech (or TL)	10:30	
24	Time SC placed part order with PS	10:35	
25	Estimated delivery time of ordered parts	11:30	
26	TIME PS texts (SC+TECH) pick up parts		
27	Text from Tech to SC additional work started		

FIGURE 11-5 This Repair Order Tracking Sheet shows the time targets the service team must meet before time runs out. The promise time is 2:30 p.m. (Row 10), the parts will arrive at 11:30 a.m. (Row 25), and the flat-rate time of the repair is 2.1 hours (Row 22). There isn't much room for error.

line for a touchdown, which is the time promised to the customer. For a coach on the sidelines, the measure of how well the team is doing is examined by where the team is at relative to the lines on the field that have numbers on it. Likewise, the critical event times tell the service consultant where the team is located in reference to the goal line, which is the time promised (touchdown) plus the other important times, such as Row 25 which concerns the estimated delivery time of ordered parts (this is like a first down and shown later). As each event time is recorded, the service consultant monitors the team's progress toward the goal. The service consultant must reach the goal before the "clock runs out" (time promised expires).

- Row 11: Records the initial hours sold (needed for scheduling purposes).

Unlike a game of football, which is played with a single football, the service consultant has several repair orders (footballs) in "play" at the same time and each one is at different phases in the shop production system. Imagine trying to play football with multiple balls and teams and players on the same field! Each player is running different plays at the same time. However, the complexity for a service consultant is not because the team members are at different phases of the Repair Order Tracking Sheet but being torn between having enough work for the service facility to earn the income needed and too much work that it cannot all get done as promised. In this case Row 11 along with Row 22 helps the service consultant determine how many hours have been sold which is compared against how many hours are available. This concept was covered in chapter 10. For example, the service consultant may need to have 50 hours to keep the technicians busy. If Row 11 is 14 hours and Row 22 is 30 hours, the sum is 44 hours. Assuming there isn't any "leftover" work from yesterday and today's work is "new" (none of the new work was started yesterday), the service consultant has six hours left in today's schedule.

- Row 12: The service consultant records the time the RO "hard copy" was given to the Team Leader and the soft copy given to the parts specialist. All of our examples assume that the team leader is a technician and will assign the RO to a technician on the team. This process may differ from shop to shop. Furthermore, when the RO "soft copy" is given to the Parts Specialist, it is assumed they will immediately pull parts for the job.

For illustration purposes, this Repair Order Tracking Sheet assumes a team or group system of management where the service consultant works with a "Team Leader" (or Group Leader); however, some shops may have another method to provide the work to the technician. This could include a dispatcher, lead technician, shop foreman, computerized RO assignment system, or even the service consultant or ASM. In all cases (except computerized RO assignment or "cueing" systems which may

> **Stop and Complete Activity**
>
> Please take the time to read and complete activity 1 at the end of the chapter. This will help reinforce your understanding of the FIRST Phase of the Repair Order Tracking Sheet.

have a form of tracking system built into it), the spreadsheet would be customized with appropriate cell descriptions to reflect the system and titles used. For an internship situation, the student typically will work in such a modification.

The Second Phase of the Repair Order Tracking Sheet: Initial Work

The second phase of the repair process monitors communication between all team members. It is assumed that communication between the service consultant and parts specialist occurred prior to the customer's arrival for the scheduled service. Therefore, the parts were ordered at least a day before, have arrived, and are waiting at the service facility. From the First Phase (Row 12), the parts specialist was notified the job has arrived and to pull the parts. The technician who is assigned the job by the team leader will be given the RO and usually the keys with a tag on them to indicate the vehicle's owner, RO number, or hang tag number (see Figure 11-6 and Figure 11-7). The

FIGURE 11-6 SUV with a hang tag number on the mirror. To identify the correct keys for the vehicle, the hang tag number is matched to the key tag number.

FIGURE 11-7 When keys are properly tagged, the technicians will be able to quickly locate the correct key for the next vehicle to be serviced, even when they are in disarray and not on a peg board.

idea is for the car to be easily found in the parking lot with minimal searching.

After obtaining the next job (RO), the technician must communicate to the service consultant that he or she has the RO and is prepared to start the job. Whether that is by a text message, email, verbal communication, or other means depends on company procedures. The method used must work and can be as simple as a sign-in sheet on the desk where the technician writes down the RO number and time before taking the keys to the car. For illustration purposes, text messages will be the means of communication. The technician sends a text to the service consultant with the RO number as the message. This will let the service consultant know the job the technician was assigned. As shown in Figure 11-8, the service consultant will record the time of the text message in Row 15 and the technician who sent the text message (assigned the job) in Row 16.

The Technician's and Parts Specialist's Role in the Second Phase

From an efficiency standpoint, the technician will perform two actions: get the parts and the vehicle. The order of what should occur first depends on logistic factors that are too numerous to list. The bottom line is: "what gets done first (get the parts or the vehicle) needs to make sense" to avoid wasting time. Obviously, to increase efficiency, the parts should be pulled by the parts specialist already from the First Phase communications (Row 12—RO is given to the TL and the PS).

	A	B	C	D	E	F
1	SERVICE CONSULTANT	PS = Parts Specialist		SC = Service Consultant		
2	REPAIR ORDER	RO = Repair Order		Tech = Technician		
3	TRACKING SHEET			TL = Team Leader		
4	**FIRST PHASE - check in**	TODAY IS:				
5	Customer Last Name	Kline				
6	RO #	77873				
7	Vehicle year	2015				
8	Vehicle Make	Chrysler				
9	Vehicle Model	Van				
10	Time Promised or Waiting	Waiting				
11	Initial hours sold	0.7				
12	Time SC gives RO to TL and PS	10:05am				
13						
14	**SECOND PHASE - initial work**					
15	Text message from Tech: TIME started job	10:15am				
16	TECH Name	Maria				
17	TIME Tech reports in person to SC with RO					
18	Additional Work (Hours) recommended?					
19						

FIGURE 11-8 This Repair Order Tracking Sheet shows the start of the second phase. Rows 15 and 16 were completed when Maria, the technician, sent her service consultant a text message with the RO number.

At highly efficient service facilities, when the parts specialist is included in communication process, such as the text message sent in Row 15, he or she can deliver the parts to the technician's bay. What is important is the technician should not stand around waiting for parts, that is costly to the service facility.

To increase technician efficiency, the technician should do the inspection and diagnosis work before any of the "sold" work, such as oil changes, brake jobs, or repair work. When the technician does not find the vehicle needs additional work, the "sold" work will be finished. The third phase—Additional Work—will be skipped and the repair order will go to the fourth phase—Check Out. If additional work is needed and parts will be required, the technician should consult the parts specialist for price and availability of the part. The availability of the parts has three options:

- The parts needed for the job are in stock.
- The parts are in stock locally and can be delivered or picked up today.
- The parts must be ordered and will arrive at some expected date or time in the future.

This is important because the service consultant needs to know what to tell the customer and how to change the time promised.

The technician and the parts specialist must work together to get the details to the service consultant about what the vehicle needs so he or she can contact the customer as soon as possible. When the service consultant gets the information needed, he or she will pivot to the Customer Service System's third stage and communicate with the customer. The service consultant will point out the features and benefits of the repair as covered in chapter 6. Presenting the repair suggestion properly (features, benefit, price, and ask for authorization) will generate an answer from the customer (yes or no). While waiting for an answer, the technician must stay efficient while the service consultant is looking up labor charges, calculating repair prices, and contacting the customer for approval of the additional work suggested. Therefore, the technician should complete any work sold during the First Phase—Check In. This could include maintenance work the service consultant sold, such as a 30,000-mile service, a special, or maybe a brake job that the customer bought. Therefore, the technician should not rush to do the maintenance or repair work; it should be done after any diagnosis or inspection work has been completed first and the service consultant has been informed.

When the customer wants the additional work to be performed (yes), he or she will authorize the repair and the RO will be updated (see Figure 11-9). The Service System will exit the Customer Service's third stage and enter the Shop Production System's Third Phase covered in the next section. When the customer declines the repair, it might be negotiated that it will be scheduled for future date (the current RO will enter the Fourth Phase). Regardless, when additional work is declined or scheduled for a later date, the suggestion of the service facility will be noted on the final invoice. This is required to protect the company from any future claims by customers that they were not told of the suggestion. It is the service consultant's job to make sure this happens.

For the technician, it is important to know the customer's answer (yes or no) as soon as possible. Either the technician will prepare to receive additional parts and the current job will enter the Shop Production System's Third Phase—Additional Work—or the technician will move on to another job after the Second Phase—Initial Work is completed. When additional work is needed and the parts required to finish the job will not arrive for a long time, the team leader may have the technician move to another job, but that decision and the reason is beyond the scope of this discussion.

Stop and Complete Activity

Please take the time to read and complete activity 2 at the end of this chapter. This will help reinforce your understanding of the SECOND Phase of the Repair Order Tracking Sheet.

	A	B	C	D	E	F
1	SERVICE CONSULTANT	PS = Parts Specialist		SC = Service Consultant		
2	REPAIR ORDER	RO = Repair Order		Tech = Technician		
3	TRACKING SHEET			TL = Team Leader		
4	**FIRST PHASE - check in**	TODAY IS:				
5	Customer Last Name	Kline				
6	RO #	77873				
7	Vehicle year	2015				
8	Vehicle Make	Chrysler				
9	Vehicle Model	Van				
10	Time Promised or Waiting	Waiting				
11	Initial hours sold	0.7				
12	Time SC gives RO to TL and PS	10:05am				
13						
14	**SECOND PHASE - initial work**					
15	Text message from Tech: TIME started job	10:15am				
16	TECH Name	Maria				
17	TIME Tech reports in person to SC with RO	10:50am				
18	Additional Work (Hours) recommended?	2.4				
19						

FIGURE 11-9 This Repair Order Tracking Sheet shows the completed Second Phase and Rows 15–18 are finished. Notice that the technician reported to the service consultant at 10:50 a.m. that the vehicle needed additional work based on her inspection and/or diagnosis. The technician got the parts price and availability before meeting with the service consultant to describe the problems found. The service consultant looked up the labor times for the additional work at 2.4 hours on top of the 0.7 hours already sold in the first phase. The technician will return to work (doing any pre-approved service work) and await an answer from the Service Consultant about whether any of the work was approved. If approved, the repair will enter the Third Phase—Additional Work. If not approved, the repair will go to the Fourth Phase—Check Out.

The Third Phase of the Repair Order Tracking Sheet—Additional Work

The Third Phase: The Service Consultant's Role

Using Figure 11-10, assume that in the second phase the service consultant obtained approval for the repair. First, the time the service consultant got approval would be entered into Row 21. Next, the number of hours approved would be entered into Row 22. Chapter 15 will discuss how the tracking sheet information is used to determine the service consultant effectiveness and efficiency. In general, the service consultant should remember, that the time between Row 17 (time the technician reports to the SC with the RO for additional work needed) and Row 21 (the time it

	A	B	C	D	E	F
1	SERVICE CONSULTANT	PS = Parts Specialist		SC = Service Consultant		
2	REPAIR ORDER	RO = Repair Order		Tech = Technician		
3	TRACKING SHEET			TL = Team Leader		
4	**FIRST PHASE - check in**	TODAY IS:				
5	Customer Last Name	Kline				
6	RO #	77873				
7	Vehicle year	2015				
8	Vehicle Make	Chrysler				
9	Vehicle Model	Van				
10	Time Promised or Waiting	Waiting				
11	Initial hours sold	0.7				
12	Time SC gives RO to TL and PS	10:05am				
13						
14	**SECOND PHASE - initial work**					
15	Text message from Tech: TIME started job	10:15am				
16	TECH Name	Maria				
17	TIME Tech reports in person to SC with RO	10:50am				
18	Additional Work (Hours) recommended?	2.4				
19						
20	**THIRD PHASE - additional work**					
21	TIME SC got Customer Approval	11:20am				
22	Hours approved (change time promised)	1.8				
23	Time SC gave updated RO to Tech (or TL)	11:30				
24	Time SC placed part order with PS	11:30				
25	Estimated delivery time of ordered parts	12:00				
26	TIME PS texts (SC+TECH) pick up parts					
27	Text from Tech to SC additional work started					
28						

FIGURE 11-10 This tracking sheet shows that the service consultant took about half an hour to get Maria's approval back by 11:20 a.m. for 1.8 hours of additional work of the 2.4 hours she suggested. Maria has earned 1.8 hours on top of the 0.7 hours already sold in the first phase for a total ticket time of 2.5 hours. The Service Consultant notified the technician and the parts specialist at 11:30 a.m. The parts specialist anticipated the parts to arrive by noon.

took to look up the labor times, calculate the price, and obtain the approval) is important to keep the technician busy. The number of repair suggestions, the complexity of each suggestion, and the ability to contact the customer are factors in how quickly the service consultant can obtain approval of all, some, or none of the suggestions. A service consultant can improve the time lapse by obtaining all of the contact information from the customer (all phone numbers and email address) as well as understanding the details of the technician's repair suggestions. Of course, to avoid a mistake, the service consultant should review the additional work suggestions with the technician before contacting the customer to assure that the proper service features and benefits are chosen.

The Third Phase: The Work Schedule and the Team Leader's Role

The complexity in the third phase is that there are other jobs that need to be done. The team leader will ultimately work with the Service Consultant to organize and prioritize the work. Discussions take place often between the service consultant and team leader to decide what jobs can wait, what jobs must wait (such as parts that are not due until later in the day), and what jobs need immediate attention. Depending on the hours approved in Row 22, the Service Consultant may need to alert the customer of additional delays depending on the situation. At the Team Leaders direction, the customer's car may need to be pulled out until approval is obtained or because the parts delivery or scope of the work will take longer than expected. This means the technician will be told by the team leader to move on to another job.

When approval has been obtained and the car has been "pulled out" of the service bay, the service consultant needs to meet with the team leader because it is likely that the technician is on to another job. The RO will be returned to the Team Leader to monitor for the next opportunity to continue working on the job. Communication must be clear between the team leader and the service consultant; otherwise mistakes can be easily made that will make the customer angry. For example, if the job is relatively complex and the team leader feels the technician will not make the "flat-rate time" on the job, this should be told to the service consultant. The service consultant must then talk to the customer and change the promise time. Other reasons for delays in the Third Phase include anticipated parts delivery delays for various reasons, other work that is backing up in the shop, unexpected problems with other jobs, technicians' illness, broken equipment, and the list goes on. The team leader will be juggling a number of different variables to try and get the work processed. He or she may decide to have a technician pull off a part and then go to another job until the part arrives. The decision may also be to pull the vehicle out of the shop and do another job such as a waiting oil change. The service consultant must share in the discussion and defer judgment about whether he or she agrees or disagrees with the team leader's decision. The idea is to work together as a team to make the shop run smooth and meet all of the customers' expectations.

The Third Phase: Processing the Additional Work

The Third Phase can be complex and eventually the service consultant will deliver the RO that was approved by the customer to the technician, if still working on the vehicle. If the technician has moved to another job, the RO will go back to the team leader to assign to the technician once the opportunity arises. The time that this occurs goes in Row 23 (see Figure 11-11). The service consultant must also tell the parts specialist what services have been approved by the customer so the parts can be ordered or pulled from inventory. The time that this information was exchanged goes into Row 24. The parts specialist should make the service consultant aware of when ALL of the parts for the vehicle will arrive and this is recorded in Row 25.

The next major event for the service consultant will be when the part specialist texts the Service Consultant and Technician (perhaps even the team leader) that the parts have arrived for RO 77873 (Kline's car). This can also be done verbally but often it is too cumbersome to tell everyone and too easy for people to forget the message that was heard. So when a text message (among other electronic means) is used, it can be more definite and efficient. The time for this is recorded in Row 26 (see Figure 11-12). Finally, the service consultant will await the text message from the technician that the additional work has started and records it in Row 27.

> ### Stop and Complete Activity
> Please take the time to read and complete activity 3 at the end of the chapter. This will help reinforce your understanding of the THIRD Phase of the Repair Order Tracking Sheet.

Catching Problems with the Tracking Sheet

The benefit of knowing the times is similar to watching a football team on the field. When you can see the numbers on the field, you know how far the players are from the goal line. The measure can help the team captain and coaches decide whether to run the ball, pass, or punt. In the same way the service consultant is watching the times. If parts are running behind schedule, the cause should be determined and any potential remedy taken. If the parts have arrived but the technician has not started the vehicle in a timely manner and the sheet shows 1.8 additional hours of work to finish (Row 22 shows additional hours), discussion needs to happen because it goes without saying that the job will be delayed. When the delay is going to cause an issue with the promised time, it should be brought to the customer's attention immediately to determine a solution. Figure 11-11 and 11-12 illustrate problems that occurred with Mrs. Kline's vehicle repair.

	A	B	C	D	E	F
1	SERVICE CONSULTANT	PS = Parts Specialist		SC = Service Consultant		
2	REPAIR ORDER	RO = Repair Order		Tech = Technician		
3	TRACKING SHEET			TL = Team Leader		
4	**FIRST PHASE - check in**	TODAY IS:				
5	Customer Last Name	Kline				
6	RO #	77873				
7	Vehicle year	2015				
8	Vehicle Make	Chrysler				
9	Vehicle Model	Van				
10	Time Promised or Waiting	(5pm)				
11	Initial hours sold	0.7				
12	Time SC gives RO to TL and PS	10:05am				
13						
14	**SECOND PHASE - initial work**					
15	Text message from Tech: TIME started job	10:15am				
16	TECH Name	Maria				
17	TIME Tech reports in person to SC with RO	10:50am				
18	Additional Work (Hours) recommended?	2.4				
19						
20	**THIRD PHASE - additional work**					
21	TIME SC got Customer Approval	11:20am				
22	Hours approved (change time promised)	1.8				
23	Time SC gave updated RO to Tech (or TL)	11:30				
24	Time SC placed part order with PS	11:30				
25	Estimated delivery time of ordered parts	(2pm)				
26	TIME PS texts (SC+TECH) pick up parts					
27	Text from Tech to SC additional work started					
28						

FIGURE 11-11 At noon the service consultant was promised the parts. When the parts did not arrive, he went to the parts specialist. The parts specialist discovered that the delivery will be delayed and the arrival time in Row 25 had to be changed to 2:00 p.m. This will mean the service consultant will have to speak with the customer who was waiting (Figure 11-10 Row 10). The customer is informed and has various alternative transportation options provided (rental, loaner, shuttle, bus, or cab) or she can continue to wait. Obviously the customer must make that decision. Mrs. Kline, the customer, decides to leave and will return at the end of the day. The time promised changes to 5:00 p.m. from Waiting (Row 10 has been changed from Waiting to 5:00 p.m.) and the Estimated Delivery of Ordered Parts changes to 2:00 p.m.

	A	B	C	D	E	F
1	SERVICE CONSULTANT	PS = Parts Specialist		SC = Service Consultant		
2	REPAIR ORDER	RO = Repair Order		Tech = Technician		
3	TRACKING SHEET			TL = Team Leader		
4	**FIRST PHASE - check in**	TODAY IS:				
5	Customer Last Name	Kline				
6	RO #	77873				
7	Vehicle year	2015				
8	Vehicle Make	Chrysler				
9	Vehicle Model	Van				
10	Time Promised or Waiting	5pm				
11	Initial hours sold	0.7				
12	Time SC gives RO to TL and PS	10:05am				
13						
14	**SECOND PHASE - initial work**					
15	Text message from Tech: TIME started job	10:15am				
16	TECH Name	Maria				
17	TIME Tech reports in person to SC with RO	10:50am				
18	Additional Work (Hours) recommended?	2.4				
19						
20	**THIRD PHASE - additional work**					
21	TIME SC got Customer Approval	11:20am				
22	Hours approved (change time promised)	1.8				
23	Time SC gave updated RO to Tech (or TL)	11:30				
24	Time SC placed part order with PS	11:30				
25	Estimated delivery time of ordered parts	2pm				
26	TIME PS texts (SC+TECH) pick up parts	Sent tech				
27	Text from Tech to SC additional work started	2:18				
28						

FIGURE 11-12 This tracking sheet was monitored by the service consultant. At 2:00 p.m. the parts still had not arrived for Mrs. Kline's vehicle. The service consultant immediately went to parts and discovered that the parts had to be picked up at the parts store—delivery was not available. The service consultant, parts specialist, and the team leader discussed the problem and decided to have Maria the technician start to disassemble the vehicle (Row 27) while another technician went to pick up the parts (Row 26). The team leader agreed to help Maria get the job done by the promise time of 5:00 p.m. if she needed assistance.

Fourth Phase—Check Out

If all goes well, the technician will give the service consultant the RO and keys with "time to spare." This time is recorded in Row 30 (see Figure 11-13). The parts specialist should provide receipts (hard copy or soft copy) of the RO with the parts section completed, or have the final

	A	B	C	D	E	F
1	SERVICE CONSULTANT	PS = Parts Specialist		SC = Service Consultant		
2	REPAIR ORDER	RO = Repair Order		Tech = Technician		
3	TRACKING SHEET			TL = Team Leader		
4	**FIRST PHASE - check in**	TODAY IS:				
5	Customer Last Name	Kline				
6	RO #	77873				
7	Vehicle year	2015				
8	Vehicle Make	Chrysler				
9	Vehicle Model	Van				
10	Time Promised or Waiting	5pm				
11	Initial hours sold	0.7				
12	Time SC gives RO to TL and PS	10:05am				
13						
14	**SECOND PHASE - initial work**					
15	Text message from Tech: TIME started job	10:15am				
16	TECH Name	Maria				
17	TIME Tech reports in person to SC with RO	10:50am				
18	Additional Work (Hours) recommended?	2.4				
19						
20	**THIRD PHASE - additional work**					
21	TIME SC got Customer Approval	11:20am				
22	Hours approved (change time promised)	1.8				
23	Time SC gave updated RO to Tech (or TL)	11:30				
24	Time SC placed part order with PS	11:30				
25	Estimated delivery time of ordered parts	2pm				
26	TIME PS texts (SC+TECH) pick up parts	Sent Tech				
27	Text from Tech to SC additional work started	2:18				
28						
29	**FOURTH PHASE - Check Out**					
30	Time Tech reports in person to SC with RO	4:45				
31	TIME PS supplies parts receipts to SC	3pm				
32	TIME final INVOICE has been completed	4:55				
33	TIME customer notified that vehicle done	4:55				
34	TIME vehicle Picked up	5pm				
35	SC was able to provide Active Delivery?	Yes				
36	Thank-you note/survey sent to customer?	Yes				

FIGURE 11-13 Maria gave the RO and keys to the service consultant at 4:45 p.m. (Row 30). The parts specialist provided the needed information for the invoice by 3:00 p.m. (Row 31), so the invoice could be partly completed. The invoice was completed at 4:55 p.m. (Row 32) and the customer was notified at 4:55 p.m. (Row 33) and she arrived at 5:00 p.m. to pick up the car (Row 34). The service consultant met with Mrs. Kline to provide her with "Active Delivery" of the vehicle and then returned to his desk where he sent her a thank-you note and a survey by email. While she may have been disappointed in the delays, for the team it required cooperation to get this job finished as promised—touchdown for the team.

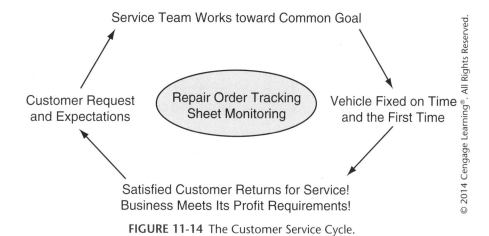

FIGURE 11-14 The Customer Service Cycle.

invoice section completed on a computerized RO well before the technician delivers the RO. This time is recorded on Row 31. When it comes to the final invoice, the paperwork "all comes together" and the hard copy of the RO reunites with the soft copies of the RO. This time the final RO is completed is recorded in Row 32 with the time the customer is called in Row 33 and the time of pick-up in Row 34 (see Figure 11-13). If Active Delivery is possible, it is recorded in Row 35 with details about whether a thank-you card, email, message, or phone call was provided with a customer survey in Row 36. This final Check-Out phase, which overlaps with the Customer Service System's Fourth Stage, should go smoothly provided the details in the previous phases and Customer Service System stages were conducted properly. More specifically, the service facility has a touchdown and the service consultant will pivot to start the business operations phase with an invoice for the customer to pay (see Figure 11-14).

Stop and Complete Activity

Please take the time to read and complete activity 4 at the end of the chapter. This will help reinforce your understanding of the FOURTH Phase of the Repair Order Tracking Sheet.

FIRST PHASE ACTIVITY: Check In

Below is an interaction between a customer, named Mr. Zell, and a service consultant. The conversation took place during the first stage of the Customer Service System. When the interaction is over, the service

	A	B	C
1	SERVICE CONSULTANT	PS = Parts Specialist	
2	REPAIR ORDER	RO = Repair Order	
3	TRACKING SHEET		
4	**FIRST PHASE - check in**	**TODAY IS:**	
5	Customer Last Name		
6	RO #		
7	Vehicle year		
8	Vehicle Make		
9	Vehicle Model		
10	Time Promised or Waiting		
11	Initial hours sold		
12	Time SC gives RO to TL and PS		
13			

FIGURE 11-15 This Repair Order Tracking Sheet goes with the First Phase Activity.

consultant will pivot and put important information in the First Phase of the Repair Order Tracking Sheet (see Figure 11-15) that is used to monitor progress in the Shop Production System. You are to pick out and write in Figure 11-15 the information from the following interaction to be entered in cells B5 to B12.

To start the Repair Order Tracking Sheet, the day of the week is Monday and is written in cell C4.

Customer Service System's First Stage Interaction:

SC: My name is John, I am the Service Consultant, (extends hand for handshake), and you are?
CUSTOMER: I am Mr. Zell (cell B5)
SC: Welcome to Auto Service World, Mr. Zell. How may I help you?
CUSTOMER: My car does not start.
SC: Okay, let me get some information. I see you are already in our system, which car is giving you a problem Mr. Zell?
CUSTOMER: It's the 2014, Hyundai, Genesis. (cells B7, B8, B9)
SC: Does it not crank over or does it crank but the engine will not start?
CUSTOMER: It does not start.
SC: Is it broken now?
CUSTOMER: No, not now.
SC: When you drive the car does the check engine light come on?
CUSTOMER: No
SC: Okay, to check this problem we charge $92.
(For this garage, $92 provides the technician with one hour (cell B11) of diagnostic time plus any other service time sold later in this interaction).

SC: I can have an answer for you by 3:00 p.m. today but depending on the problem may not have it fixed today. Is that okay with you?

CUSTOMER: That is okay as long as I have an answer by 3:00 p.m. so I can make other transportation arrangements. (The time promised for the Repair Order Tracking Sheet is 3:00 p.m. and goes in cell B10) . If the customer approves any work, the SC will change the promise time based on parts arrival and time to make the repair).

SC: Yes, Mr. Zell, I will call you by 3:00 p.m. Now I see from your service file that the front brake lining was low and we suggested replacing them during your next service. Would you like for me to schedule your brake replacement now?

CUSTOMER: Okay, may as well get it done how much is that and how much will that delay fixing my car?

SC: The charge is $129 and the technician will probably have it done soon after we diagnose the vehicle for you. It should not delay the diagnosis process.

(The replacement of the brakes is 9/10th of an hour or 0.9 hr. This goes in cell B11 along and is added to the diagnosis time above).

CUSTOMER: That is fine.

The Service Consultant prints out the estimate.

SC: Mr. Zell, here is your estimate for the diagnosis and the work; I need your authorization to start the work.

This ends the Customer Service System's Stage 1. A Service System review of what happens next:

- The Service Consultant hands the estimate to Mr. Zell.
- Mr. Zell examines it and then signs the estimate authorizing the repair.
- The estimate document (Customer Service System's Stage 1) turns into a repair order document with the customer's authorization. The authorization signals the start of the Customer Service System's Stage 2.
- The Shop Production System's First Phase starts when the service consultant starts to fill in the Repair Order Tracking Sheet after Stage 2 of the Customer Service System ends.
- The Repair Order number assigned to Mr. Zell's job is #3087 (cell B6).

SC: Do you need our courtesy shuttle?

CUSTOMER: No, I have a ride waiting; I have to get to my 9:00 a.m. meeting.

SC: Thank you for your business Mr. Zell. (The SC shakes the customer's hand).

This ends the Customer Service System's Stage 2.

The Service Consultant pivots and fills in the Repair Order Tracking Sheet. To help you understand the process, fill in Figure 11-15. Find in the interaction above the following information:

- Customer last name for cell B5
- The repair order number in cell B6
- Vehicle year, make, and model for cells B7, B8, and B9
- The time the SC has promised to get back to Mr. Zell in cell B10
- The hours sold for the diagnosis and the brake pad replacement in cell B11

The SC takes the RO #3087's "hard copy" to the Team Leader to assign to a technician with the soft copy to the Parts Department as per this company's procedure. The SC notes that the time is 8:52 a.m. This time is recorded in cell B12 of the Repair Order Tracking Sheet.

SECOND PHASE ACTIVITY: Initial Work

Examine Figure 11-16. This is a blank copy of the Second Phase the Repair Order Tracking Sheet with the first phase completed. You are to pick out and write in Figure 11-16 the information needed to complete cells B15 to B18.

	A	B	C
1	SERVICE CONSULTANT	PS = Parts Specialist	
2	REPAIR ORDER	RO = Repair Order	
3	TRACKING SHEET		
4	**FIRST PHASE - check in**	TODAY IS:	MONDAY
5	Customer Last Name	Penton	
6	RO #	3456	
7	Vehicle year	2015	
8	Vehicle Make	GMC	
9	Vehicle Model	Yukon	
10	Time Promised or Waiting	Waiting	
11	Initial hours sold	1.3	
12	Time SC gives RO to TL and PS	11:15	
13			
14	**SECOND PHASE - initial work**		
15	Text message from Tech: TIME started job		
16	TECH Name		
17	TIME Tech reports in person to SC with RO		
18	Additional Work (Hours) recommended?		
19			

FIGURE 11-16 This Repair Order Tracking Sheet goes with the Second Phase Activity.

Look at the First Phase information:

- Look at cell B5; the customer's name is Mr. Penton.
- cell B10 indicates that the customer is waiting for the repair to be done.
- The RO was given to the team leader at 11:15 a.m. (cell B12).
- 1.3 hours of work (cell B11) was sold on the customer's 2015 GMC (cells B7 and B8).

Using Figure 11-16, fill in the information for the Second Phase:

- At 11:35 a.m., the service consultant got a text message from Jerry the technician that RO #3456 was being serviced (cells B15 and B16).
- After working on the truck, Jerry reported to the service consultant at 12:10 p.m. that the vehicle was finished (cell B17) and required no additional work (cell B18 = 0 hours).

ADDITIONAL ANALYSIS:

- Looking at cells B12 and B15, the customer waited around 20 minutes for Jerry to start the job.
- From the tracking sheet information (B15 and B17) it took the technician about 0.6 hours to do the work requested for a job that paid 1.3 hours (cell B11).

THIRD PHASE ACTIVITY: Additional Work

Examine Figure 11-17. This is a blank copy of the Third Phase the Repair Order Tracking Sheet with the First and Second Phases completed. You are to pick out and write in Figure 11-17 the information needed to complete cells B21 to B27. Mr. English's Dodge required 2.2 hours of additional work (cell B18) in addition to the 0.7 hours of work that was sold initially (cell B11). At 10:15 a.m., Todd the technician reported to the service consultant the truck needed work (cell B17 and B16).

- At 12:05 p.m. the service consultant was able to contact the difficult-to-reach Mr. English (cell B21) and performed Stage 3 of the Customer Service System (Additional Work Stage). He updated the repair order and pivoted back to the Shop Production System's Third Phase to complete the Repair Order Tracking Sheet.
- Mr. English approved all of the 2.2 hours of suggested repairs (cell B22).
- The service consultant negotiated a new promise time of 5:00 p.m. (This would change cell B10 from 3:00 p.m. to 5:00 p.m.).
- At 12:06 p.m. the repair order was returned to Todd (cell B23).
- At 12:07 p.m. the repair order was given to the parts specialist (cell B24).

	A	B	C	D	E	F
1	SERVICE CONSULTANT	PS = Parts Specialist		SC = Service Consultant		
2	REPAIR ORDER	RO = Repair Order		Tech = Technician		
3	TRACKING SHEET			TL = Team Leader		
4	**FIRST PHASE - check in**	**TODAY IS:**	Thursday			
5	Customer Last Name	English				
6	RO #	8900				
7	Vehicle year	2014				
8	Vehicle Make	Dodge				
9	Vehicle Model	Charger				
10	Time Promised or Waiting	3 pm				
11	Initial hours sold	.7				
12	Time SC gives RO to TL and PS	8:00				
13						
14	**SECOND PHASE - initial work**					
15	Text message from Tech: TIME started job	9:45				
16	TECH Name	Todd				
17	TIME Tech reports in person to SC with RO	10:15				
18	Additional Work (Hours) recommended?	2.2				
19						
20	**THIRD PHASE - additional work**					
21	TIME SC got Customer Approval					
22	Hours approved (change time promised)					
23	Time SC gave updated RO to Tech (or TL)					
24	Time SC placed part order with PS					
25	Estimated delivery time of ordered parts					
26	TIME PS texts (SC+TECH) pick up parts					
27	Text from Tech to SC additional work started					
28						

FIGURE 11-17 This Repair Order Tracking Sheet goes with the Third Phase Activity.

- The parts specialist said the parts would be in after lunch at around 1:00 p.m. (cell B25).
- At 1:00 p.m., the service consultant got a text message that the parts arrived (Cell B26).
- The service consultant obtained a text at 1:10 p.m. from Todd the technician that he started the repairs (cell B27).

The Fourth Phase of the Repair Order Tracking Sheet will be entered when:

- The Parts Specialist provides the Service Consultant with parts invoices. These are needed to check the repair order against the final invoice and update it if necessary. The service consultant wants to assure all parts needed for the repair(s) have been charged out on the invoice at the correct price.
- The technician finishes the job and provides the repair order to the service consultant. If there are any questions about the parts used in the repair process or important notes that should go on the invoice, this will be discussed at their meeting, This will help the service consultant create an accurate invoice for the customer.

FOURTH PHASE ACTIVITY: Check Out

Examine Figure 11-18. This is a blank copy of the Fourth Phase of the Repair Order Tracking Sheet with the first, second, and third phases completed. You are to pick out and write in Figure 11-18 the information needed to complete cells B31 to B36.

The service consultant is out sick today (Friday) and you have to fill in for him. To get up to speed on Mr. Plessinger's repair request, you examine the First, Second, and Third Phases of the Repair Order Tracking Sheet. From cell B12 you see the job came in on Thursday at 9:30 a.m. (cell B12) and today is Friday (cell C4). It is promised for today (Friday) at 4:00 p.m. (cell B10). You notice that the Toyota Prius (cells B8 and B9) arrived on Thursday for one hour of diagnosis (cell B11) that required an additional 7.3 hours of work to repair and that the customer approved (cell B22). You notice that the technician has the parts (cell B26) and started the job yesterday at 1:10 p.m. (cell B27).

Since the repair was longer than could be completed on Thursday, the job went into Friday. You can see Thurdsay's B5–B36 cells were copied into Friday's Repair Order Tracking Sheet and the day (Thursday) was added after the time to avoid confusion. You are waiting for parts invoices from the parts specialist and the RO from the technician in order to start the Fourth Phase of the Repair Order Tracking Sheet—Check Out.

Fill in the following information on Figure 11-18.

- Shortly after the day starts at 8:30 a.m., the Parts Specialist provided all of the receipts (cell B31).
- At 12:00 p.m. the technician reported that the job was completed (cell B30).
- As the substitute service consultant, you pivoted into the Customer Service System Stage 4 and completed the invoice at 1:00 p.m. The fourth stage of the Customer Service System "Time the final invoice has been completed" is tracked on the Repair Order Tracking Sheet (cell B32).

	A	B	C	D	E	F
1	SERVICE CONSULTANT	PS = Parts Specialist		SC = Service Consultant		
2	REPAIR ORDER	RO = Repair Order		Tech = Technician		
3	TRACKING SHEET			TL = Team Leader		
4	**FIRST PHASE - check in**	TODAY IS:	Friday			
5	Customer Last Name	Plessinger				
6	RO #	7589				
7	Vehicle year	2010				
8	Vehicle Make	Toyota				
9	Vehicle Model	Prius				
10	Time Promised or Waiting	4pm Fri				
11	Initial hours sold	1				
12	Time SC gives RO to TL and PS	9:30Thur				
13						
14	**SECOND PHASE - initial work**					
15	Text message from Tech: TIME started job	9:45Thur				
16	TECH Name	Brian				
17	TIME Tech reports in person to SC with RO	10:15Thur				
18	Additional Work (Hours) recommended?	7.3				
19						
20	**THIRD PHASE - additional work**					
21	TIME SC got Customer Approval	10:30Thur				
22	Hours approved (change time promised)	7.3				
23	Time SC gave updated RO to Tech (or TL)	10:45Thur				
24	Time SC placed part order with PS	10:45Thur				
25	Estimated delivery time of ordered parts	In stock				
26	TIME PS texts (SC+TECH) pick up parts	11 Thur				
27	Text from Tech to SC additional work started	1:10 Thur				
28						
29	**FOURTH PHASE - Check Out**					
30	Time Tech reports in person to SC with RO					
31	TIME PS supplies parts receipts to SC					
32	TIME final INVOICE has been completed					
33	TIME customer notified that vehicle done					
34	TIME vehicle Picked up					
35	SC was able to provide Active Delivery?					
36	Thank you note/survey sent to customer?					

FIGURE 11-18 This Repair Order Tracking Sheet goes with the Fourth Phase Activity.

- You called the customer at 1:15 p.m. (cell B33).
- The customer arrived at 3:45 p.m. to pick up the car (cell B34).
- The customer saw the cashier, paid the invoice, and picked up the car keys. You took the time to provide Active Delivery (cell B35) and after the visit generated a thank-you note and sent a follow-up survey to the customer (cell B36).

Even though you had to substitute for the service consultant, the Customer Service System and Shop Production System's Repair Order Tracking Sheet allowed you to pick up where the sick service consultant left off. A seamless service system that can continue to serve the customer regardless of which trained employee does the job. This allowed you to, without confusion, help reach the goal: a customer who had his or her vehicle completed on time. As you can see, a service system approach can help stay on top of repairs in a busy shop where there are often several cars in different phases of the Shop Production System.

Review Questions
Multiple Choice

1. What phase is the time located when the service consultant receives a text that the technician started additional work?
   A. First Phase
   B. Second Phase
   C. Third Phase
   D. Fourth Phase

2. What phase is the time located when the service consultant meets with the technician to reviews the additional service needs of the vehicle?
   A. First Phase
   B. Second Phase
   C. Third Phase
   D. Fourth Phase

3. What information is usually NOT part of a text message between team members?
   A. Repair Order Number
   B. The time of the message
   C. The team member who sent the message
   D. The cause of the repair problem

4. If the parts deliveries are delayed, what row would the service consultant examine and compare against the current time of day?
   A. Row 24
   B. Row 25
   C. Row 26
   D. Row 27

5. What phase are details about the customer, vehicle, and repair order information found?
   A. First Phase
   B. Second Phase
   C. Third Phase
   D. Fourth Phase

6. The total time sold for a job is the combination of which rows?
   A. Rows 11 and 22
   B. Rows 11 and 18
   C. Rows 18 and 22
   D. Rows 11, 18, and 22.

Short Answer Questions

1. Explain each of the four phases and how they work together to help meet customer expectations.
2. Why is the third phase sometimes "skipped" and only three of the four phases completed?
3. Explain what a service consultant must do with the customer and the Repair Order Tracking Sheet when a repair is delayed and the time promised cannot be met.
4. Explain what the service consultant must do when the hours approved by the customer is greater than the difference between the time promised and the time the technician texts the service consultant that the additional work is started.
5. Sometimes the numbers in the additional work recommended (Row 18) and the hours approved (Row 22) are not always the same. Why are they different and also should the number in Row 18 be larger than Row 22?

CHAPTER 12

CUSTOMER SATISFACTION AND MARKETING

OBJECTIVES

Upon reading this chapter, you should be able to:

- *Demonstrate how to greet customers and respond to angry customers (A.1.6).*
- *Explain the importance of identifying and prioritizing customer concerns (A.1.3, C.2).*
- *Give examples of how to promote the procedures, benefits, and capabilities of the service facility (A.1.8).*
- *Identify methods to communicate the value of performing related and additional services (C.4).*
- *Describe methods used for customer follow-ups (A.1.5).*

CAREER FOCUS

When developing your career, you cannot be the best in the business if your customers do not tell you they are happy with the service you provided to them. As the best you know how to make known your success, identify your shortcomings, and challenge yourself to improve. Criticism is solicited because it is the breakfast of champions and you like to be fed. More importantly, if you cannot improve yourself, you cannot help improve the business. Knowing the content of this chapter goes beyond helping the business; it applies to improving yourself. You know the best aren't born the best, they work to earn it every day starting with motivation and a "can-do/will-do" attitude when they get out of bed in the morning.

Introduction

The previous chapters cover in detail the processes and procedures to be followed when working with customers. The purpose of this chapter is to focus on customer relations and satisfaction. Specifically, this chapter discusses marketing the business, the different advertising techniques, basic sales techniques, follow-ups, methods used to promote sales, and handling angry customers. It is an important chapter to master because it contains the practical information needed to obtain new customers and retain existing customers needed to operate a successful service facility. In other words, if not mastered, the technicians will not have any work because there won't be enough customers. The shop does not make money unless there are customers that need work done and are satisfied with the work performed. Failure to understand the basic principles presented in this chapter has ruined more than one business.

Selling Services

As explained in previous chapters, service consultants have several opportunities to promote sales when they come in contact with customers. These include the initial contact on the telephone, when the customer comes into the shop, after the technician's initial inspections and diagnosis, and during active delivery. During these contacts, the job of the service consultant is to sell what the service facility offers. In time, service consultants develop their own style of selling; however, there are several basic principles to be recognized.

The first principle, as noted previously, is to always offer customers a pleasant and friendly greeting. When a person calls on the phone, the caller must be treated with courtesy and the conversation should be given proper attention. The service consultant must speak clearly and distinctly. To ensure that phone conversations are handled properly, the phone should not be located where there is a lot of background noise. This can cause communication problems for both the caller and the service consultant. If cordless phones are used, the service consultant should not attempt to carry on conversations in the shop area where technicians are working on automobiles.

When customers enter the facility, service consultants must offer an appropriate greeting as follows:

- "Welcome to Renrag Auto Repair."
- "My name is _____, and your name is?" (if possible, extend the right hand to offer a firm and friendly handshake while waiting for the customer's name)
- "How may I help you?"

After a proper welcome, customers will state what they need, which could include the repair of a problem, maintenance, or just general information. When customers state their concerns, service consultants must be good listeners and give them their full attention. Service consultants must listen carefully to the customers' problems, what the customers think, and what has been done (or what the customers think has been done) to their automobiles. Service consultants must ask "when," "how often," and "where" questions. Most importantly, sales cannot be made if service consultants are not aware of what the customers need, and, more importantly, wish to buy!

Selling and Angry Customers

A critical sales technique for service consultants to develop is the ability to handle angry customers (a more in-depth discussion about dealing with angry customers is included later in the chapter). First, the service consultant must always treat the customer, including angry ones, with courtesy. Second, service consultants must keep in mind that the objective is to close and complete a sale. This interaction should be handled like a business meeting! Third, the service consultant must realize that when a customer is angry, it is typically about the inconvenience and unknown costs (anticipating the worst) associated with a repair. It is not personal.

Service consultants must realize that customers rely on their automobiles to meet their personal needs. When their automobile cannot be used, even for a brief period of time, customers and their families will experience transportation problems and will not be able to take care of their daily obligations as easily. This may cause feelings of helplessness, agitation, and frustration that cause some people to be angry. Because these encounters are not common, service consultants must learn to interact in a disciplined and positive business manner.

Customer Personalities and Sales

After service consultants learn what a customer's problems and concerns are with respect to his or her automobile, they should prioritize the work as discussed in chapter 8. In addition, they should focus on the features of the service and benefits received as also discussed in chapter 8. Service consultants must provide customers with accurate, detailed explanations of the service to be purchased and answer all questions. The focus should be on the capabilities of the facility to take care of the customer's needs.

When discussing a service with a customer, service consultants should attempt to adapt their discussion to the customer's personality. For example, if a person seems to be:

- Anxious, cynical, and/or distrustful, the consultant should allow plenty of opportunity for the person to ask questions, give thoughtful answers, and permit the customer to have more time than usual to make a decision (in other words, the customer should not be pressured to give an answer).
- Not interested or in a hurry, and/or seems bored with the description of the service, the service consultant should provide explanations that are brief with limited details.
- Irritated and/or upset about the problem, the service consultant should use the business approach that revolves around the features of the product, benefits of the service, and repair priorities.

Overselling and Up-Selling

There is often confusion about when up-selling becomes overselling. Overselling means a customer is sold a service that is not needed. For example, a manufacturer's manual may recommend a maintenance service and part replacement at 50,000 miles. If a new part is sold to the customer at 25,000 instead of 50,000 miles as recommended, it is likely not needed and the customer has been oversold. However, if the customer knows the recommended replacement interval is 50,000 miles and understands that early replacement is not needed but wants it done anyway, then that is the customer's preference. The reason a customer might want it done early is not significant; rather the principle to remember is the service consultant is an advisor to the customer. This means that while the business needs sales, it is just as important not to oversell with an understanding of how to balance customer preferences with service recommendations.

Up-selling is when a customer needs service and has an option on the quality of the parts used or the amount of labor to perform the repair. For example, assume that a customer has to have new brakes. The choice is between less expensive, lower-quality brake pads and more expensive, higher-quality pads. The customer would obviously save money by buying the cheaper pads; however, the person would have to pay the same labor and also to have them replaced sooner. The cheaper pads may also have different warranty coverage. If the higher-quality pads are purchased, the pads will perform better, the customer would save money in the future and may have better warranty coverage plus the facility would make more profit on the sale of the better set of pads. In the long run, the customer will be more satisfied with the repair. Hence, the service consultant should try to up-sell the customer on the purchase of the higher-quality brake pads.

Up-selling also occurs when a customer's automobile is having a maintenance or repair service and the technician sees a maintenance or repair that should (not must) be performed. For example, assume that an automobile is having an oil change and the technician sees that the front tires are wearing poorly. If the service consultant sells a tire rotation and alignment check, it would be an up-sell. However, if the rear tires are worn worse than the front tires, then a tire rotation may not be the best idea and if sold , it would be an oversell. Oversells are how facilities get a bad reputation, eventually losing customers, and they may break a state's consumer protection law!

Customer Disagreements with Sales Proposals

If a customer does not agree with a recommended repair or set of repairs, the service consultant should ask for his or her recommendation. In some cases, the customer may want to purchase more than what the service consultant is trying to sell. This is okay and why recommendations must be based on manufacturer service information, technician's expertise, or industry standards!

In a few situations, the service consultant may not agree with a customer's repair request because it will not solve the customer's problem. If the customer continues to insist, and the service staff does not feel comfortable performing the repair, the service consultant should suggest that the person seek another opinion.

In one situation, a customer wanted organic brake pads when his car required semi-metallic pads. However, the organic pads would not last as long or bring the automobile to a stop as well as the semi-metallic pads. The service consultant refused the customer's request. The customer then offered to buy the pads and bring them to the facility for installation. The service consultant again refused and the customer left. This proved to be a good decision because after the customer had the pads installed at another facility, he rear-ended another automobile when he could not stop as quickly as he did before the replacement.

Refusing to make an improper or inappropriate repair at the request of a customer might result in a lost sale, but it is preferable to risking a comeback, a breakdown on the road, an accident, or a lawsuit. Any of these would cost a facility more money than it would make on the repair.

Factory and Franchise Representatives

Service facilities may be part of a larger network of businesses. For example, a car dealership may be owned by Ms. Brown who has contractual rights to sell new cars in a specific region for a vehicle manufacturer. In other cases, a service facility may also be part of a larger network of independently owned service facilities under a franchise agreement with a franchisor, such as

Midas Muffler. At times regional businesses may be part of a chain owned by the same company; such as Sears Automotive Centers or a multi-franchise dealership chain that has dealerships in several locations. Any time there is a service facility that is part of a larger network, then the regional business must be overseen at least in part by the vehicle manufacturer, franchisor, or the parent company's upper management. The reason is to make sure the operations of the regional service facility are conducted properly and to promote the improvement suggestions of the franchisor, vehicle manufacturer, or larger chain operations' home office.

To assure the regional service facility gets the help needed, the franchisor, vehicle manufacturer, or chain operations' home office will employ a representative to visit their regional service facilities. The job title of their representative varies but the duties have some common characteristics. Basically, the representative's job is to promote the franchisor's or manufacturer's initiatives (often called programs) as well as to protect their interests. The franchisor, vehicle manufacturer, or "home office" wants to make sure retail customers are treated properly and this means the warranty information, as well as services and products sold, meet customer and "home office" expectations. For example, when a product warranty is involved, such as replacement of a defective module, which is ultimately paid by the vehicle manufacturer, the work must be done correctly by following the manufacturer's guidelines. This was, covered in earlier chapters. The representative ultimately wants to help the regional service facility management improve their performance so that the dealer can obtain good customer satisfaction surveys. The representative must also look out for their employer's interests, in this example the vehicle manufacturer. The representative's job gets challenging when processes aren't followed, customers are unhappy, and he or she must troubleshoot the problem.

Service Facility and New Car Dealer programs

The initiatives or programs offered by auto, parts, and other manufacturers, as well as franchisers, are usually optional and are used to encourage the business to do better. Some are incentive based (do this and you will get that), but some are a process change that will help the business and the manufacturer. Some new car dealer programs, such as the Toyota Signature program, the Ford Blue Oval program, or the Chrysler Five-Star program, are easily recognized by the consumer because of advertising. The purpose, reason, and establishment of these programs are beyond the scope of this book but generally involve manufacturer-level negotiations with outside providers to help keep their local dealers competitive. At the dealer level when the program fits the business owner's needs, then it can have a very positive impact for everyone at the dealership. For example, a service manager was offered a program to compete against other dealers in his region to increase sales and Customer Satisfaction Index (CSI). The challenge was to improve service and parts sales by 10% each as well as improve CSI scores over last year's score (see Figure 12-1). At the end of the year, the successful service managers were given a trip to Hawaii with all expenses paid and any employees who helped with the program were provided with rewards.

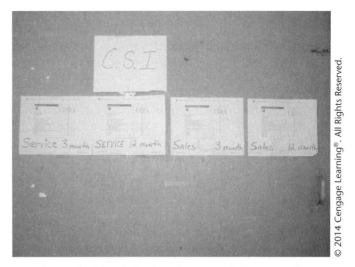

FIGURE 12-1 The goal of the dealer is to have good CSI scores that improve over time. These CSI scores are posted so employees can track the results. This type of feedback is helpful when an objective is established and the employees can see the results. Similar to bowling, fallen pins coupled with the resulting score provides feedback about a bowler's performance for each frame. From this feedback, bowlers know what they need to do to improve their score in the next frame. The same is true for CSI, when it is used to improve employee performance.

CAREER PROFILE

New Vehicle Manufacturer: The Factory Representative's Job

Factory representatives work for the manufacturer, or vehicle distributor. Their responsibility is to their employers and their job is to help their dealers. They are given company cars, laptops and cell phones, an expense account, as well as support and training. Many are required to hold a B.S. degree and have held jobs in the company that helped them prepare to become a representative. Their B.S. degree is usually in a related area such as Automotive Technology Management, but some hold degrees in other programs of study as well. The representative will have exceptional people skills, always speaks highly of dealers or manufacturers, and constantly promotes the manufacturer's products and services.

The factory representative's job, however, is more difficult than simply promoting his or her company and its initiatives to sell more cars or services though various programs. Each day the representative must take phone calls from dealers and answer questions about policy, procedures, and various business-related questions. At times, they need to give authorization to override a warranty claim issue. For example, assume a new vehicle is out of warranty by a few hundred miles and requires a repair. The claim may not be processed because the mileage is too high. Also the repair may be too expensive for the service manager to override. Therefore, the factory representative must be contacted for approval. The factory representative will consider several factors before providing approval such as:

- Is the concern common for the year and model of the affected vehicle? Also is approval commonly given to other dealers?
- Was the vehicle purchased from this dealer and, therefore, an approval can help foster a positive relationship between the customer and the dealer for future business?

- Does the customer have a long history of loyalty and repeat transactions at the dealership?
- Do the records show that the customer has completed all the required maintenance requirements and the concern is not a result of the customer's failure to service the vehicle?
- Was the vehicle concern the result of the customer damaging or modifying the vehicle?

The factory representative will use these and other various criteria to make his or her decision about whether to approve or reject the claim. When the representative decides to pay a non-warranty claim, it is called "goodwill." Sometimes a "goodwill" claim will require a company manager who is above the representative to approve it.

Factory representatives handle a multitude of other concerns and must perform a variety of other duties. At some manufacturers, however, there are different representatives who specialize in certain jobs, such as only handling fixed operations (service and parts), while other manufacturers may have their representatives handle multiple issues and duties, such as variable operations (new vehicle sales). On the fixed operations side, they may be involved with factory promotional events, special service tool needs, warranty claim audits, parts inventory review, technician training, college automotive programs, dealer meetings, charitable donations, lemon law actions, and so on.

On the variable operations or retail side, the representative may have to be involved in vehicle wholesale issues to convince a dealer to take his or her fair share of a product that may not be selling very well or helping to adjust inventory to keep an adequate number of products on the lot. Dealers also have issues such as not receiving enough "hot-selling products" that have a limited supply. These and other concerns mean the representative may have to handle disgruntled dealers in person or at meetings. Resolution is as much of an art in communications and interactions as it is working with reports and data to link performance issues to the dealer's bottom line.

To address a dealer's service concern, a representative might show a dealer how he or she performs against other dealers in the district. However, the review should go further than just a score in a report. For example, the survey question (covered in more detail later) that asks customers, "Was the service a good value for the money?" offers insights on pricing. Unfortunately though, poor satisfaction measured by consistently lower CSI scores often is a communication issue between the service consultant and the customers over basic performances, such as the work was not explained thoroughly relative to the warranty policy. To help with dealer improvement, a factory representative using report information might mention to the dealer that by increasing the service department CSI by just 4pp (percentage points) it can equal X dollars in improved sales. In other words, the idea is to use all of the data and resources available to help the dealer make decisions.

As a career path, the concept idea to understand is that the list of duties performed by a factory representative is very long and cannot be presented here in detail. The point to be recognized, however, is that a factory representative handles many different assignments during his or her career making him or her invaluable and a good person for service staff to get to know.

Effective Communication Advice to Promote Future Sales

For effective communication one company executive explained, that to maintain a proper attitude, "there are no ugly babies; every mother thinks their baby is beautiful." This means a factory representative and dealership employees must be positive at all times and focus on what needs to be accomplished. They must communicate effectively to avoid issues because a careless comment or decision can lose customers and destroy a career. The following suggestions apply to all service employees even though the examples involve the factory representative.

The first suggestion involves a factory representative who discovered that a dealership service consultant was upsetting customers with slow, unprofessional, and rude email responses to service inquiries. For manufacturers, the dealer is their customer. The factory representative works to help the dealer become a leader in its retail market area. So in this case, the representative could have gone to the dealership owner and commented on the incompetence of the service consultant and the inability of the manager to oversee the employee. Instead the representative approached the problem by asking about the email process at the dealership relative to answering customer's questions. More specific questions were then asked about the response times and communications along with other questions. As these questions were answered, the representative discovered that the service consultant in question was the dealer's son. Further, he was in charge of the electronic communications (Facebook, email, and internet communication) for service customers. Therefore, it was wise that the problem was handled as an effective communication issue before pointing out concerns. A direct approach in this situation could have put the owner in a defensive position, and this example would have ended poorly had the representative not been cautious. Therefore, when trying to fix a problem, it is best to ask questions and understand the entire situation before getting into the details of how to fix it.

The second suggestion involves a new model that was reviewed by selected factory representatives with the engineering design team present. It was observed that the switch to turn on an accessory was in a hard-to-reach place on the dashboard. To say that is a stupid place to put a switch would clearly offend the engineers who put their best work into designing the system. Rather than to risk a fight with engineering resulting in the switch not being moved, the senior factory representative pointed out the positive attributes of the vehicle. At the end of the discussion he followed with a question, "do you think it would be easier to turn on the accessory if the switch was further to the right?" That comment was received well even though there was a "push back" by the design team about the idea. This is, of course, human nature; however, when the vehicle was manufactured, it was noted that the switch was moved to the right as pointed out by the senior representative. Sometimes a well-phrased question to a boss, coworker, or customer at the right time can make a point that will result in a needed change without offending them.

The third and final suggestion concerns dealership personnel who serve the retail customer. At times a factory representative may help dealership employees with a customer, although this is not common. In one situation, a customer thought his car was broken because he could not integrate his handheld device to the new vehicle's remote access systems. The dealership personnel asked the factory representative for help because they did not have the recent training needed to understand its function or setup. The factory representative went with the service manager to look at the car with the customer. With the customer and dealership staff watching, the representative showed them how to load the correct apps, and then how to push the right buttons to make the system work properly. Collectively, the group reacted to the results and then asked for the customer's approval of its function. The factory representative made sure the customer was pleased and during the process never made the customer feel that he was at fault or "stupid" because he could not make it work. He also helped the dealership staff save face by not making them admit they did not know how to make it work either. Taking the time to look, explain, and teach others goes further than trying to make a short comment that could make the customer or others "feel stupid" and lead to a disagreement that is difficult to resolve later.

While these examples did not result in additional sales, they do explain how a future sale of a service or product will never occur when past problems are not handled properly. In other words, a service facility will never get a chance to sell anything to a customer they have offended. Effective communication must be practiced when working with customers, other businesses, and associates to solve problems; otherwise the opportunity to engage in future business may be limited.

Marketing the Facility

Service consultants must constantly promote the facility and the personnel who work there. They must have a positive attitude (we can fix your car) and must be the company cheerleader (we are the best). In other words, they must always be selling the facility and the technicians' ability to fix automobiles. They must tell customers about the good reputation of the facility, customer compliments, how their technicians go to extra lengths to solve complicated problems, the quality of the parts put on customer automobiles, and the benefits of allowing the facility to keep customer automobiles in top shape. Service consultants must also promote the capabilities of the facility, the expertise of the technicians, and the use of expensive equipment to maintain and repair automobiles.

In addition, when the technicians have ASE, factory, and other training certifications, they must be placed on the wall near the service counter for customers to read. Any signs such as the ASE certification sign should be placed in the front window or on the building for people to see when passing the business. Technician certification information and symbols should appear in newsletters and specials sent to customers and found, on professional business cards, uniforms, and in all advertisements. At Renrag Auto Repair, the ASE training certification displays impressed customers, many of whom commented or asked questions.

Other displays that attract attention are a clean and organized "bulletin board" and a professional display of products, such as tires or service "posters" near the business entrance and customer waiting area. Point of sale information, such as product or service brochures, may be placed in the customer waiting area, near the service consultant's desk, or when appropriate near the cashier. Whatever the display, it must be attractive, clean, kept updated, and promptly removed when out of date or no longer not effective at generating sales. When a bulletin board is used, it should have copies of all service facility specials, advertisements, and newsletters presented attractively. Furthermore, it provides a place to post thank-you cards, publicity, photos, items of information, and results from customer surveys. When customers are waiting for their automobiles, most take the time to look at this board; the service consultant should encourage those who do not.

Sales Follow-Ups

The service consultant's job is not over after the customer's automobile has been serviced and the bill has been paid. Rather, another phase of the job has just begun. This is the "follow-up" part of the job and is closely related to the promotion of sales. Service consultants must be concerned about the customer's satisfaction if they wish to sell customers services in the future.

The primary objective of the follow-up, however, is to identify as soon as possible those customers who are NOT satisfied with their service or treatment. If a customer is unhappy about a service, the likelihood of making a future sale to that customer is not good. It is therefore important to find out why the customer is dissatisfied and to correct the problem if possible, and then to avoid creating the same problem in the future. Furthermore, if a dissatisfied customer tells other people about the bad experience, serious damage to the reputation of the facility is likely to occur.

When service consultants realize that a customer is not satisfied, they must collect as much information as possible about the complaint. They must check if other customers have the same complaint and then investigate the matter. If there are any weaknesses, errors being made, or employees not doing their job properly, then corrective action must be taken as soon as possible. After actions are taken to correct a problem, the service consultant should call the customer, or customers, who made the complaint to inform them of the corrections, express appreciation for their interest in assisting the facility, and invite them to return for service (possibly with a discount).

First Follow-Up

Customer follow-ups begin with the "thank you" during the active delivery of the customer's automobile. At this time, the service consultant might say:

- "Thank you for allowing us to work on your automobile. If there is anything that was not satisfactory about your experience or service, would you please let me know?"

- (Pause to permit the customer to respond.)
- Then continue with, "If you would be willing to permit us to service your vehicle in the future, please call me."
- (Then hand the customer a business card and possibly a self-addressed, prepaid postage customer satisfaction survey card, described next.)

Second Follow-Up

After a week or so, a follow-up "thank you" is recommended. This form of appreciation may be: (1) sending a thank-you card by mail (see figure 12-2), email, or text message to the customer, (2) mailing or emailing the customer a satisfaction survey form with a thank-you card, or (3) making a personal phone call or sending an email or text to ask about the automobile's performance. In addition, regular reminders should be sent to customers inviting them to schedule an appointment for a future service, especially seasonal specials. The means of delivery is unimportant but must be on the customer's terms; if U.S. mail is desired, then use it.

Several companies sell attractive thank-you and service reminder cards to send to customers see Figure 12-2). At Renrag Auto Repair, because these cards were quite popular, they were changed from season to season. Most importantly, however, many customers expressed their appreciation for the attention when they returned to the facility.

Second Follow-Up—Web-Based Services

While thank-you cards will appeal to certain customers, some service facilities are part of larger networks. Mechanicnet.com provides

FIGURE 12-2 This thank-you card was sent by U.S. mail. On the back is contact information for the customer to visit a website to view service history and sign up for reminders. It is a version of the second follow-up.

marketing tools such as allowing customers to access their service history on line, providing service reminders, and the ability to leave feedback for a service facility among other customer tools. Links to Facebook, Twitter, and other social networks make a service facility's job easier to stay in touch with customer and find new ones.

The Personal Touch

Another follow-up involves personal notes to the customer from the service consultant who fills out the thank you card. An example was a customer who shared she was going to Disneyland for vacation. The Service Consultant wrote "Thanks for your business and have fun in Disneyland, Susan". These personal touches help build an ongoing relationship with customers so that they feel quite comfortable when they come to the facility. They also help to make conversation, but they require service consultants to remember customers.

One suggestion is for service consultants to keep a notebook on their customers. The notebook contains the name of the customer with notes about conversations, names of children, special interests, and so on. For example, when customers return to a facility, a service consultant may mention how much he or she enjoyed a story told at the last visit, ask them about their children by name, or wish them luck in a bowling tournament or fishing trip. Some computer systems allow notes to be added to a customer's file. However care is required because a service consultant who never served the customer before would not want to ask a personal question or engage in a personal discussion.

Advertising and Sales Promotion

The primary objectives of advertising and **sales promotions** are to keep regular customers and attract new ones. Different types of advertising are discussed in the next sections with some being more effective promotions of a service facility than others. In addition, two important points to be recognized when advertising are (1) a method may not show any results until it has been used several times and (2) the program's design and the mediums selected (type of advertising) must "fit the service facility's needs." This means that when paying for these services, such as advertising, it must make sense. When a service facility's target customers live within 10 miles of the service facility, then advertising that targets customers in a 125-mile radius or further may not make sense. The service facility owner must access the promotional needs of the business. Also, the owner must have a reasonable budget to get the message to the customer, always remembering that the advertising must bring in more money than is spent. There is no single effective method or the same target for every business. A tire dealer in a city has a different customer target than a small service facility that performs Rolls Royce mechanical restorations and draws customers from a tri-state area (figure 12-3).

FIGURE 12-3 This Rolls Royce restoration shop draws customers from several states.

Internet Advertising

Internet "website" advertising, like most other forms of advertising, is a one-way communication that can focus on a single message, such as "get your car fixed at Bill's Garage." The success of an Internet marketing campaign, however, should start with a company website and company ads strategically placed on other websites. There must be an effective message, appealing design, and placement that will reach a specific target audience. Often Internet advertising for a service facility is most effective when coupled with other advertising to be part of a campaign to promote the service facility and its services.

Social Networking

Social networking developed as a two-way communication medium that connects individuals together, so they could interact much like a service consultant communicates with customers when they enter the service facility as discussed in chapter 6. Messages are exchanged on a variety of issues but hopefully focused on a topic, such as vehicle repair, with efforts directed to engaging new automotive service customers. However, a concern for automotive service facilities is that those who are engaged in the social network don't live near the service facility and will not become customers. This means that a service facility in Colorado may have a wonderful relationship with a potential customer but since the customer lives in Mississippi, it is unlikely this person will patronize the service facility.

Facebook is a good example of how a social network can be used to market to customers, even automotive facility customers. It has millions of users and can target specific populations because users voluntarily provide information about themselves, such as their location and vehicle information. The social network sites can also allow the service facility to access what customers are saying about its services. With positive comments, it is like good "word-of-mouth" advertising that can take a business a long time to create. However, the possibility of getting negative feedback can be scary. However, it is better to participate in the discussion because a negative comment may turn into gaining a positive reputation for the service facility; plus it may even generate ideas for new services from the comments received.

The potential greatest advantage of social networking to a service facility is the possibility that it can quickly send messages as well as offers and news that will generate enthusiasm. Since the service facility may have a day without enough work, messages to customers can turn a slow day that may ruin the week into one that is profitable. For example, on one day a service consultant did not have enough alignments because a couple of customers cancelled their appointments. Knowing this early in the day, the service consultant sent out a message that the alignment machine had some unexpected open appointment times and a discount was being given to the first customers to fill the schedule. While this example has potential problems that must be considered, customers will rally to take advantage of a special.

A social network is not just an opportunity to connect to customers but to create "followers" through the company's social networking site. While this is very powerful for marketing, the rest of this book's concepts must be in place first. For example, the service systems must function properly to take care of the customer's transportation problems. If the service system is not set up or the social network site has not been monitored frequently before using it, the results can be a nightmare. This means that among the tasks for a service consultant or owner, is to each day monitor the social network site just as someone must turn on the air compressor each morning.

Social Network Advertising

A study of Facebook as a social networking site will show it to have two main products. The first is the infamous Facebook page, which is a web page within the Facebook community that, like a "traditional" website, describes businesses and services. Many companies use this to direct potential customer to their "traditional websites." A new service facility may even start a Facebook page before it has a website host for its traditional internet site.

The second way for companies to utilize Facebook is to purchase and place advertising on Facebook. This is similar to web advertisements but when a company runs an advertisement on Facebook, the company can select what category of viewers it wants to see its ads. A marketing phenomenon, "Like Us on Facebook," has convinced Facebook users to disclose what they "LIKE" so advertising can focus on the users' interests. For example, a service facility may want people over

30 years old who like classic cars and live within Austin, Texas. Facebook has information about how many of its users meet that criteria, frequency of login to Facebook, and the time spent on Facebook pages. The purpose is to place all promotional efforts effectively so more business is generated.

Internet Search Engines Advertising

People want answers fast and want the web to find answers for them immediately. Google is the most popular search engine because Google can give users the best match rather than providing matches for businesses that paid the most money for the best listing position. To make money, Google started to place small text ads in the margin of the search results page. A business can buy "Google Ad Words," such as "Replace Brake Pads" or "Transmission Repair" by placing a maximum bid amount for those words. If the bid wins, then it has the ability to associate the title to the search result with text description or advertisement. The service facility can control the exact days or months the Ad Words campaign runs and is charged the bid amount each time a viewer clicks through to the business's site in addition to a fee for every search viewer who sees the text description or advertisement. Of course, the business text description on Google always includes a web link so the user can click to visit the business website.

For the owner to research the best words to bid on, Google Ad Words tools estimate how many times the specific word combinations are searched each day or month. This can be broken down into "hits" in the geographic regions the owner selects and so forth. While the Google information on users is not as detailed as Facebook's user profiles, it can be useful when targeting larger geographic regions.

The Advertising Message

Advertisements provide information to tell potential customers about the business and attempt to convey the service facility's ability and desire to serve them. Whether they are a first-time customer (never been to the service facility and therefore need to be enticed) or a repeat customer who needs to be reminded to come to the service facility again, each requires a message to be directed to them. There are many direct advertising messages; this textbook will present three that can be targeted to the people the owners feel will be best served by the business. They include newsletters, customer specials, and service reminders.

The first to be discussed are **newsletters** that are sent to customers usually by mail or email. They can contain articles of interest and useful information about automobiles. The purpose of newsletters is to convey information and keep the "service facility name" in the forefront to help develop a relationship with customers so they will remember the service facility when they need service.

Newsletters are a benefit because some customers may forget where they had their automobile serviced last. Newsletters, therefore, can serve as nice reminders. Studies have shown that it is less expensive to get a previous customer to return to a service facility than to attract a new one. As noted earlier in the book, repeat customers keep a facility in business and a regular newsletter can help. Newsletters can also be provided to prospective customers, but care must be taken to identify them and deliver the newsletter in an appropriate manner.

Service Reminders

In addition to newsletters, facilities should send out by mail, email, or text message a service reminder to inform existing customers of maintenance services due on their automobile (see Figure 12-4). The creation of a reminder is a fairly simple task that can be done using computer systems, provided the database is up-to-date with current customer contact information. When the facility signs up to be part of a service "network" or it is part of a computerized service system package, these notices can be set to be generated automatically as monthly reminders or for customers whose annual safety inspection is due. In some cases, a special or discount might accompany the reminder to further entice the customer to return for a service.

Before sending out service reminders, service consultants should review their content as well as the names of the people to whom they will be sent. For example, a customer whose check bounced (meaning the account did not have enough money in it to pay the bill) should not be sent a reminder with a discount. Also when reviewing the names, the service consultant can handwrite personal notes on some of them.

Customer Specials and Coupons

Customer specials and coupons are sent to regular as well as potential customers, usually by mail, and are posted at the service facility. **Specials** should be attractive, have a catchy heading, and serve a particular purpose. For example, a special may be offered because of a change in weather (see Figure 12-5) or because of a large inventory purchased at a reduced price, such as a large, bulk delivery of oil. A special may be offered on a particular day of the week, week of the month, or month of the year when business is slow. Another special may promote a new product line or service.

Because some specials may have a low profit margin, they should be promoted inexpensively. Emails, the company website, a social network, or perhaps mail can be cost-effective. For example, a business can buy a carrier route from the post office to deliver the specials to specific neighborhoods and areas of towns. Other methods are to print the information on colorful paper to create flyers that are handed out, placed on windshields, sent through bulk third-class mail to a targeted group

SERVICE REMINDER

...that will ring your bell

Dear Valued Customer,

Our records show that your car is due for its **annual Pennsylvania state safety and emission inspection** in the upcoming month. We understand how much you depend on your car and how difficult it is to fit service into your busy schedule. But if you call us now, we will help to make this <u>state mandated inspection</u> service as convenient as possible. So, call 555-5555 NOW to make an appointment:

Schedule before the 10th of the month
and as a special "thank you"
we will take $6 off any other regularly priced service
(oil change, coolant flush, alignment)*

*If performed at the same time. Applies only to regularly priced "menu" services, not valid with any other specials or discounts.

If you have a friend or relative who also needs this service, tell them to call us and we will gladly extend the special "thank you" discount to them, as a courtesy to you. Again, thank you for being such a good customer and we hope to see you soon.

THANK YOU
Put in the owner's, manager's, service writer's, or technician's name

FIGURE 12-4 An example of a service reminder.

of potential customers, placed in high traffic areas, enclosed as circulars in newspapers, and so on. The objective is to gain high visibility at the cheapest price. Another method to advertise specials is to place them on large advertising signs in front of the facility. The special announcements, of course, must be short and to the point; for example, "Oil Change, Lube, and Filter—$29.99."

A more recent means to promote specials is through email. Many businesses ask customers for their email addresses to send them specials

Hot Summer Special

Dear Valued Customer,

Because (**insert Business Name**) values good customers like you, we wish to say "thank you" by giving you a Summer Special. Summer weather will soon be upon us and we want to help you and your car stay cool. As a result, we are offering you this special. CALL (**555-5555**) NOW to make an appointment before (**MONTH 1, 200X**) and we will perform our:

HOT SUMMER SPECIAL

- Check ALL of your A/C system's parts
- Check A/C system for leaks
- Check refrigerant level and add up to one pound
- Perform a # point summer inspection of your car
- Inspect your car's cooling system
- Give you advice about problems your car may have

Normally $ _____ NOW $ ____ **for R-134 systems**
Normally $ _____ NOW $ ____ **for R-12 systems**

Most Cars and Light Trucks
Referigerant dye extra if required

If you know someone who also needs this special, tell them to call us at (555-5555 by Month 1, 200X) and we will let them have this summer special at this price as a courtesy to you. Again, thank you for being such a good customer and we hope to see you soon.

THANK YOU,
(Put in the owner's, manager's, service consultant's or technician's name)

FIGURE 12-5 An announcement of a special.

and reminders. This is the cheapest means to promote specials and maintain contact with customers. The facility can use its computer system to prepare and send out the announcements.

Public Advertising in Newspapers, Radio, Television, and Solicitations

As many service consultants soon learn, advertising sales representatives are regular visitors at a service facility. They want to sell advertisements in newspapers, radio stations, television, restaurant paper placemats, church bulletins, public school and college sports programs, high school yearbooks, special charitable events, car racing, billboards in stadiums, and so on. Advertising is an absolute necessity for

ADDITIONAL DISCUSSION

Use of the Telephone and Other Electronic Mediums

A service facility owner or service consultant attempting to get more business on a slow day may try to randomly call potential customers to tell them about a special or advertise the service facility desire to do work for them. In other words, the purpose for this call is to solicit business. The people called may have been a past customer or a referral from another customer perhaps; even a random phone number might have been taken from the phone book. While this may seem that the consultant is showing initiative, telephone and other electronic communications (email, text, etc.) of this nature may violate federal and even your state's laws. According to the Federal Government FCC website (accessed 2012);

> A telephone solicitation is a telephone call that acts as an advertisement. The term does not include calls or messages placed with [the customer's]. . . express prior permission, . . . , or from a person or [company] with which you have an established business relationship (EBR). An EBR exists if you have made an inquiry, application, purchase or transaction regarding products or services offered by the person or entity involved. Generally, [a customer]. . . may put an end to that relationship by telling [the service consultant]. . . . or [service facility management]. . . not to place any more solicitation calls to [the customer's]. . . home. Additionally, the EBR is only in effect for 18 months after [the customer's]. . . last business transaction or three months after [the customer's]. . . last inquiry or application. After these time periods, calls placed to [the customer]. . . home phone number or numbers by that person or entity are considered telephone solicitations subject to the do-not-call rules." (http://www.fcc.gov/guides/unwanted-telephone-marketing-calls)

The federal "do-not-call rules" and a specific state's law, which may require obtaining a special license in some states, are beyond this textbook's purpose. Rather this is an area of concern and a service consultant should seek additional information from management before attempting to make random calls or other electronic communications to customers advertising the business. In addition, the laws and requirements change over time because of new laws and court decisions. So, keep up to date with the regulations.

a business; however, it is expensive, complicated, and may not be effective unless done properly.

Because of the many types of public advertising methods, a facility should consider employing an advertising consulting firm. First, it can put a balanced advertising plan together. A good consulting firm also has or can obtain the number of people and their demographics (e.g., age, gender) that an advertising source reaches. Second, an advertising firm can screen all advertising sales requests made to a service facility. This removes the service consultant and managers from dealing with advertising representatives during business hours. It also keeps them from choosing one business over another and possibly losing a customer. In these cases, the representatives are simply referred to the advertising firm. Third, the advertising legal disclosures, consultants can prepare and review all public advertisements to ensure proper content and appearance. For example, the size of a newspaper ad and the words, pictures, diagrams, and placement in the paper are critical to the success of the promotion.

Without an advertising consultant, an automobile service facility must make all of the decisions about where, when, and how to advertise. Public advertising must be seen by potential customers and fit the needs of the facility. For example, an independent repair facility may be more interested in advertising to people who own automobiles that are under warranty. In addition, there is no need to advertise if a facility is operating at 100+% of capacity for a particular service.

When public advertising is being considered to attract new customers, then the design of the type, mix, and amount of advertising must be carefully prepared. Each type has its advantages and disadvantages. If service consultants are responsible for public advertising, they must carefully study the types of advertising available to the facility and then prepare an advertising campaign.

Customer Satisfaction and Surveys

Customer feedback is an important assessment tool often used to determine if customers are pleased with the services they received and the work performed on their automobile. As noted previously, some feedback should be gained during the active delivery process; however, many facilities use a formal customer survey. For example, some service facilities:

- email the survey, (or mail a paper survey with prepaid return postage) to the customer. (This method is preferred by many automobile manufacturers.)
- leave the survey in the customer's car to be mailed back to the service facility or other location. (This method may be used by organizations who partner with the service facility, such as AAA.)

- call the customer on the phone to ask survey questions. (This is often used by larger dealerships or companies that perform market research.)
- give the survey form to the customer to fill out and place it in a locked box before leaving the facility. (This is often used by independent service facilities and dealership service departments for internal use and feedback on the customer service experience.)

Survey designs and the method used to obtain useful information may be complicated; however, most have a number of common features. In general, the questions in a survey look at three basic questions: (1) whether the repair bill was more or less than what was expected, (2) whether the automobile was finished on time, and (3) whether the services met expectations.

Manufacturer surveys (see figure 12-6) are typically much longer, often in excess of 20 questions that ask customers to rate the facility on a scale of 1 to 5 (1 being the least satisfied and 5 being the most satisfied). The survey may also ask a series of yes or no questions in order to look for a pattern or obtain greater detail. When responses on a satisfaction scale (positive) do not agree with a yes or no answer (not positive), a problem may exist and the customer deserves some attention, such as a follow-up call to extract more information. A dedicated service manager will monitor the customer surveys and when a low score is found or inconsistent information is submitted, he or she will personally follow up with the customer to determine the problem and try to rectify it.

Analysis of Survey Information

After the surveys are collected, a customer satisfaction index score can be calculated from a single question response, such as "Based on your service visit, how satisfied are you with XYZ dealership" to a series of questions. When

FIGURE 12-6 Customer Service Satisfaction survey being filled out by customer.

multiple questions are considered in the CSI score, the average score for each question, the average of all of the questions, the range of answers for each question, the number of choices for each number in the scale, and the number of yes or no answers could all be considered. The CSI score is then compared to previous surveys for the facility, in the past 90 days for example, and to those of other service facilities in the region. CSI can also be calculated for individual employees such as the technician who performed the job and the service consultant who served the customer.

In some cases, advanced statistics are used by a market research company to identify patterns. For example, automobile manufacturers can compare their dealers' scores with each other and with competitors' service departments. In some cases, questions and their answers may reflect on a specific employee or group of employees. In some instances, this score is so important that service consultants may be tempted to ask customers, "if you cannot give me an excellent rating what can I do to meet your expectations in the future?" While this seems like an innocent question, caution is suggested because any employee who tries to influence a customer's response may receive a reprimand or worse. The reason is because, larger companies (such as automobile manufacturers) track customer surveys over a period of time to determine if any changes in perceptions of a facility occur and how they compare to those of the facility's competitors.

To get accurate information, some companies collect multiple surveys from the same customers to determine if a change in marketing emphasis or service procedures is warranted. For example, a company may survey customers shortly after visiting the service facility for the first time, then again several days, weeks, or months later. The purpose is to see if the customers' perceptions have changed after they had more time to think about their responses. Employees who attempt to influence customers initial responses ultimately undermine efforts to improve the customer's experience.

While customer satisfaction ratings are an important assessment tool for a service facility, they are not necessarily the magic bullet for success. There are many variables that go into customer service to produce superior satisfaction ratings as well as higher profits. The idea that "doing anything the customer wants" or offering gimmicks typically erodes profits that must be paid through higher labor rates and markups on parts. In these cases, a service facility may find that it cannot compete in the marketplace. Therefore, caution is necessary when a person believes there are no limits on how far a company should go to improve customer satisfaction.

Customer Satisfaction, Behavior, and the Psychology of Problem Resolution

The psychology of customer satisfaction and future behavior to become a repeat customer is a very complex subject. This textbook does not explore the most recent ideas surrounding this issue but must point out that there is a lot of thought and research currently conducted on this topic. The automotive industry tries to obtain customers' feelings after service with a survey and then would like to predict future customer behavior from the results. Although surveys are easier to distribute and analyze than in the past, they have not necessarily led to an accurate prediction between satisfaction and future behavior even though logic might dictate that a relationship should exist.

Regardless of the research, customer satisfaction survey scores are viewed as a diagnostic tool that are important in attempts to judge the quality of a service facility and its employees. So CSI scores must be taken seriously especially when they are connected to employee raises, promotions, or continued employment. While a relationship between satisfaction and future behavior may not exist, the reason CSI scores have value is because it helps management identify unsatisfied customers so they can be contacted and the problem rectified as soon as possible. However, this is not necessarily done for the customer's future business. Rather it is to improve the business and its reputation by showing the customer that even though they may not have been satisfied, the service facility cared enough to contact the customer to understand his or her feelings.

Walking in the customer's shoes can help all employees relate to the customer's frustrations. However, when analyzing why the customer was not satisfied, it ultimately comes down to the fact that satisfied customers require three elements: *trust* (the customer feels the service consultant is truthful and looking out for customer's interests), *respect* (the service consultant communicates at the customer's level and is not perceived as arrogant), and *feeling of being in control* (the customers feels they have a voice and can help shape the outcomes; they are not dictated by the service consultant). To illustrate these principles, a second chance for the customer to communicate with a manager about a poor survey response, allows the dealer to rise to the challenge. At a minimum, productive communication with management gives the customer some control by allowing him or her to express the problem for fair consideration. Trust can possibly be renewed with a resolution of the concern for the customer in a respectful manner.

When possible, the unsatisfied customer's problem should be resolved and the service facility will benefit with knowledge of how to better help future customers. When the problem cannot be resolved to the customer's satisfaction, then at least the customer may remember that the service facility personnel tried to be understanding and helpful. The basic idea is that the automotive industry is very competitive and it is up to the trained service facility personnel to use the service systems in place to do a "good job" for the customer.

Unfortunately, trying hard does not always result in a "good job" in the customer's opinion because customer expectations are different for different service facilities. Therefore, a customer who has a vehicle repair made at a salvage yard with used parts will have different expectations of a "good job" than a customer who went to a high-end luxury dealership for a similar repair. Measuring each customer's satisfaction in reference to his or her expectations and then trying to predict future behavior (repeat customer) is a very complex subject.

Dealing with Angry Customers

Some customers express their anger at the time they pick up their automobile, whereas others pick up their car and drive it home and then unleash their anger the next day. Their anger is typically because they believe the service facility did not meet their expectation. More specifically, the customer may believe that the automobile was not serviced properly, that the invoice was too high, that the service facility forgot a request that was important to him or her, or that the automobile was not returned in the same condition as when it was left at the facility. The service consultant must understand that is it okay for the customer to be angry and that the customer's anger must be resolved before the issue that caused the anger can be fixed, if possible.

First, the service consultant must determine if what the customer believes is true. When a customer is rightfully angry, the service consultant should immediately apologize. The service consultant should follow up with explanations (but not excuses) and/or immediate action to determine why the problem occurred and what can be done to correct it. One action may be to bring the automobile back to the service facility to examine why the service did not meet the customer's expectation. The problem may be as serious as a part that was not installed correctly and has to be reinstalled, or it may be as simple as a grease spot that must be removed.

The service consultant should consider the following steps to get through an inherently bad situation when a customer is clearly angry:

1. Welcome and greet the customer in the same manner as a person who is not showing signs of anger.
2. Allow the customer to vent his or her frustration and do not make an attempt to interrupt.
3. Do not take anything the customer says personally. Remember, the problem is frustrating the customer, not you.
4. As the customer begins to exit the stage where his or her frustrations have been vented, examine the situation and assess the real problem in a professional business manner. Do not get side tracked into unrelated issues and the following suggestions may help when a customer's frustration cannot be controlled:
    A. If a customer becomes abusive, the service consultant must stop the interaction and ask the customer to refrain from using abusive language.
    B. If after an appropriate period of time, a customer refuses or does not seem to be able to exit the frustration phase so the issues at hand can be addressed, the customer must be asked to leave. The customer may be invited to return when they are not so angry or set a time and date for a meeting to resolve the problem, perhaps with management and even the technician present.

C. If a customer does not leave when requested, possibly because of being under the influence of drugs or alcohol, then local law enforcement should be called to handle the situation before it escalates out of control.
5. Identify the real problem and focus statements and the conversation on the real problem.
    A. Start this process by summarizing why the customer is angry.
    B. Ask the customer to agree with your assessment of the problem so that he or she knows you understand.
6. If the problem was in some way caused by the service facility, offer an apology immediately. For example, if a stranded customer needed a tow truck and it arrived late because the service consultant forgot to call the towing company after the customer called, an apology is necessary.
7. Understand company policies and how far a service consultant can go to help the customer. Watch what is said and how it is said. Do not make statements that are outside the authority of the service consultant's position.
8. Provide a solution or several options to rectify the problem within the authority of the service consultant's position.
    A. Many service facilities give service consultants a variety of tools to help in a situation. At some facilities, a **policy check**, which is an account with "make believe" money in it can be used, for credit toward a future purchase (see Figure 12-7). Often it is used to compensate a customer for inconvenience, a mistake, or other circumstance where the service consultant felt "money off" the current or future invoice was required.
    B. Another option is to ask the customer to suggest a solution by saying "how would you like to see this resolved" or let the customer choose an option from a couple of suggestions.
9. After the situation is rectified, the customer may be embarrassed by his or her behavior. This could cause the customer to be too embarrassed to return for future service. The service consultant wants repeat business and appreciation for the customer's time for bringing this problem to the service facilities attention. Followed by a meaningful "thank you" for the opportunity to rectify the problem is a positive way to end the process.

If a customer does not accept any of the solutions or suggestions offered to correct the problem or requires financial compensation for the trouble and anguish caused, then the problem must go to the next level of authority. When management has to be involved, service consultants must document everything that occurred, and the conversation with the customer must be reconstructed as accurately as possible. This documentation will be critical if the complaint should escalate to the point that legal actions will be taken by either the service facility or the customer.

In some cases, a complaint may not be the fault of the service facility and would not require an apology. This sometimes occurs because

Policy Check

Renrag Auto Repair
(Address)
555-555-5555

No: _____

Date: _____

Pay to the

order of _____

$ _____ Dollars

Good only for credit toward a repair performed at Renrag Auto Repair.
Void without proper validation and authorized signature. (No cash value)

_____

Authorized Signature — RAR

Reason _____

_____

R/O number _____ Date _____

FIGURE 12-7 A voucher for a policy check.

customers refuse the services recommended. For example, a customer at Renrag Auto Repair had an antifreeze leak that was coming from a loose bypass hose at the front of her engine. The hose was relatively new and the clamp simply needed to be tightened to fix the problem. The repair was performed and recommendations of the technician presented to her that the heater hoses were very old (original by all appearances) and worn in some spots. The technician recommended replacing the hoses, but the customer refused the suggestion. She did not feel that it would be a problem even though her car was over 10 years old and had over 120,000 miles on it. A couple of weeks later, her heater hoses blew apart on the freeway. The customer continued to drive her automobile until the engine seized. The angry customer demanded a new engine from Renrag Auto. After a lengthy discussion with her and her attorney, which required an extensive review of the notes on her invoice and all conversations, she agreed that the service facility was not in the wrong. Although unfortunate, the problem occurred because she did not take the recommendation of the service consultant at the time of her service.

Serious Issues and the "Policy Check" Concept

When an invoice charge is greater than expected (possibly justified) and disputed, service consultants may be able to use a policy check with the "make believe" money in it. This permits a service consultant to adjust the amount of the invoice by writing out a policy check (see Figure 12-7) for the amount of the disagreement. Either the amount on the policy check is deducted from the invoice in dispute or the customer presents it for payment toward a future service, depending on management policy.

Serious Issues

While the purpose of a policy check is to improve customer relations, it should not be construed as admitting fault in the repair process. Therefore, it must be clear that the policy check is like an "open-ended" coupon the service consultant fills out and provides as a courtesy to the customer. When there is a serious "repair process" failure at the service facility, such as a tire fell off while the customer was driving the vehicle resulting in damages or injury, the service consultant should say little to the customer other than apologize. A policy check should not be offered to the customer. Investigation by the insurance company, law enforcement, or the owners will be required to determine the exact details of the accident. Until the investigation is over, the service consultant should not discuss the situation until prompted by the service manager, owner, or company attorney.

Inappropriate use of Policy Checks

In some cases, a policy check may be effective to appease an angry customer who wants compensation. This concept is easy to understand when a customer feels he or she was over charged, inconvenienced, or financially harmed and the policy check will help to "make it right." However, there are times when the policy check will not make a difference. For example, a customer needed an oil change and a wash because she was going to a funeral the next morning. The service facility made the oil change and found a recall that needed to be done. The recall did not pose an immediate threat and could have been scheduled at a future date. The service consultant forgot, or disregarded the customer's request and told the technician to perform the recall. The service facility finished the wash and the recall by closing but ran out of time to wash the automobile. The customer was angry because her car was still dirty. Neither a policy check, nor a free wash, nor a review of the services performed would make this up to the customer. Therefore, an apology was all that could be offered. The service consultant had to recognize that listening to the requests of the customer is the most important part of the job. When dealing with the disappointed and angry customer, the service consultant has to think and avoid answering or reacting too hastily. The policy check is a mere tool that might be used and not a solution to all problems encountered.

Review Questions

Multiple Choice

1. Which of the following is considered an up-sell?
   A. Higher-quality brake pads that will last longer
   B. Replacing all fuses in a fuse box with ones that are not blown
   C. Removing air from the tires and replacing it with new air
   D. Installing a new battery because the old one is bad
2. Two service consultants are discussing customer follow-up. Service consultant A says that their purpose is to identify customers who are NOT satisfied with their service or treatment. Service consultant B says it is important to find out why the customer is dissatisfied, correct the problem if possible, and then avoid the same problem in the future. Who is correct?
   A. A only
   B. B only
   C. both A and B
   D. neither A nor B
3. The primary objective(s) of sales promotions is/are to:
   A. attract new customers
   B. keep regular customers
   C. both A and B
   D. neither A nor B
4. Which of the following is NOT a method to obtain customer feedback?
   A. Mail the survey to the customer.
   B. Leave the survey in the customer's automobile (hung on the mirror) to be filled out and sent in later.
   C. Hand the customer the survey when he or she pays the bill.
   D. Have the technician call the customer.
5. When discussing a service with a customer, Service consultant A says that the discussion should be adapted to fit the customer's personality. Service consultant B says that the customer should always be pressured to give an answer. Who is correct?
   A. A only
   B. B only
   C. both A and B
   D. neither A nor B

Short Answer Questions

1. Explain how to greet customers and respond to those who are angry.
2. Explain why a consultant must identify and prioritize customer concerns.
3. Give examples of how to promote the procedures, benefits, and capabilities of the service facility.
4. Identify methods to communicate the value of performing related and additional services.
5. Describe the different methods used for customer follow-ups.

Activity

Activity 1: For a general repair facility, research different advertising mediums that would help attract new customers. Determine which ones appear best at promoting the business for a region served (you select the region). Summarize the information found and justify which mediums are best at attracting the customers desired. Include in the report a recommendation to the owners about which ones might be good to use.

Activity 2: Explore the Internet for sites that help promote the service facility to current customers. Consider sites that send reminders to customers, perhaps access to their repair records among other features, such as mechanicnet.com. Research how the features of the sites function. Determine which ones help the service facility stay in touch with customers. Summarize the information in a report with a recommendation to the owners about which ones have the most appealing features.

Activity 3: Find examples of social networks where a service facility is promoted. Examine the information available and make a judgment about what you like and dislike. Make suggestions that if you were the facility owner, what you might change.

Activity 4: Prepare the following for your service facility.
   A. A one-page flyer for a service special.
   B. An email announcement that is to promote services at the facility.
   C. A questionnaire to be given to customers with their invoice inquiring whether or not their experience was satisfactory.
   D. An email asking for feedback from customers about the work done on their car.

CHAPTER 13

THE ASSISTANT SERVICE MANAGER'S AUTHORITY AND RESPONSIBILITIES

OBJECTIVES

Upon reading this chapter, you will be able to:

- *Compare the job duties and tasks of the service consultant and the Assistant Service Manager.*
- *Classify employees as staff exempt or staff nonexempt employees based on their management authority.*
- *Explain the difference between management authority and responsibility.*
- *Describe "implied responsibility" as it affects the Assistant Service Manager's job.*
- *Describe Assistant Service Manager's job relative to leading a team of technicians.*
- *Describe the purpose of OSHA and how the standards improve workplace safety.*

Introduction

The information presented in the past chapters can be applied to the job of either a service consultant or an Assistant Service Manager. This chapter continues the discussion but focuses on the additional tasks and responsibilities assigned to an "Assistant Service Manager," or ASM.

Some consider the positions of "Service Consultant" (SC) and "Assistant Service Manager" (ASM) to be essentially the same and only the names are different. This is not correct because while the basic tasks and duties may be the same, such as preparing a repair order, the ASM has additional job responsibilities. A major factor for the difference comes from the federal government's labor standards for staff "exempt" employees that are not paid overtime because they are in a "management" position. This law is explained in more detail below as well as the authors' second book on *Managing Automotive Businesses: Strategic Planning, Personnel, and Finance*.

Beyond the law it is difficult to specifically define the Assistant Service Manager's job because the exact job duties vary from one service facility to another. In general, the Assistance Service Manager has more operational authority and responsibility. Furthermore, they are expected to provide leadership in their work area and meet the expectations of the Service Manager. Often, differences beyond this general statement depend on the ownership and size of the service facility. For example, the position of a single ASM at a small general repair facility will be different than at a large dealership where there are several ASMs and teams of technicians.

Three Types of Assistant Service Managers

While not a rule but based on the authors' research and experiences there are three types of Assistant Service managers whose tasks, responsibilities, and authority vary by size of the service facility. Specifically, a single ASM at a five-bay shop will have different job expectations than an ASM who is one of three ASMs who oversee 15 bays. In addition, differences in ASM tasks can be examined by the type of ownership at a service facility, such as a proprietorship versus a corporation with a board of directors. To recognize these influences on the ASM's job, service facilities have been divided into three groups for further discussion.

The first of the three groups, which may or may not have an ASM, is small- to medium-sized service facilities with 3 to 12 employees. The business is often owned by a single person whose family members may help run the business. The second of the three groups consists of medium to larger repair shops with 8 to 20 employees that typically operate under a strict "formatted" service system, such as a franchise, large dealership, or chain of service facilities. These usually employ an ASM. The third group is typically the largest service facility with over 20 employees and operates with a management team. They may have multiple operations or profit centers as discussed further in the author's next book. Examples of the profit centers at these service facilities could include a collision repair center, parts department, used car sales center, vehicle reconditioning operation, fast lube lane, and a service department, among other operations. The members of this group include larger dealerships, chain operations, and multiple franchises often owned by the same people. They have a large customer base, possibly over 20 technicians, and employ several ASMs. The size of these businesses requires a formal business approach, including an elaborate chain of command.

ASM Positions within the Three Groups

The ASM's position in the first group of small to medium sized service facilities will likely have considerable independence to "just get the job done." Therefore, the ASM will be paid a salary and have a workweek that is over 40 hours. Since the ASM is a "staff exempt" employee with a salary, he or she will not receive overtime pay like a service consultant after 40 hours of work. At these service facilities the ASM will perform all the tasks of the service consultant, as described in earlier chapters plus other duties as explained later. However, because the service facility is smaller and the business has been established to benefit the owner's family, the owners may decide that some ASM tasks will be performed by the owner, owner's friend, or family member. For example, if the ASMs have the technical skills to fix cars, they may be told to repair a customer's vehicle or oversee or help a technician with his or her work.

An ASM working at a service facility in the first of the three groups will find the work environment to be more informal with fewer "systems" in place for daily operations. Policies, rules, and procedures may not be in writing and may change as conditions warrant. This means the ASM's job expectations may be less defined in writing and more at the will of the owner. For example, when the workload becomes heavy or a person is out sick, the ASM will have to cover the extra duties. Also, at the owner's discretion, the ASM may have duties taken away and assigned to the owner's family member, friend, or the owner him- or herself. If the owner wishes to do some tasks such as greeting customers and writing up repair orders, then he or she can do it until it is decided to reassign it to the ASM again; it is the owner's business. Such shifts in assignments are not uncommon and also not permanent as the next day may find another change as needs arise. As a result, "cross-training" of employees to do each other's jobs is often necessary and conducted informally when the need arises.

The point to this discussion is that among the first group of service facilities, there is no "right" or "wrong" job description for an ASM. Further, Repair Order Tracking Sheets, formal job descriptions, and other "system" approaches to doing business are usually not found. Running the business, the work environment, and employee interactions are usually informal and the climate can be stressful because operations revolve around the owner's personal life and preferences. Since the owner is an integral part of the business, when the owner is unable to run the business any longer, it may close.

In the second group, the most important difference between it and the first group is that there are a larger number of customers and this requires a larger facility and more employees. There may be one or more ASMs (staff exempt from overtime pay) and an array of other staff, such as warranty specialists, clerical personnel, and even service consultants (staff nonexempt employees who must be paid overtime). In some cases, the ASM position is awarded to the "senior service consultant." While the ASM will have job duties similar to the service consultant, he or she is usually on salary and expected to work more than 40 hours per week as well as provide some leadership. The leadership responsibilities might include such tasks as training new employees, reviewing Repair Order Tracking Sheets, oversight of technicians, assisting with business performance analyses, and even being the contact person for an absent owner or service manager.

Because the facilities in the second group consist of larger operations, a system that consists of operational procedures written in an "operations manual" is typically in place. A franchise service facility will have the equivalent called a franchise manual. When the second group does not have an operations manual to assist employees, it may be due to the inability of the owners to create a service system. A system may not be desired because the owner wants to be the "boss". Like a small sized business, a medium size business also may close when the owner is no longer able to run the business. In a well-organized second group, when the owner, who is the boss, cannot work any longer, the systems are designed so the managers can continue to operate the business indefinitely. Therefore, a Service System with an orderly arrangement of procedures that are linked together to form a process followed by employees and used by management to control shop production, customer service, and business operations is important for a business to continue. This is called a "going concern" and a term used by accountants and other business professionals.

The third group includes larger dealerships, chains, and same owner multiple franchises that typically have formatted business systems that in many respects appear more like a "corporate structure." This means they have various departments with profit centers in each one. In the service department area there may be multiple ASMs assigned to lead smaller work groups of technicians (see Figure 13-1). For example, a facility could have over 35 technicians and seven ASMs who are divided into seven teams with five technicians each. There is not a "rule to follow" with respect to the number of technicians an ASM can have on a team or how many teams a business can effectively use. How the teams are created and the job description of the ASM will depend on the owner(s) and service manager.

FIGURE 13-1 This dealership has multiple Assistant Service Manager (ASM); each is responsible for a team of technicians.

ADDITIONAL DISCUSSION

An Example of ASM Job Demands

An ASM at a very large dealership found he had a distressed customer who felt her repair should be covered by warranty. She had authorized the repair and knew the charge before the work was performed. However, when she picked up her car she demanded that the repair be paid by warranty and she should not have to pay for it. She became very upset. This was the ASM's problem to handle, not the Service Manager or any other employee at the dealership.

The ASM followed company procedures and relied on his training to work through her frustration with great calm and dignity. He took time to explain a second time why the warranty would not pay for her problem. As required by the company's procedure, the technician who worked on the vehicle was interviewed by the service manager to provide details about the situation in a formal statement that was added to the record. Shortly after the customer left, as a team, the dealership employees had to meet to review the situation. They wanted to learn from the problem and determine any additional steps that could have been taken to work with customers in the future. This entire process was disciplined and professional because of the systems that were in place and employee training that included in-house training. Credit for handling the situation went to the ASM and the Service Manager who trained him. Unfortunately, not all service facilities operate this way and often the larger businesses provide management teams who can concentrate on company policy and business systems development. The larger size also allows more opportunity for employee training so they are able to handle tough situations more consistently. As the ASM position is discussed in this chapter, keep the differences of three types of service facilities in mind because the job of an ASM will vary considerably from one type of service facility to another.

The ASM at a large service facility is expected to follow the company policies, procedures, and rules. They are expected to know the organizational management chart, job descriptions of employees, and any relevant contract the business uses, such as the manufacturer's Policy and Procedures manual. To keep their job the ASM must be "on top of their game." This means they know the performance of each member on the team as covered in the second text on *Managing Automotive Businesses*. The business environment is more formal in terms of customer, subordinate, and higher-level manager interactions. Commonly, the ASM must be a master at the company computer systems, maintain and access a variety of service facility information data bases, and participate in management initiatives, such as employee performance evaluations, training, and marketing efforts. At these service facilities, there are business systems, constant training initiatives, and rules that must be followed such as dress code among many others. The ASM job at the largest service facilities is demanding, and as a result the pay and benefits are usually higher than at smaller service facilities.

Federal Job Classification: Staff Exempt and Nonexempt

As discussed in more detail in the authors' second book *Managing Automotive Businesses, Strategic Planning, Personnel, and Finance* the federal government's Fair Labor Standards Act requires employers to classify their employees as staff exempt or staff nonexempt. This means employees are either exempt by definition from the act or not exempt from the entitlements in the act. For instance, staff nonexempt employees are entitled to a minimum hourly wage established by the federal government or state government if it has a higher minimum wage requirement. Staff exempt must be paid a minimum salary.

Staff nonexempt employees do not have any executive, professional, or administrative duties, such as technicians, custodial employees, secretaries, clerks, service consultants, and so on, and they must be paid overtime. Staff exempt employees do not have a right to be paid for any overtime hours. At most businesses, policy manuals specify that nonexempt employees must gain permission to work overtime from the owner or a manager. This is typically written and initialed by the manager on their time card (see Figure 13-2).

Staff nonexempt employees also have other rights under the law, such as the right to bargain collectively as a union. They also may not have any personnel supervisory duties, meaning they do not hire, fire, evaluate, assign, direct, or create working terms. These duties are performed by owners or managers (staff exempt employees) who are representative(s) of the owner.

Staff exempt employees (managers) do not receive an hourly minimum wage but must be paid a minimum "weekly" salary established by the federal or state government. Therefore, these employees are "exempt" from receiving overtime pay. In other words, if a service facility

FIGURE 13-2 This time clock is where nonexempt employees "punch in" to start work and "punch out" when they are done. The payroll time cards that are "punched" are on the left. The "time tickets" on the right are used by technicians to keep track of the work completed and record the time it took to complete the job. The time tickets are used for other recordkeeping as well, such as a record on the flat-rate hours charged to the repair order.

opens Monday through Friday at 7:00 a.m. and closes at 6:00 p.m. a staff nonexempt employee, such as a service consultant, would have to be paid overtime after performing 40 hours of work (assuming overtime was approved by the owner or manager). A staff exempt employee would not be entitled to overtime. Consequently, an ASM may work a 50-hour workweek and be paid a weekly salary as opposed to a Service Consultant, who would receive 10 hours of overtime pay. Both staff exempt and nonexempt employee jobs may provide additional compensation such as bonuses, commissions, and other incentive payments in addition to their hourly wages or salary.

Staff exempt employees, such as a service manager, represent the owner. The ASM, as a staff exempt employee, represents the Service Manager and is considered to be part of the company's management team. As a result, both have some level of autonomy to decide how to "get the job done" as opposed to working under the strict oversight of the owner or manager. Consequently, an ASM, as a manager, usually has some authority granted to him or her to make some decisions, which are often guided by company policy. An example would be to approve overtime for technicians or other staff nonexempt employees under his or her supervision. Policy may even give the ASM the authority to sign certain business documents in a manager's absence, such as a payment voucher. Of course, in addition, as a staff exempt employee the ASM would be responsible for overseeing the work and conduct of technicians, as well as for assigning work to them. However, an ASM may delegate some authority to perform some tasks to a service consultant, or lead or senior technician as shown in Figure 13-3.

FIGURE 13-3 This service consultant has been authorized by the service manager to assign the next job to a technician.

Management Authority and Responsibility

The number one management rule is that a manager may delegate authority to act on his or her behalf but not the responsibility for errors or mistakes. This means a manager may give (delegate to) a subordinate the authority to do something, but if it is not done correctly, the manager must assume the blame from his or her supervisor. For example, a Service Manager may tell the ASM to oversee the toolroom reorganization (see Figure 13-4). The ASM may delegate the job to a technician. Remember, the ASM can delegate the authority to do the job to the technician but cannot delegate his or her responsibility that the job will be done correctly. If the technician does not get the job done properly, he or she is responsible for the failure to the ASM. The ASM bears the responsibility for the failure to the service manager, who gave the ASM the job to perform. However, if the job is done correctly, the ASM may offer praise to the technician. In turn, the ASM may receive praise for a job well done from the service manager or owner and should say in return, "happy to lead the effort. I am glad you are pleased and it was a success because of (name of the technicians) hard work." As a leader, it is good to recognize others who helped with a successful project but never try to defer blame.

An ASM, as a manager, may be given the authority and responsibility by the service manager (higher-level manager) to manage a team of technicians. In turn, the ASM may delegate some authority to one of the technicians (staff nonexempt employee) to hand out work. When this occurs, the technician will serve as the lead technician or team leader. The concept is the same as when a service manager delegates authority to a service consultant or dispatcher (both are staff nonexempt

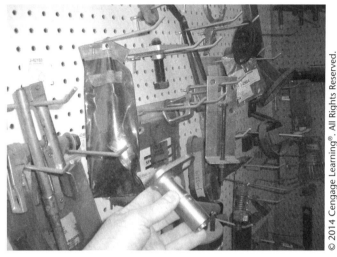

FIGURE 13-4 Special Service Tools need to be organized. These tools are labeled with the tool number and easy to find when needed.

CAREER PROFILE — The Job: Assistant to the Service Manager

Often there is confusion when the title of "Assistant to the Service Manager" is used. This is often classified as a staff nonexempt position and is not to be confused with an ASM. The Assistant to a Service Manager is not a manager but a staff nonexempt employee. A person in this position may be given considerable authority by the Service Manager but the authority cannot be of a supervisory nature.

To illustrate this point, an Assistant to the Service Manager collects, processes, and sometimes analyzes service department data for the Service Manager, so he or she can make a decision. Examples of the data include technician time tickets, subcontractor invoices, repair order to final invoice inspection, and customer complaint details among other information. This position also oversees the work of the warranty clerk but is not the clerk's supervisor. When the manager is gone, the Assistant to the Service Manager may be given "temporary" or "limited" authority to make some decisions; however, the decisions may be reversed by the Service Manager upon his or her return. In other words, the Assistant to the Service Manager does not have authority to direct the work of the warranty clerk because the delegation of authority is limited and temporary. Also decisions made by the Assistant to the Service Manager, or other employee who fills in for the service manager, need to be reviewed by the Service Manager because the responsibility for any actions taken by the Assistant cannot be delegated.

Therefore, a Service Manager must seriously consider what and how much authority to delegate to others, such as the assistant to the service manager. The extent of authority given to a subordinate requires good judgment and experience. Managers are hired to look out for the best interests of the company's owners and they are assigned a level of trust. At the same time, subordinates should think about how much authority to request and accept.

employees) to hand out work. When a lead technician (team leader) is given some management-level duties by the ASM, the ASM must remember the concept of authority and responsibility. Therefore, delegation of authority requires an ASM (as a manager) to be certain that the technician assigned the authority understands and is qualified to do the job. The ASM should recall the "Number One Rule" in Management: authority may be delegated but not his or her responsibility to assure the job is done properly (see Figure 13-5).

FIGURE 13-5 This Service Manager meets with the technician to review the repair progress. The Service Manager knows the job was assigned to a qualified technician by the Service Consultant. While the Service Manager gave the Service Consultant the authority to assign work to the technician, the Service Manager is taking responsibility by checking with the technician. The Service Manager knows he is responsible for every job. He makes sure the technician is not having problems with the repair and the job will be done on time.

CAREER PROFILE — Team Leader: Responsibility and Authority

Selection of a technician to lead a team can be one of the most important management-level decisions an ASM will make. The technician selected to be a lead technician or "team leader" may not be the most technical person in the shop. Rather, the technician should be the most dependable to lead a group of technicians. At some service facilities, the team leader may be elected by the other technicians on the team. The ASM must assure the team leader understands and can follow management procedures, can be seen as a person with authority by the other technicians, and will treat all technicians fairly. For instance, assigning work properly requires a team leader who understands how the service facility's systems function to process work. The team leader also needs to know how data, such as the Technician's Flat Rate Objectives, are used to help the technicians meet their goals (this was discussed in chapter 3). More importantly the team leader must be knowledgeable about the technicians on the team, their abilities, attitudes, and work ethic.

CASE STUDY: Unqualified Technician Repairs

John is a Team Leader (staff nonexempt employee) and he also understands that his authority to hand out work comes from the Service Manager (higher-level manager) via the ASM (lower-level manager). To do his job, John has been given access to the records of each technician's licenses and manufacturer's certifications to repair certain systems. He knows that if he allows a technician to do a repair that he or she is not certified to perform, the claim will not be paid and the business will have to accept responsibility that may include fines and other penalties. At the same time, John knows that in his state, technicians must be licensed to perform emission inspections.

John's team is backed up with work that includes a vehicle waiting for emission inspection. The workday is coming to an end and there is pressure to inspect the vehicle. John cannot get his work done and is tempted to permit an unlicensed technician to perform the inspection by using his inspection license number. If the inspection is not done, the customer will leave without the inspection and report being dissatisfied with the service. If the inspection is performed and the illegal use of his license number discovered, the results would be disastrous to the service facility and both technicians could lose their jobs. Furthermore, John's emission inspection license would be revoked (taken from him) by the state authorities. In addition, the service facility would lose its state-issued license to conduct emission inspections and legal charges may be filed by the state that would include fines to be paid by the business plus the people involved. This means the ASM and Service Manager who had authority over and responsibility for both emission inspections and each technician's work will be held accountable. Obviously, John must withstand the pressure as the team leader when the inspection is not completed as promised.

Delegating authority to a Team Leader is most important when it comes to assigning work to qualified technicians. An ASM must trust that the team leader understands and follows all of the shop's management procedures. If an unqualified technician happens to perform a job he or she is not qualified to perform, there can be legal ramifications as well as payment issues such as warranty repairs.

ASM Workplace Responsibilities

As noted in the previous section, the ASM and Service Manager are both managers and staff exempt employees. They have workplace authority and responsibility, as well as other job performance expectations. The "manager" title indicates the person is part of the business management team with a duty to oversee the conduct of all employees, even if they are not his or her direct subordinates. Managers carry the burden of the success or failure of the company unlike staff nonexempt employees, such as the Service Consultant or technicians.

From the standpoint of workplace responsibility, the Service Consultant, as a staff nonexempt employee, has the same employment status as a technician. They have an obligation to perform the job assigned and neither officially supervises the other. The following example comes from a composite of cases to illustrate the point regarding the differences in management responsibility. Assume that Bob, a technician, was hurt on the job by a defective brake lathe. He knew that it was defective before he used it to machine a customer's rotor but used it anyway. Since Bob knew it was defective, a claim for personal injury beyond his actual medical bills would not likely be recognized. His medical bills will be paid because employers are required by state law to have workers' compensation insurance. Depending on factors beyond this example, Bob may or may not be compensated for wages lost as a result of the injury.

To continue the example, assume that Bob asked a fellow technician, who was not "in charge" of the equipment or its maintenance, whether he should use the lathe. Since the technician was not part of the management team, like an ASM or Service Manager, nor Bob's supervisor, the response would be considered an "opinion." It is an "opinion" and not an "order" because the response has little or no consequence to Bob's future employment or his job performance evaluation. The coworker does not have authority to supervise Bob's work nor can he reprimand Bob when he used the equipment or refused to use the equipment to finish the job assigned.

However, had Bob asked the service manager or any Assistant Service Manager, the response (yes or no) would constitute an "order" not an "opinion." The circumstances change because an ASM or Service Manager is in a management position with management authority and responsibility. As a member of the management team, the ASM would be seen to have the authority of a service manager with a level of supervisory responsibility. Consequently, when Bob reported that the brake lathe was defective and whether he should use it, the expectation is a manager will understand the risk to the company. As a member of the management team, the ASM should order the machine to be "tagged" as being defective and "locked out" as per OSHA regulations (see Figure 13-6). Further, the ASM has the authority to order Bob not to use the machine until repaired and then immediately rectify the issue related to completing the job for the customer.

The concept of risk must be considered as a member of the management team. Had the ASM told Bob to finish the job and use the defective lathe, even if Bob thought he could use it without getting hurt, this would constitute management approval or even an order. This is because of the ASM's managerial authority and responsibility for safety. Had Bob been hurt, the business as well as the ASM would be at risk for compensation beyond Bob's medical bills and lost wages for his injury. It must be understood that managers, such as the ASM, represent the owner. Their decisions such as permission to use the defective lathe that result in injury, or worse death, could make everyone face serious legal charges brought by Bob as well as other government agencies concerned with worker safety. In other words, the ASM and service manager would face more than owner discipline for the action, and the business could even be shut down. Members of the

FIGURE 13-6 This machine is broken and an OSHA "Do Not Operate" tag, written both in English and Spanish, is ready to be attached to the equipment with a wire so it can be seen easily. The machine is unplugged and the electrical plug should be "lock boxed" so that it cannot be plugged in again without management opening it. More OSHA rules and procedures may apply depending on the equipment and situation. Managers should consult the OSHA website for details concerning defective equipment and safety of employees.

ADDITIONAL DISCUSSION

OSHA as a Regulatory Agency

According to the Occupational Safety and Health Administration (OSHA) website at http://www.osha.gov/, "The Occupational Safety and Health Act of 1970 (OSH Act) was passed to prevent workers from being killed or seriously harmed at work. The law requires employers to provide their employees with working conditions that are free of known dangers. The act created the OSHA, which sets and enforces protective workplace safety and health standards. OSHA also provides information, training, and assistance to workers and employers. Workers may file a complaint to have OSHA inspect their workplace if they believe that their employer is not following OSHA standards or that there are "serious hazards."

Discussion of the OSHA regulations that govern shop operations and management decisions is beyond the scope of this textbook.

OSHA's oversight extends from literally the shop floor to ceiling and every corner of the business. It includes all the equipment, tools, storage, and chemicals the technician uses in addition to work procedures. OSHA information is not only helpful to run a safe shop but it is the law. The regulations, updates, forms, recordkeeping, and information provided to employees (access) are mandates that a manager is responsible to know.

The best way learn these regulations is to visit the OSHA website and spend time under the website regulations tab reading the rules that pertain to the equipment and type of work the automotive service facility performs. In addition, there is training available to help managers understand OSHA regulatory information and a small business section to assist owners with information.

management team are not only responsible for worker performance with respect to productivity but they are also held legally accountable for worker safety.

Real and Implied Manager Responsibility

Managers have real responsibilities and expectations assigned to them that are found in company documents that include the manager's job description, policy manual and update memos, and operations manual. The company documents are influenced by both internal and external information that includes agreements, contracts, and government laws. Example of external agreements include dealer contracts and the manufacturer's warranty policy manuals, franchise contracts, union agreements, insurance company policy requirements, and government laws.

Government laws include state regulations often found in handbooks, such as those for state emission inspection among others. External information includes trade and industry information, such as repair specifications, market competition, and customer expectations. Internal company information that influences company documents are owner's expectations, decisions, and instructions (verbal or written) as well as proprietary processes that give the company market advantage. These processes may range from how customers are processed to how the shop area is set up and operates.

All managers, including the ASM, must read and understand the company documents related to their job. The three basic documents include the policy manual that governs rules all employees in the business must follow. The operations manual that explains the business system of a work area and directs how work is performed that includes area employee job duties. Since the operations manual gives the business competitive advantage, it may contain proprietary processes that are unique to the business and not necessarily an industry standard. The final document is the manager's job description that defines the tasks to be performed. Since the company documents are based on external and internal information, the manager must understand what influences each document. To illustrate the hierarchy of these relationships relative to a manager's job, below (Figure 13-7) is a table to help visualize some of the internal and external influences that define the company documents a manager will use to meet company expectations.

Implied Expectations

When a task or duty of an ASM is not specifically stated in a job description or other company documents, then the expectations are implied. In chapter 2 of this book, the job description for a service consultant includes the tasks that might be assigned to the position. The tasks listed in these descriptions would be considered "real" expectations and responsibilities. However, there are also "implied" expectations that are a

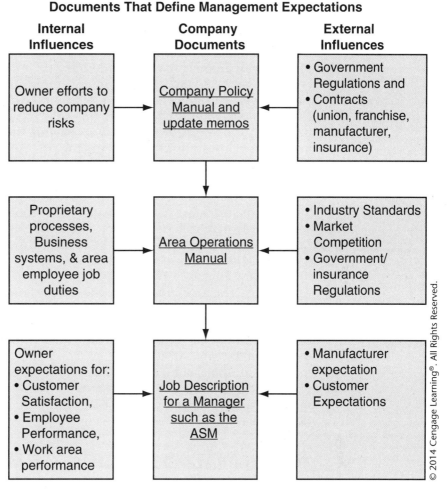

FIGURE 13-7 This diagram shows the documents that the owners expect a manager to be responsible for, enforce, or perform. Examples of the internal or external influences that shape the documents' policies, procedures, and duties are illustrated to either side of the documents listed.

"gray area"; meaning they are learned by reading company policies and talking with immediate supervisors. Therefore, some implied responsibilities for an employee, such as an ASM, would come from the Service Manager, or a higher manager, perhaps the owner. Over time, changes to company documents will come from internal and external influences. Instead of rewriting an entire document, such as policy manual, there will be update memos. Furthermore, there can be changes to real and implied tasks and duties as well as responsibilities when there is a change in managers or owners.

To provide a concrete illustration of implied responsibilities, one of them is for an ASM to inform the service manager and owner aware of any situation that would put the business or its systems at risk. For example, the ASM must always be on the lookout for employee drug abuse, unauthorized changes in procedures, downtime created by service facility equipment, building problems, safety issues, plus numerous other daily operational concerns. Problem identification is likely not found in the ASM job description but is an implied responsibility. The ASM, as part of the management team, must report these problems to managers or owners otherwise the ASM can be criticized for ignoring them. On the other side of the coin, the ASM can be criticized for reporting situations and taking on "too much" authority, being a "busy body," "cry baby," "whistleblower," or not being a "team player" for reporting it. This criticism can come from employees under the ASM's supervision or even the service manager or owner. In some cases, the ASM will be in a "no win" situation or worse.

The best advice to an ASM new to the job is to first learn the job thoroughly. The ASM must earn a track record of success and trust. This can be accomplished by developing relationships with experienced managers and to ask them for guidance when needed. ASMs must understand how to be patient when it comes to being a member of the management team. All teams are different and there are always informal and unofficial rules and practices that vary from one team to another. The worst thing is to be assertive in an attempt to impress team members and to suggest changes that imply criticisms of current operations or managers.

Team and Lateral Support Group Management

This text does not attempt to explain the many team and group variations used by repair shops because there are too many configurations. For instance, there are shops where the service manager or ASM is hired as the shop leader for the technicians. There are cases where management appoints an ASM, shop foreman, or dispatcher but relies on a technician to serve as the shop leader. There are also group systems where the technicians elect the group leader to head small groups of technicians and the ASM's job is to represent the interests of the management team. In some team system arrangements the members on the team split their flat-rate hours with each other so there is incentive to "work together."

Regardless of the design, for numerous reasons a "team approach" (not to be confused with a "team system") makes the service facility shop operations easier to manage. This can benefit company performance and allow expansion of the service facility when it works as designed. The common element among any team or group, regardless of design, is the ability of the members to work together to serve the best interests of the business. When a person cannot work on a team, then he or she needs to be placed in a job that does not require him or her to work with others to attain common objectives.

Teams are commonly composed of technicians connected to management through an ASM or an appointed nonmanagerial (nonexempt) employee. Examples of appointed or possibly "elected" nonmanagerial employees who would head a team include a lead technician or team leader, shop foreman, or service consultant. The choice of the person selected would depend upon the willingness of team members to follow the leader while at the same time adhering to management directives to reach a predetermined goal. Technicians are selected for a team because of their technical skill sets and ability to work together toward common goals. Too often, businesses have gone into a decline or failed because management and production employee attitudes did not support the interests of the company. Teams ideally are organized to prevent this from happening.

The Assistant Service Manager's Role as a Team Member

The Assistant Service Manager could potentially be a part of three teams. The first is the team that consists of the ASM and the technicians on his or her team. In cases where a repair facility, such as a large dealership, has multiple ASMs, each one may oversee a team of technicians. Therefore, a second team is the service department team leaders that consist of the other ASMs and technicians who are Team/Group Leaders as well as the Parts Department Specialist(s) who work(s) with each team. These employees directly work with each other on a daily basis. At one dealership, this team meets briefly every morning and at the end of the work day. The third team the ASM would be part of would be the management team for the business. In a small shop this may consist of the owner and managers (service manager, ASM) plus perhaps staff members, such as a bookkeeper or warranty specialist. This team may even include outside contractors who support operations, such as a factory representative and parts suppliers. At a dealership the third team could be extended to include the sales manager, used car manager, operations manager, business manager among others.

The coordination of work between teams and team members is crucial to owners and vary from one service facility to another. Achieving positive interactions among these different employees and operations is important and requires good leadership and positive member collaboration. Basically, when an ASM is hired to work on a team, an orientation may require exposure to the different job duties of the members. In other words, the ASM must understand the nature of the different jobs and processes at the service facility.

CAREER PROFILE: The Automotive Business Consultant

Some owners often seek out business advice from outside consultants; such as how to establish Service Systems and the related procedures. Owners may want help in creating and utilizing teams, improving facility utilization, and so on. This is because automotive business operations have become complex. Automotive business consultants are hired to assist an owner to maintain profitability, to expand, or to keep up with changes in technology, laws, practices, among many other reasons. These consultants are often professionals educated in business management with expertise and experiences working with automotive businesses. They will not be influenced by personal relationships which can be a problem when owners and managers try to make changes. Business consultants do a good job when their client has achieved improvement after they have completed their work.

Business consultants are not employment agencies or "head hunters" or human resource contractors; therefore, they are not the type of consultants who will take over management's responsibility for hiring or firing. They are trained to collect data to solve problems and provide recommendations. Unfortunately, managers may not know the role of the business consultant, how to work with them, and may resent their intrusion. This seriously hampers the business consultant's ability to find ways to improve a business that may be needed for the business to survive, especially when past "hit" or "miss" improvement efforts were used to try and fix the business resulting in new problems.

How a Business Consultant helps a business is beyond the scope of this book, knowing when to ask for help is the first step toward improvement. The second step is for all managers to work with consultants as professionals with a common purpose. The goal is for all parties to enter the process with an open mind and work with each other to solve the business problems. When the process involves a business consultant and others outside of the business such as a factory representative, vendors, unions, accountant, lawyer, and so forth, the role of each must be defined. As input is obtained, it must be balanced by the owner so decisions can be reached so the business can reach its objectives and targets.

Review Questions

Multiple Choice

1. What is it called when an owner gives a manager "power" to oversee something?
   A. Authority
   B. Responsibly
   C. Team player
   D. An order
2. What can be delegated by a manager to others?
   A. Authority
   B. Responsibility
   C. both A and B
   D. neither A nor B
3. When "things don't go right," who must ultimately accept responsibility?
   A. Management
   B. Customer
   C. Employee
   D. Coworkers
4. Which employee is staff exempt?
   A. A Service Consultant
   B. The Assistant to the Service Manager
   C. The Team Leader
   D. The ASM

5. Service Consultant A says a technician doing manufacturer warranty repairs must be certified in the services being performed? Service Consultant B says an unqualified technician doing warranty repairs can result in not being paid by warranty? Who is correct?
   A. Service Consultant A
   B. Service Consultant B
   C. both A and B
   D. neither A nor B
6. Which company documents define the owner's expectations of the manager?
   A. The Company Policy Manual
   B. The Area Operations Manual
   C. The Customer Satisfaction Surveys
   D. The Job Description

Activity

Part 1: Visit the OSHA website and write a brief description of the information found in each tab.

Part 2: Return to each OSHA website tab and compile a list of information that you think would be relevant to a service manager.

Part 3: Pick out three items from the list compiled in part 2 and read the information. Summarize the OSHA information in a memo and explain why each item might be important for a service manager to know. Suggested topics to review in more detail include personal safety protection, Material Safety Data Sheets (MSDS) or chemical safety, as well as equipment such as hoists (automotive lifts).

Part 4: Visit the following website: http://www.health.state.mn.us/divs/hpcd/cdee/occhealth/wsws.html

Scroll down to "Individual Lesson Plans and Curriculum Components" and complete any lessons or activities your instructor feels will complement your understanding of OSHA.

DISCUSSION QUESTIONS

In the following cases, the ASM could be seriously criticized for ignoring a problem, not being a team player, not assuming leadership or management responsibilities, stepping out of line, or being too ambitious. These can be "no win" situations unless the ASM has a clear understanding of the implied expectations of the service manager and the owners. The answers to these problems do not appear in the ASM's job description tasks. At your instructor's discretion either as a class or as an individual, determine what might be expected of the ASM and how he or she might handle the problem.

1. With respect to people, indirect supervisory responsibility and implied expectations: Sally, the ASM, observes a technician, not under her direct supervision, acting "strange." Other technicians suspect drug use may help to explain the behavior and poor job performance. Recognizing that the technician is not a direct subordinate, what action should be taken? After taking action, should the ASM follow up to determine what was done and if it was effective?

2. With respect to process observations: Work orders stopped flowing as usual. The ASM recognizes the "clog" in the service process is caused by a change in the existing computer system by a new software vendor. The changes were made with the promise that the work would flow more efficiently. The clog in the process will impact the performance of the ASM's team. The manager promised these changes will not cause production problems, but they are a concern to the ASM. Should the ASM ignore these observations? If not, how should the ASM proceed to make his concerns known?

3. With respect to facility operations: When the ASM opened the facility in the morning, he found that the lights in one bay did not work. For safety reasons this would mean the bay was "down" until repairs were made to the lights. The ASM asked a technician to make the repair. The technician told him that it is not his job to make building repairs. The ASM does not have the authority to call for facility repairs as it is the sole responsibility of the Service Manager. The Service Manager cannot be contacted and is not expected to come into the building until later in the afternoon. Should the ASM ignore the problem until the manager comes into the shop? Should he assign the technician to another bay that is temporarily available? Obviously shop performance will be negatively impacted and how might the service manager perceive a phone call to an electrician to fix the problem?

4. With respect to operations: The Service Manager, who has been drinking heavily over lunch, orders a technician to replace a plastic fuel line with special fittings that are broken with a rubber hose and clamps. He wants the vehicle finished immediately. Parts are not available to do the job correctly. The technician fears being terminated for not obeying the Service Manager and tells the ASM that the repair request is unsafe. The technician feels the rubber hose and clamp will not hold together under high fuel pressure and will likely come apart causing a fuel fed fire. The ASM will have to answer to the service manager and customer if the vehicle is not done or to investigators if there is an accident. Explain the different options that the ASM has available and whether each could cost the ASM his or her job? What option should be chosen?

CHAPTER 14

MANAGEMENT OF OPERATIONS, CONTRACTS, AND INSURANCE

OBJECTIVES

Upon reading this chapter, you will be able to:

- Understand the basic elements of a contract and management's role in their execution
- Outline the procedures for making arrangements for a sublet sale (D.4).
- Describe the various "other duties" an Assistant Service Manager may have to perform.
- Explain the purpose of a petty cash account.
- Explain the reason for liability insurance, garage keepers, property, and workman's compensation.
- Describe the security systems that can be implemented at a facility to reduce loss.

CAREER FOCUS

You know that you are paid to watch out for the best interest of the owners and the owners admire you for your management abilities because you pay attention to the details. The owners don't have to worry about giving you the authority to run their business because they know you take responsibility for and can handle contracts, systems, customers, vendors, and employees. When you are given new management opportunities by the owners, you respond by taking control and leading the employees after you organize the procedures and plan how to integrate it into the Service Systems. Specifically you implement system changes that are designed to handle many different types of problems. This is why the owners employ you as the manager and trust you with the money they invested in the business. This chapter will discuss insurance and other contracts a manager must understand.

Introduction

Up to this chapter, the text has pointed out the primary responsibilities and expectations for Service Consultant (SC) and Assistant Service Manager (ASM). Those who have worked as a SC or an ASM can testify that there are a number of "other duties" that are unofficially assigned and change over time.

In many cases the "other duties" are assigned because the SC/ASM's workstation is conveniently located to the duty performed or because of the SC/ASM's working relationship to other employees such as the technicians. By design, the SC/ASM's workstation is located "on the front line" where customers who enter the service facility can be met as they come through the door (see Figure 14-1 and 14-2). The SC/ASM is there to help people with questions, requests, problems, and sometimes even deliveries to the business. This leads to the variety of "other duties" such as accepting packages, taking messages, and referring people to someone else in the business for assistance. Often the SC/ASM must personally take care of the request or problems because they by default are the "go-to" person for solutions. Consequently, when the ASM/SC work location changes, their "other duties" will change as well.

In addition to people entering the service facility seeking the SC/ASM's help, other employees, such as the parts specialists among others, go to the SC/ASM when they have a question or need a solution to a problem. The problem may be as simple as a new light bulb to replace one that has burned out to asking where the extra rolls of duct tape are stored. The reason is because the SC/ASM is known to them and can easily be contacted. So they are often seen as the person who knows where "things" are kept and what is happening. Thus, the SC/ASM's are often referred to as the "go-to" person, a common attribute associated with employees who are in charge and starting to take on management or leadership duties.

FIGURE 14-1 This Service Department counter is at the center of the dealership activity. The ASM/SC helps and answers questions for a variety of departments.

In many cases these "other duties" make the "go-to" person indispensable and valued colleague. Employees and customers know that a top-notch "go-to" SC/ASM gets things done and this makes them important as compared to the ones who are not inclined to readily offer assistance. Unfortunately, it is impossible to list all of unassigned "other duties" because they vary from one facility to another. For example, although the tasks may be the same, the ones performed by an SC/ASM at

FIGURE 14-2 At many dealerships the Service Department has a separate entrance. The ASM/SC will primarily meet customers who need service but employees will still seek out advice.

a small independent repair facility are different from those performed at a large dealership, chain, or a franchise. The purpose of this chapter is to introduce contracts that make up some of the common "other duties" that may be assigned to the SC/ASMs at different types of service facilities along with important considerations when performing them.

Management Authority to Enter Business Contracts

Because a number of the "other duties" discussed below involve working with business contracts, several important legal considerations must be understood. To begin with, SCs and ASMs for their own personal protection should not enter into any business contracts for the service facility unless given a direct order to do so by management or the owners, preferably in writing. For instance, it is not unusual for a sales person to ask an SC/ASM to approve something, such as putting a soft-drink vending machine in the waiting room. Before it can be installed, the SC/ASM is asked to sign a permission form that includes details about the services and payments. This permission is likely a contract and cannot be signed by the SC/ASM because it includes more than just putting a vending machine in place and plugging it into an outlet. The agreement has four parts that make it a valid contract. The parts are:

- What each party will provide or do,
- Consideration: What "of value" will be exchanged between the parties (typically money for something in return).
- Capacity: Both parties are competent by legal definition
- Legality: the contract is not against public policy or illegal

In terms of who can enter a contract for the business, it is without argument that the owner may enter the contract. In a corporation, the board of directors can vote to enter the contract or give authority to a manager to enter certain contracts as the board's agent. With a power of attorney document, the manager can enter all contracts on owner's behalf including signing legal documents and deeds.

In some cases, the authority to enter a valid contract may be implied because of a person's position. It is customary in the Automotive Repair business for someone with a manager's title to enter certain contracts, especially for operational necessities; such as equipment repair or purchase, and contracted services, trash removal, and so forth. One reason Service Consultants cannot enter contracts is because it is not customary in the automotive repair business for the SC to have that authority without manager's approval or other approval mechanism; such as obtaining a PO (purchase order) number. An ASM, as a manager, might be able to legally enter into some business contracts that might have to do with his or her work area; such as setting up an agreement with a towing service (contracting for) or booking a lunch meeting at a local restaurant.

However, it is the ASM's and the third party's responsibility to know the limit and what is customary. Ultimately, it rests with the third party or contractor to know the implied authority before delivery so the contractor can be paid for the goods or services according to the contract.

At the same time, the SC and the ASM must know as a third party what they can legally ask a customer to enter. More specifically, as studied in an earlier chapter, a mechanics lien cannot be enforced (keeping the vehicle until the customer pays) unless certain information is on the invoice that includes the customer's signature. In some states the lien is not enforceable unless all owners of the car have signed the invoice or provided approval by some other acceptable means, such as telephone authorization. So for a very expensive job, a husband and wife, who are both listed on the vehicle title, must sign or give approval to enforce a lien. If the leasing company owns the car, they must provide written approval for a repair on a leased vehicle, not the person who is the leasee and/or drives the vehicle. Fleets require the fleet manager approval before working on the vehicle, not the driver of the fleet vehicle except in emergency circumstances, such as towing the vehicle when disabled on the highway. Furthermore, when an aftermarket (non-factory) warranty is involved, the vehicle owner must understand that when the warranty does not pay the entire repair bill, he or she is responsible. If an SC/ASM or even service manager enters into a contract without having the proper authority, the contract would not be valid. In addition, the SC/ASM could be personally in jeopardy with both businesses and all parties to the contract.

Execution of Business Contracts

Even though an SC/ASM may not legally enter into a contract, they are often asked for advice involving a contract. For example, an SC/ASM may be asked how to handle an aftermarket warranty company's contract for the customer especially when it is unclear whether the repair will be covered. Further, upon delivery of a contract's goods or services it is expected that the contract is understood well enough to assure the goods or work meets the contract specifications. This means the contract must be studied (not just read) to know what goods or services are to be exchanged; the terms of delivery; as well as how and when payments are to be made between the parties. Contracts may specify how, where, and when the work is to be done, how long it will take, and possibly even the person or business that is permitted to perform the work. There can be no deviations from a contract unless all parties agree in writing. The amendment to the contract is the only option to modify the existing contract; otherwise a new one must be prepared.

A good example of a contract problem was at a repair facility contracted to provide oil changes for a fleet. Other work could be provided but only upon the authorization of the fleet manager. Every Monday morning the service manager of the repair facility had to deliver and review the invoices of the work conducted the previous week with the fleet manager. To show how well the facility could wash and detail vehicles, the owner decided that one of the vehicles should be washed and detailed. Thinking the fleet manager would be impressed with the quality of the work, a charge for the service was entered on the weekly invoice. The following Monday the manager of the fleet saw the charge and refused to pay. A charge could not be made because the fleet manager did not give permission for the wash and detail as per the service contract. The irritation and loss in confidence by the fleet manager eventually cost the repair shop the contract to service their vehicles.

External and Internal Contracts

At a service facility there may actually be two types of contracts: external and internal. An external contract is made with another business as studied up to now. An internal contract is made with a separate operation in the same business, such as the "Service Department" making a repair on a car that came from the inventory of the "Used Car Department." Under an internal contract, the Used Car Department would pay the Service Department and possibly the Parts Department for the repair. At some businesses the charges are based on cost with possibly a small markup to cover indirect costs. Therefore, internal contract negotiations are often between managers of the different departments regarding terms, such as costs and markup as well as priority given to the work. These internal contract discussions or agreements are often tense because each manager is watching his or her area's profit margin.

External contracts are, of course, quite formal. They include insurance policies, service facility estimates and invoices, and most vehicle warranty-related repairs. Some contracts might involve an attorney, even when the contract may be a "standard contract;" such as a franchise contract, building improvement contract, or even equipment purchase agreement. A contract is without a doubt the safest method for one business to engage in work for another. Personnel and even business ownership changes do not necessarily void a contract unless specified in the contract, therefore business continues as usual. In all cases, the contracts must be carefully reviewed by owners and perhaps managers and an attorney before it is signed by all personnel involved, including the SC/ASM if appropriate.

Service Facility Insurance Contracts

Assumed by customers and the ASM/SC is that the service facility has insurance and that a policy will cover possible accidents. However, this is an assumption because some service facilities don't have insurance or may not carry enough of the proper type! When this occurs, not only will there be an unhappy customer but there will be civil law suits as well as potentially other undesirable results. Although review and selection of liability and workman's compensation insurance are not a responsibility assigned to ASM/SCs, they must understand what they are and how they work.

Business Liability Insurance

Business Liability Insurance or Business Operations Policy (called a "BOP" in the insurance industry) covers a customer who may be accidentally hurt while at the facility. Other possible liability claims include damages from accidents caused by the negligence of the employees, such as faulty work that results in an accident with damages. Excluded in the liability policy are damages to the customer's vehicle when the service facility has possession of it. Also excluded are defective parts sold to the customer. However, if the part's failure caused property damage, then the damage the part caused would be covered but not the part itself.

Garage Keepers Insurance

To insure against damages to the customer's vehicle while road testing, parking, or storing, as well as when working on it, the service facility will buy **Garage Keepers Insurance.** If a customer's automobile is damaged while at the service facility, the ASM/SCs must be aware of the procedures the insurance company expects to be followed. This usually includes pictures and written notations about what happened.

Property Insurance

Property Insurance is required by many commercial building leases and banks that loan the owner money to buy the service facility building. If the building is damaged in a storm, fire, or other accident, the building damages will be covered. There are many details with respect to this coverage but of importance to the ASM/SC is following the procedures to document a loss. Pictures, notations, and lists of items damaged should be recorded immediately following the loss.

In some cases inventory, shop-owned tools and equipment, and technician tools or toolbox may be damaged in the accident. The business policy must have loss limits that are high enough to cover these items including the employee tools. If employee tool coverage is not within the policy or the limits are too low, it should be purchased. To help determine coverage limits and evaluate a loss, an inventory of tools, equipment, and parts for sale should be kept and regularly updated. In the event of a loss, this will help an insurance adjuster quickly assess the claim and make payment. Of course, a copy of the inventory list should be kept off site in a safe place so it is not damaged in an accident.

Of special interest is when a service facility technician might leave the premises to repair a vehicle. Most policies will provide a small amount of coverage for off-premises tool and equipment losses. If the manager plans to have more tools/equipment off site than the insurance limit, then the owner should consider "InLand Marine" coverage. This coverage is designed to provide higher coverage limits for tool or equipment loss when assets are "off premises;" such as a pit crew at a race track or mobile fleet repair operation.

ADDITIONAL DISCUSSION

Additional Garage Keepers Policy Details

Some customers may ask about the coverage of their automobile when they leave it at a service facility. The ASM/SCs must be able to explain the coverage. The Garage Keepers Policy coverage is for physical damage and includes collision with another object and comprehensive for any other damage such as theft or vandalism among other perils while at the service facility. However, Garage Keepers Insurance can be complicated because the service facility owner must select whether to purchase Legal Liability, Direct Excess, or Direct Primary. This affects the ASM/SC job because if there is damage to the customer's car, a Legal Liability policy pays for a loss only when the business is liable for the damages. For example, service customers had their vehicles damaged due to a hail storm. The business was not liable for the hail damage unless management agreed to put the vehicle inside and they did not. Therefore, the customers who had comprehensive coverage on their vehicles were covered by their vehicle insurance policy. Those who did not have compensation insurance on their car insurance policy did not get paid for the loss and were perhaps unhappy with the business owner. However, the business owner was not liable for damages caused by the storm and Garage Keepers' Legal Liability insurance did not have to pay the claims.

To perhaps improve customer relations, an owner may select Garage Keepers coverage that is Direct Excess or Direct Primary. Direct Excess will pay above what the customer's car insurance policy covers and Direct Primary pays a loss regardless of liability. From the storm example, a service facility with Direct Primary Garage Keepers Insurance will pay the storm damage of the customers' cars regardless of whether the customer even had car insurance. Needless to say, Direct Primary is the more expensive of the three types of coverage and commonly dealers have this coverage. While it may seem that this is the "best coverage" with the "least hassles" to the ASM/SC and customer when there is a loss. Some service facility owners object when the insurance company pays some claims because it implies they were negligent; such as a broken rear truck window caused by the customer's unsecured cargo that came loose or a customer who did not pick up her vehicle as promised and it was damaged in a storm.

Commercial Auto Coverage

Vehicles owned and operated by the service facility employees must have **Commercial Auto Coverage.** Examples of vehicles owned by a service facility include courtesy shuttle, tow trucks, owners' company vehicle, parts delivery vehicles, and so forth. Coverage is similar to a personal auto policy. However, within the Commercial Auto Coverage is "Employee-Owned Auto" coverage that is not found in a personal auto policy.

Employee-Owned Auto Coverage is important for an ASM to understand because when an employee uses his or her vehicle to pick up parts or "run an errand" the business has liability if there is an accident. This coverage is needed if an employee is traveling in his or her vehicle and has an accident which harms another person or the person's property. Without this coverage, the business can face significant legal challenges when an employee uses his or her vehicle and has an accident, even if the employee is "off the clock" or doing the errand as a "favor." This should be remembered when an employee is asked to use his or her vehicles for work-related activities.

Worker Compensation

Workman's compensation is an insurance policy that covers the costs of injuries received by the employees of the facility while performing their job. This cost may include the pay lost by the employee due to the injury. For example, an employee injured his eye while making a brake repair. The emergency room medical bill was paid by the insurance company under the worker's compensation portion of the policy. The service facility owner decided to pay the employee for the one-half day of lost wages due to the injury. In this case, the employee was injured because he did not have safety glasses on while making the repair. After the accident report was filed, the insurance company instructed the owners to require all employees to wear safety glasses when making repairs. In addition, safety slogans were provided by the insurance company who also required that they be posted throughout the facility. After several weeks passed, the insurance company representative checked to see if their requests were enforced.

ADDITIONAL DISCUSSION

Dealership Insurance Issues

Dealers often have different insurance needs than an independent service facility. They require "Dealer on Lot" coverage to insure the vehicles held for sale (called a floor plan). Further, they have "Drive Away" coverage (Garage Keepers) when the vehicle is taken for a test-drive by the customer. The dealership owner is of course covered for theft of the vehicle; however, they are not covered for theft if it is due to "Voluntary Parting." This is when the customer is given the keys to a new car for a test-drive and the customer never returned the car. There is also no coverage for "False Pretenses" such as not having a clear title for a vehicle the dealer recently traded or when a personal check did not clear (the dealer did not get paid—the check bounced) before the vehicle was titled to the customer. Voluntary Parting and False Pretenses are not an insurance claim but rather a civil court or perhaps criminal issue. For managers, there must be clear procedures for employees to follow when dealing with customers who want to test-drive the vehicle to avoid "Voluntary Parting" issues and when to title vehicles to prevent "False Pretenses."

At the same time, the dealership insurance plan for vehicles held for sale (floor plan) limits the dealer to a maximum range of 50 miles. Therefore, when a dealer must deliver a vehicle more than 50 miles, such as an Internet sale or driving the vehicle to/from an auction, it needs "Dealer Drive Away" coverage. This will extend the range of transport over the 50-mile limit so the dealer's vehicle, intended for resale, will remain covered by insurance. Failure to have this coverage will mean the damage claim will not be paid if it occurs past the maximum range of the policy.

Unwritten Agreements with Other Businesses

A common alternative to a written contract is for a service facility to have a general agreement with other businesses, such as a towing company (see Figure 14-3), to provide goods or services. When a call is made for a tow, a representative of the facility obtains an amount of the charge after a phone call. For this discussion, the PO or purchase order system that tracks these business transactions by issuing a PO number after approving the charge (PO numbers are obtained from business operations), is not used. While the following procedures in the next case study example may vary from one business to another, it illustrates the importance that certain procedures must be followed.

CASE STUDY: Business Agreements and the Customer

Assume Renrag Auto Repair calls Ace Towing to tow a customer's broken car because the customer is not a member of AAA and does not have towing insurance. Ace Towing provides an estimated cost. Next, the SC/ASM must call the customer to approve the charge after marking-up the subcontractor's (towing company) charge to cover the time to make arrangements, accepting delivery, storage, and so on. If approved by the customer, the vehicle can be picked up and delivered to Renrag. The SC/ASM would accept delivery by signing a delivery receipt.

If the customer feels Ace Towing is too expensive, the SC/ASM may inform ACE Towing to see if they wish to change their charges. If not, then the Renrag SC/ASM must call another towing company until they can obtain a charge the customer approves. If a customer will not approve any charges, then the person must make his or her own arrangements or the SC/ASM must consider it a lost sale. The basic rule is that the ASM/SC may not authorize any subcontractor services that the customer will not agree to pay.

For this reason, it is best when a customer approves a charge, the customer must sign the repair order indicating approval. When done over the phone, the SC/ASM must keep a record on the work order of the amount approved, date and time of phone approval, name of business providing the goods or services, and then initial the work order. Basically, the customer signature or a record of a phone approval provides a contractual agreement. Therefore, the SC/ASM must be very clear about the terms (cost and delivery) by repeating everything to the customer.

FIGURE 14-3 This tow truck picks up the customer and his vehicle.

Employee Bonding

For some agreements, the service facility may have to identify the person who has permission to make a call to authorize the service or perhaps part order. This is for internal control purposes. In addition, the service facility may have to identify the employee who may accept delivery. For instance, a new car delivery from the manufacturer may require a service manager to sign and check in the delivery.

The reason for the formality is because such a person may be bonded (insured) so that if the person should commit any theft, the business would be covered for the loss and the bonding company will process legal charges. Of course, in the midst of a busy day with usual complications that occurs, the temptation is to take short cuts and not worry with signatures, approvals, procedures, and so on. When the process is ignored, which is discussed in the book on *Managing Automotive Businesses*, the consequences can be unfortunate.

CASE STUDY: Theft and Bonding Contracts

An ASM (who was not bonded for theft as a result of job performance) was responsible for ordering and receiving all parts from four different auto parts stores and several dealerships. With parts and core returns, the ASM was so overloaded that he became careless in making sure all deliveries were received and recorded. Consequently, when the owner of the service facility compared the invoices for parts received to customer invoice receipts, discrepancies were discovered. Parts were ordered and charges were paid to the parts supplier but there was no record of them being sold to a customer.

After a couple of weeks of listening to the excuses of the ASM, the owner found they involved only one parts supplier. Upon visiting the manager of the parts store, the owner discovered that the parts in question were never delivered to his service facility. Instead, the parts were picked up by the service facility's ASM or delivered to the garage that the ASM owned and operated part time. Clearly the parts were used by the ASM in the repair of vehicles in the evenings and weekends at his garage. The service facility owner was sure, but could not prove that some of the customers who contacted his facility for services were redirected to the ASM's side business. The ASM was fired and since he was not bonded, the owner suffered the loss. Additional legal charges could have been filed against the fired employee; however, the owner decided it was not in his best interest given the time and effort required. The parts store owner, knowing he had responsibility under his contract with the Service Facility for delivery of the parts to authorized personnel at the service facility (this contractual requirement was not bonded), paid for some of the loss and then reviewed his employee's conduct as well as his business procedures.

Internal Contracts and Managers

Internal contracts are like external contracts except they are used when a facility has several operations within the company and the owners treat them as sub-businesses or departments. In these cases the sub-businesses are designed to show the owners that they are profitable and often have their own department managers. For example, at a dealership there is often a parts department, new car sales department, a collision repair shop, and so on. Each department manager must show his or her financial profitability yet they are encouraged by the owner to cooperate by working with each other as subcontractors. Therefore, they treat each other as a business unit or profit center. This requires that the work be done on a contract basis with a formal transfer of money internally between departments. For example, if a body/collision department has to have mechanical repairs made on a car it is repairing, then the service department prepares an estimate for the manager of each shop to review and approve. When the work is completed, the service department sends an invoice for payment to the body/collision department. In turn the body/collision department will charge its customer. This can work but negotiation is required when it comes to the details of how much the department managers will charge each other. The owner may get involved because in the end the dealership may make the required profit; however, the department managers have to cooperate and "share the profit." When a department discounts its parts or services to each other, this can affect the department's profit and even the department manager's incentive pay.

Internal Users of the Dealership Service Department

At large dealerships there could be a variety of departments using the service department. The services they would require is shown in the following examples.

- New cars sales department—The service department is required to perform new car preparation by completing the factory's inspection process as well as state safety/emission inspection of vehicles before they are sold. Technicians must remove shipping covers and install parts shipped separately, such as installing the license plate brackets. Furthermore, the technicians may be required to install any accessories requested by the customer before putting the dealer logo on the vehicle.
- Used car sales department—The service department will perform mechanical checks and repair any vehicles traded for a new vehicle or bought at an auction to be resold. These cars must be state safety and emission inspected. Some dealers have their own service bay and technicians to service used cars instead of contracting services with the service department or outside service facility.
- Body/collision shop—The service department will perform some mechanical repairs and computer-controlled diagnostics.
- Car rentals department—A service department will perform periodic maintenance on the rental vehicles.
- Car reconditioning department—The service department will perform repairs when the detailer finds a problem, such as head lamp

replacement. In some cases the Recon department will perform these "minor" repairs and use the service department to do more "major" repairs. For example, a power window that will not roll back up when the reconditioning has been completed.
- Parts department—The service department may be asked to replace a defective part that the parts department sold. This will usually be paid under a "part warranty" where repairs charges are paid to the parts department by a manufacturer of the parts. For example, a part was sold to the customer and the part was later found to be defective. The defective part was originally installed by the service department, an authorized installer. Therefore, the defect fell under the part warranty and a replacement part was provided as well as the labor was paid to replace the defective part.

As explained previously, internal contracts are and should be negotiated between the department managers. This is necessary because each is trying to make a profit. If the owner does not have negotiations, then the business may not be competitive compared to using an external contractor. This means that without any negotiations, a department may make attempts to have service department repairs made by an independent repair facility. As witnessed at one dealership, this ended the careers of at least one manager and harmed the owner's relationship with others as internal conflict caused issues that could not be resolved.

Other Duties of the ASM/SC

The types of "other duties" assumed by an SC/ASM vary depending on the size of the business and the business systems in place. A small repair facility will have different subcontractor needs compared to a large dealership. The following examples are limited to the more popular ones handled by both SCs/ASMs. They do not cover all of those performed by SCs/ASMs but offer a general representation.

Sublet Sales

The first agreements with other businesses to be discussed are arrangements for **sublet sales**. This is because service facilities are often quite dependent on the maintenance of good relations with a business to whom they sublet sales as well as a business that sends them sublet sales.

There are many types of **sublet sales** arrangements. A facility may make arrangements through a written contract or an unwritten contract, sometimes known as a gentleman's agreement. This means they will do business with each other as long as both are satisfied. For example, assume a service facility does not have alignment equipment and not enough business to support the high equipment cost and space required. Rather than buying one and causing a loss of profit, the service facility makes an agreement to have alignments performed by another business. When the service facility requires an alignment after replacing steering parts or at a customer request, the ASM/SC calls the subcontractor who can perform the alignment. The service facility will charge the customer for the alignment and will pay the subcontractor's fee. In many, but perhaps not all cases, the service facility that serves the customer and makes the arrangement for the subcontracted service adds a fee to the charge. For instance, if the alignment charge is $60, the service facility may add $20 to the fee and charge the customer $80 on the invoice.

Under an unwritten agreement, if one of the two businesses is dissatisfied with the arrangement, they simply end it. For example, assume a towing service decided to double its charges to the service facility. The service facility could simply make arrangements with another towing service. If the two businesses had a legal contract to do business with each other, the charges for towing would be in the contract and could not be increased until the contract expired.

Sublet sale arrangements may be made for a variety of services, such as windshield replacement (see Figure 14-4), muffler replacement, radiator repair, sound system installation, alignments, tires, body repairs, computer diagnostics, installation of accessories such as a sunroof, and others. Often customers are not aware that a repair was made as a sublet sale unless a warranty agreement was a part of the sale. In most of these cases, however, the warranty is provided to the service facility who arranged the sublet sale. So, if a warranty is provided, customers will return to the business who sold them the service and do not go to the business that performed the repair. The service facility that sold the repair to the customer would then go to the business that performed the repair to have the warranty work done.

FIGURE 14-4 This business was contracted by the ASM/SC to install a new windshield into a customer's car. The windshield installer drove to the service facility to install the part.

Building Repair and Shop Equipment Maintenance

Maintenance of the building and shop equipment is an ongoing challenge that cannot be avoided. For example, maintenance contracts for the service of some equipment, such as the update of the alignment rack, must be done at a time when it will not disrupt business. The ASM/SC, of course, is probably the only person aware of the time when this service can be performed. When a piece of shop equipment breaks down, such as the air compressor, the ASM/SC must be directly involved because it can cause delays and appointments may have to be cancelled.

When equipment and facility maintenance and repair service personnel come to a facility, they often go to the ASM/SC for instructions. The ASM/SC must try to insure the maintenance and repair work is done as quickly as possible and does not disturb management and shop employees (see Figure 14-5). When the work has been completed, the ASM/SC must often verify that the job was completed. When the ASM/SC is the contact person, they must understand the maintenance agreements and contracts. Examples of contracts/agreements at a facility include the communications systems, garbage collection, recycling of waste materials, the security system, TV cable for reception area, magazines, heating and air conditioning, computers, shop equipment, and lawn and landscaping care.

ADDITIONAL DISCUSSION

Independent Contractors Tests

A subcontractor may arrange for a service facility to perform jobs using their employees who may not have the time or expertise to do the work. Also some subcontractors are not really "contractors" but rather "employees" "paid under the table," often to avoid taxes. How management "handles" the contract will dictate whether they are really a contractor or an employee subject to numerous regulations and tax withholding requirements.

First, all employees are subject to state law that requires they be covered by Worker Compensation Insurance. Second, as per Federal IRS definition, an individual is an "independent contractor" if the service facility can only control the result of the work (accept or reject the work as being acceptable) and cannot tell the contractor "how" or "what" will be done. Otherwise the contractor is considered an employee to be covered by Worker Compensation Insurance and subject to tax payments.

Under some state laws, such as Pennsylvania's, the criterion is stricter, and there are more "tests" such as a written agreement between the parties for the specific work to be completed. Furthermore, contractor must have their own tools and own their business (by state definition this means the contractor can suffer a loss as well as earn a profit), and the contractor's business office (location) is not the "payee's" (or service facility) business location (essentially the same location). In terms of insurance, the contractor must have his or her own liability insurance and produce an insurance certificate before work begins.

If the person doing work for the service facility is not a contractor by the federal and the state's law definition, then he or she is an employee subject to minimum wage and workers compensation laws among other regulations. Fines can be significant for the violation of these laws and regulations. Also, from a legal liability standpoint, the risk if the "fake contractor" is hurt, does the work improperly, or a customer is injured can be staggering.

FIGURE 14-5 These technicians perform service to the drive-on lift after the daily work has been completed.

Uniforms and Cleaning of Uniforms

Uniforms for technicians and ASM/SCs may be purchased or rented by the facility from another business. When they are rented (and in some cases purchased), they are picked up and cleaned once a week. The arrangements are usually clear-cut and may go through the ASM/SC. As a result, when problems occur, the ASM/SC may become involved, for example, if a technician does not have the right pair of pants delivered, the ASM/SC is often the one who has to "keep an eye out" for the delivery truck. Likewise, if the cleaner is short a pair of pants, the delivery person must go to the ASM/SC to ask the technician where the pants are located. As a result, the ASM/SC usually has to be aware of the agreements made for the ordering and cleaning of uniforms.

At the same time, the presentation of the employees to the public at a service facility is important. At some service facilities, the ASM/SCs are provided with a uniform or shirt with the name of the facility, the name of the employee, and, sometimes, the employee's job title. While this may not seem important, helping customers recognize an employee's position can avoid confusion. For the ASM/SC, being supplied with uniforms and laundry services can be a financial advantage, especially since they will likely get oil/grease on them from handling parts.

Suppliers of Parts and Goods for Resale

Service facilities must make arrangements with other businesses from which they purchase parts and a variety of goods, such as oil and tires, to sell to customers. These arrangements are usually made by the managers, owners, or business representatives of the company and typically require an approval for credit purchases. In return, the suppliers typically sell their goods to a service facility at wholesale, as opposed to retail, prices.

The suppliers of parts and goods to a facility are referred to as **vendors**. When a state levies a sales tax, the tax is not charged for parts and goods to be resold by another business. Rather, the sales tax is collected when the service facility charges the customers for the parts and services sold to them.

An ASM/SC must understand the business relationship and policies the facility has with different vendors. For example, ASM/SC must know the return policy of the vendor and how old parts removed during service (called **cores**) are to be returned to the vendor (see Figure 14-6).

In addition, since the inventory of parts kept at the facility (such as oil filters, see Figure 14-7) may be important for production, the ASM/SC must know about the resupplying arrangements made with each vendor. While a parts specialist may order the parts, in some cases the inventory is restocked automatically. If the restocking of the parts is not done by the time the ASM/SC makes a sale, problems will occur. With respect to parts and supplies at a franchise service facility, ASM/SCs must be thoroughly informed of the arrangements setup for the purchase of parts and goods by the franchise corporation.

ASM/SCs must also learn how to obtain parts from new automobile dealerships. This is because the auto parts stores often do not have all of the parts required to fix newer automobiles. As a result, a service facility must have arrangements with the different automobile manufacturer dealerships for new automobile parts.

It is not always common for dealerships to deliver parts so the service facility must be able to pick them up. For example, if a new automobile is being repaired, the ASM/SCs must order the part (that is hopefully in stock at the dealership), and then make arrangements to

FIGURE 14-6 This engine core is ready to be secured to its skid so it can be returned for a credit from the parts vendor.

FIGURE 14-7 This cabinet in the oil-change bay contains oil filters that will be sold to customers. As filters are used, the technician rips off the box top with the part number on it. The box tops will be given to the ASM/SC at the end of the day so new filters are ordered to replace the ones that were used.

pick it up. In some cases, this affects the estimated time needed for a repair.

Petty Cash

An operational responsibility that may involve ASM/SCs is the **petty cash fund**. Although ASM/SCs do not usually manage the petty cash fund, they have to know how it works and when to use it. **Petty cash** is kept on hand to pay for items that must be purchased from a vendor that does not have a credit arrangement with the service facility. For example, a service facility may need to purchase a fluorescent light from the hardware store. As a result, the employee making the purchase can obtain the cash needed from petty cash, make the purchase, and return the receipt to petty cash. Another option is for the employee to pay for the light and then submit the receipt for reimbursement.

Petty cash accounts are not to be used for major purchases. As a result, the amount of cash placed in the account is usually $50 or less. The money should be kept in a locked drawer or special locked box and only one employee and a manager or owner should have access to it. The drawer or box should be replenished periodically, such as at the end of each month, or when the cash available gets below a set amount, such as $10. When replenished, the amount of cash left in the drawer or box plus the total receipts must equal the amount placed in the account, such as $50.

Petty cash can be a nuisance but is invaluable when a small purchase is needed or a delivery must be paid for in cash. For instance, when a delivery is made to the ASM/SC and the person making the delivery needs to be paid for postage or a fee before the package can be left at the facility, cash is needed. Without a petty cash account, the ASM/SC will either have to personally pay for the delivery or not accept it. Either option is not desirable.

Utilities

Water and electricity are essential to an automotive service facility. Usually an ASM/SC does not have any need to be involved with these services until they have a problem. In these cases, it is important to know where the water shut-off valves and electric circuit breakers are located. When a utility service is disrupted or any building repairs are made, the people sent to make a repair must be shown where the control valves and panels are located. Again, because the ASM/SC is usually the first person to be met inside a facility, they are asked for assistance.

Banking

When ASM/SCs receive payments from customers, they should know how to make deposits to the bank account. There should be two rules regarding the handling of cash at a service facility. First, there should be a limit as to the amount of cash kept in the register. Once the amount is exceeded (let's say $1,000), then a certain amount should be deposited to leave an acceptable balance (let's say $200). In the case of a large company, the money may be placed in a vault. Second, at the end of the business day, only the amount of cash needed for opening the next day (let's say $200) should be left in the register overnight. This usually requires a night deposit into a bank's drop box.

To follow these two rules, ASM/SCs must fill out bank account deposit forms and a daily earnings report, and then make deposits into the facilities account. When a manager or cashier is employed by a facility, they usually handle the reports and deposits. However, cross-training is a key so employees can cover for each other. Therefore, the ASM/SC is often the backup for the cashier.

Advertising and Promotions

When advertising and promotions are prepared, ASM/SCs are often asked for input. An ASM/SC will be asked whether he or she thinks customers will buy a combination of specials at a certain price. The ASM/SC must be included because a customer will not just buy a special because they saw an advertisement. The ASM/SC must know the special and take time to explain the benefit of the special (see Figure 14-8) to the customer before asking if they wish to buy it. In addition, when specials and advertisements are being prepared, the ASM/SC may also be asked for his or her opinions on the presentation of the

FIGURE 14-8 As customers enter the service facility, they see the specials advertised. Since the specials were advertised using several advertising mediums during the week, this may be a reminder to them. As the customer meets with the ASM/SC, the specials have already been introduced before the ASM/SC discusses them with the customer. For some "in-store" advertisements, smartphone technology allows customers to learn more about the offer before they even reach the service desk.

information; such as the layout and the words being used. An ASM/SC may even be asked to check the original before it goes to print. Finally, after the advertisement or promotion is released, phone calls from customers asking about information in the announcement will come to the ASM/SC.

Since ASM/SCs answer inquires about the services offered by a facility, their opinion and observations are logically sought out. As a result, to make these promotions successful, ASM/SCs should always make notes about what customers want, like, and dislike. Further, to successfully answer a customer's questions about a particular advertisement, the ASM/SC should always have a copy of the ad handy.

General Supplies

ASM/SCs must keep an inventory of office and other supplies on hand and they may be required to order them from a supply store. For example, they must be sure to have extra appointment books on hand, estimate-repair forms for the computer printer, printer cartridges for the computer, pens (sometimes with chains so they do not "walk away"), paper, credit card tapes (when they collect money from customers), phone message pads, tags to put on automobiles waiting to be serviced, key tags, business cards, and on and on. In addition, at some smaller independent and franchise facilities, the ASM/SC may be expected to monitor the first aid kit as well as supplies needed in the restrooms.

In some cases as per the last task in the job description shown in Chapter 2 ("other duties as assigned"), ASM/SCs are responsible for maintaining the customer waiting area. This may include making coffee and even, in some cases, having pastries on hand in the morning. When the coffee detail is assigned to the ASM/SCs, they must make sure the proper supplies are on hand and that the room is neat and clean.

In other words, one of the first things ASM/SCs must do after being hired is to determine what supplies they must monitor and how they order them. Next, they must check the inventory to see how much is available. Finally, they must bring their inventory supply up to an appropriate level. As the supplies are being used, they must then monitor their use. Running out of some supplies can be terribly disruptive.

Security

Most automotive repair shops have some form of a security system. At Renrag Auto Repair, located in a small city, there were two security systems. The first was on the doors and windows and was monitored by a private security firm. The second was a set of video monitors at the ASM/SC's work area connected to cameras in the service bays. The ASM/SCs had to know how to turn on and off the door alarms and TV system when opening and closing the facility. They, as well as the other employees, also were expected to keep an eye on the monitors to identify when non-authorized people were in a service bay.

Security is important to a service facility for several reasons. One is that they have money, expensive tools, equipment, and supplies to protect. Protection is needed both during the night when the facility is locked up and during the day when it is open. Automotive facilities are particularly vulnerable to theft because the large doors to the service bays are often open (see Figure 14-9). In addition, a number of emergency exits must be located throughout the building. On hot days these emergency doors may be opened for ventilation. At some facilities, a plastic yellow chain is placed across open doors to keep unauthorized visitors out of restricted areas.

Another reason for a good security system is to reduce insurance costs. For example, customers are not permitted in service bays. Insurance companies usually require signs to be placed on all bay doors stating that customers are not permitted in the work areas. In addition, to protect the insured equipment and tools, they like to have alarms placed on the doors and windows. Further, they are quite pleased when the bays are being further protected through the use of a TV security system. In some cases, a good security system can reduce insurance premiums.

FIGURE 14-9 These large open garage doors are inviting for customers to come and "do business" at this service facility. However, the ASM/SC must be vigilant to watch for unwelcome visitors who desire to cause problems such as stealing tools and equipment.

A note on the TV monitors for the service bays regards the interest they generated from customers. At Renrag Auto Repair, the TV security monitors were located at the ASM/SC workstation. Because the service bays could not be seen from the waiting area, many customers appreciated the opportunity to watch the technicians working on their automobile. They could watch it being brought into the service bay, how the work was proceeding, and when it was completed. They knew the repairs and maintenance work being conducted was, in fact, taking place. Given the popularity of web monitoring, these systems are now connected to the Internet. Managers and even customers can log into a service facility's website to "watch" their car being repaired no matter where they are located.

Customers under Contract

Some service facilities have contracts to service the automobiles of other businesses. These contracts vary in terms of the different services the facility provides to the business. In any case, the type of services directly impact the "other duties" assigned to the ASM/SC.

One of the services that an automobile facility may agree to provide to another business may be to pick up, service, and deliver the company vehicles. As a result, the ASM/SC may have to call the business to schedule the maintenance appointments. In some cases, the maintenance services may be provided in the evening or a weekend. This requires the ASM/SC to make arrangements for drivers and technicians to work late or on the weekend.

Contracted services can also require ASM/SCs to make sublet sales arrangements with another business. For example, if a service agreement includes emergency repairs, the ASM/SC may have to call a towing company to pick up a vehicle. In other words, service contracts with other businesses can add another dimension to an ASM/SC's position. ASM/SCs must know what services are to be provided and then have a plan to provide them. Most importantly, service facilities and ASM/SCs must not promise something they cannot deliver.

Review Questions

Multiple Choice

1. In most general repair shops, which of these is *least* likely to be a sublet operation?
   A. Windshield replacement
   B. Spark plug replacement
   C. Automatic transmission overhaul
   D. Drive shaft balancing

2. A vehicle is being serviced under an aftermarket warranty contract. ASM/SC A suggests the contract should have been read by the customer and to just do the work without the ASM/SC reading it. ASM/SC B asks the customer to bring in the warranty contract to be read carefully as well as any related repair records to help the shop get an idea of the vehicles history before starting the repairs. Who is right?
   A. A only
   B. B only
   C. both A and B
   D. neither A nor B

3. ASM/SC A says that the liability insurance covers mistakes made by the technicians that caused a customer to be injured. ASM/SC B says that Garage Keepers insurance is for damage to the customer's vehicle when test-driving it. Who is correct?
   A. A only
   B. B only
   C. both A and B
   D. neither A nor B

4. ASM/SC A says that like automobiles, maintenance schedules for shop equipment are printed in the "owner's manual." ASM/SC B says that like automobiles, maintenance schedules should be selected based on the use of the equipment. Who is correct?
   A. A only
   B. B only
   C. both A and B
   D. neither A nor B

5. Which of the following is true?
   A. ASM/SCs must know the return policy of the vendor and how old parts removed during service (called **cores**) are to be returned to the vendor.
   B. ASM/SCs must know how to obtain parts from new automobile dealerships. This is because the auto parts stores often do not have all of the parts required to fix newer automobiles.
   C. Petty cash is kept on hand to pay for items that must be purchased from a vendor that does not have a credit arrangement with the service facility.
   D. All of the above

Short Answer Questions

1. Provide examples of sublet sales and the different methods to arrange for a sublet sale.
2. Describe the various "other duties" an ASM/SC may have to perform.
3. What is petty cash used to purchase?
4. Provide an example when a service facility will use each of the following:
   a. liability insurance,
   b. garage keepers,
   c. workman's compensation.
5. Explain the purposes of security systems at a service facility and how they might be designed.
6. Explain the difference between an internal and external contract and provide examples.

Activity

Activity 1: Obtain a vehicle warranty contract from a dealer or an example of an aftermarket warranty from the Internet. Read it and determine what the ASM must know to provide high-quality service to the customer. Write a memo to a fictitious service manager about the "most important" contract terms the business needs to know to do the work for the customer. Important contract terms might include authorization procedures, payment, any limits on the types of repairs that can be performed, deductibles, labor rate "maximum" allowances, and so forth.

Activity 2: Obtain a vehicle lift "owner's manual" for your school lift(s). If not available, get one from the lift manufacturer's website or lift vendor. Read it and determine what inspection and maintenance services are recommended and how often. Research the OSHA website (type in Automotive Lift Inspection into the search box) and also examine the Automotive Lift Institute (ALI) website for lift inspection and maintenance. Determine what they recommend for lift inspection. Summarize your findings in a report with a plan for a "shop maintenance program." Conclude your report with an explanation about why the plan would be important for management to implement and whether a contract might be used with an independent inspector to do the work. If a contract is part of the plan, what would management be advised to put in the contract terms. Think about when they would inspect (after closing?), how frequently to inspect, what is inspected (which items and to what standards), and so forth?

Activity 3: Repeat activity 2 for other shop equipment your instructor assigns such as compressors, alignment equipment, frame "straightening" equipment, spray booths ventilation systems, and so forth.

CHAPTER 15

SERVICE MANAGEMENT: TRACKING EFFICIENCY AND IMPROVING EFFECTIVENESS

OBJECTIVES

Upon reading this chapter, you should be able to:

- Explain how tracking efficiency and improving effectiveness is related to productivity.
- Outline steps to set completion expectations (A.2.4).
- Define effectiveness and efficiency.
- Explain how effectiveness and efficiency are related to productivity.
- Explain why customer expectations, effectiveness, and efficiency are critical to a business (A.2.8).
- Describe the basic requirements needed to establish and maintain a positive work environment.

CAREER FOCUS

Your management skill goes beyond planning, organizing, leading, and controlling the service facility. You strive to create a "culture" of communication and respect. You admire hard work and praise innovation. Even though the technicians' job is to fix the vehicle, when they are "stuck" you already thought about it and have a plan to help them solve the problem. Likewise the service consultants' job is to work with the customer and make the service system function. When they have problems, you already identified the resources that can be deployed to help make a bad situation better. Therefore, you know your job as manager is not to fix the cars or make the service system function. Rather, as the manager, you take responsibility that "the customer car gets fixed on time and on the first visit." Anything less is your fault!

You know it is hard to take responsibility for mistakes because you can't know every detail that happens during a workday. However, as a manager, you are the leader and no excuses are acceptable. The only acceptable answer for a mistake is to apologize, learn from it, and then figure out how to fix the problem. Therefore, every problem is an opportunity to get information and use it to figure out how to solve the issue so it doesn't happen again.

When the culture of the service facility is based on communication and respect with systems for employees to follow, problems should be fewer and easier to solve. Also systems for employees to follow can give you time to solve problems, plan, and improve systems. Specifically, you obtain information and develop plans about how to help your employees succeed. You know that if you fail to plan, your service team will fail and there is never an excuse for management failure. However, there is always a reason for team success—great team members who prepared to meet a challenge and executed a solid plan!

Introduction

Managing a Customer Service System and Shop Production Operations that is effective (meets customer expectations) and efficient (enough profit is earned to stay in business) depends upon the work environment. If the work environment is positive, it is one of the reasons why people will keep working for an employer; if not, the employees will leave.

One of the responsibilities of an ASM/SC is to create and maintain a positive environment with the objective of generating quality work and high productivity. The purpose of this chapter is to discuss the connection between the work environment and service facility performance and how management can influence them.

Productivity, Effectiveness, and Efficiency

When a skilled and experienced technician works on an automobile and all goes well, he may do the work faster than estimated by the labor guide. This causes productivity to go up! When a job does not go well and problems arise or when a less-skilled and experienced technician works on a job, it may take longer than the labor guide allows. Productivity declines! When a job is done in less time than estimated, the profit for a business will increase. A job that takes longer than estimated will reduce profits, possibly to a point where a loss occurs.

Whether a service facility is productive depends on the effective and efficient performances of the ASM/SC, parts specialist, and technicians.

- **Effectiveness** requires the estimates, ordering of parts, communication among team members, and the work conducted to be done correctly. Errors reduce productivity, cause customers to be unhappy, and cost the business money!
- **Efficiency** regards whether or not resources of the facility (including labor) are used properly and to their best advantage. For example, too much time to process repair orders, order and receive parts, and complete a job reduces productivity, which reduces the profit of the business.

Efficiency and effectiveness are connected. When a job is done effectively but it takes too long (not done efficiently), it will cost the business more money to do the job than it should, resulting in lost profit. If a job is done quickly and efficiently but not effectively because it has to be done again, the business will lose money. Therefore, jobs have to be done both effectively and efficiently. For a service facility to make a profit, more jobs must go "right" (effectively and efficiently) than go wrong. Essentially as a concept:

Efficiency + effectiveness = required profit (objective target)

Tracking Effectiveness and Efficiency

Effectiveness (job is done correctly) is often tracked by customer satisfaction (survey rating) after the repair, whether the vehicle had to return for a second attempt (comeback), and whether the job was done on time at the price promised. These are common measures of effectiveness. Efficiency is commonly tracked by comparing the billable hours of the job (flat-rate hours) relative to how long the technician took to complete the job. This common measure of efficiency will be discussed later in this chapter.

To track efficiency a shop may use nothing more than a piece of paper with a column for the repair order number, billable hours (flat-rate hours), and clock hours (actual time on the job). Then in the final column is the difference between the two times (figure 15-1). Either it took longer than expected to do the work or less than expected.

Tracking Technician Efficiency

Some shops may use a paper tracking system called **time tickets**. The following illustrations show the sequence of "how they are used." Once the paper system is understood, it is easy to understand how shop software systems perform these basic functions automatically. To start a time ticket, the technician writes his or her name and number on the top of the ticket, such as seen in figure 15-3. The time tickets are located next

Technician:	Hank			
Day:	Wednesday			
Job	Billable	Clock	Time	
R.O. #	Hours	Hours	Difference	
3458	4	3	1	Job took less time than expected
3562	1.5	1.5	-0.5	Job took more time than expected
3565	0.3	0.2	0.1	
3569	1.6	2	-0.4	
3577	1.7	1.3	0.5	
TOTAL	9.1	8	**0.7**	Hank made more than his clock hours

FIGURE 15-1 This is an example of how efficiency can be tracked for a technician using a spreadsheet. More detail can be added to the sheet such as the type of repair performed for each job among other information a manager will find important to help improve a technician's performance.

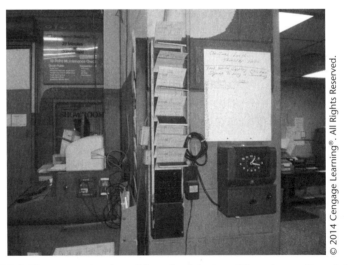

FIGURE 15-2 Shown are the time tickets for this shop. There is one time ticket for each technician and a new one is started each day.

to a time clock (figure 15-2) so the technicians can "punch on" when they start a job and "punch off" when they finish a job (figure 15-4). The series of figures from 15-3 to 15-9 found below show the sequence of events for processing time tickets.

FIGURE 15-3 This technician is assigned a repair and will write on the "**time ticket tag**" his technician number, the repair order number, and customer name. He will start with the tag at the bottom then move upward to the next tag when he starts the next job.

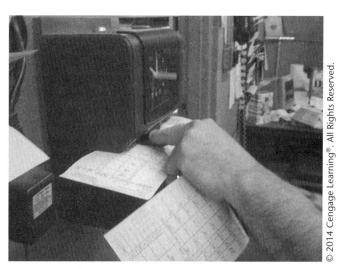

FIGURE 15-4 After filling in the time ticket tag, the technician will punch "on" the time ticket as well as the repair order before he starts the job. When finished, he will "punch off" the time ticket and the repair order using the time clock.

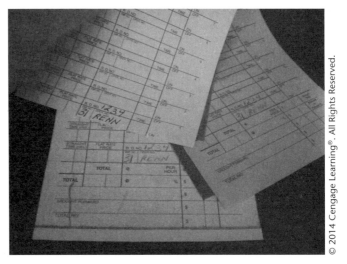

FIGURE 15-5 The time ticket has three layers. A top layer has "tags" that are peeled off later by the ASM/SC and stuck to the back of the repair order. A second layer has carbon paper on top to record all of the information written on the tags. The bottom layer has carbon paper on top of it to record all of the information written on the tags.

FIGURE 15-6 This time ticket has the punch times as well as the billable (flat-rate) hours recorded for this job.

CHAPTER 15   SERVICE MANAGEMENT: TRACKING EFFICIENCY AND IMPROVING EFFECTIVENESS   333

FIGURE 15-7 This service consultant has removed the top layer time ticket tag and is pasting it to the repair order hard copy.

FIGURE 15-8 This service consultant prepares to review and total a technician's time ticket.

FIGURE 15-9 Example of a completed time ticket. When a technician works eight hours and has flat-rate hours greater than eight hours, his or her efficiency is over 100%. When a technician has flat-rate hours that are under eight hours, the efficiency is under 100%. This technician earned 9.8 hours for working 8 hours for 123% efficiency.

After the technician finishes the job, he or she will punch off the time ticket as shown in Figure 15-4. The time ticket and the repair order are given to the ASM/SC. The ASM/SC will record the billable hours for the entire repair order on the top layer tag (figure 15-6). This of course is recorded to the lower two layers for the service manager and the technician to use as feedback (figure 15-5).

After the billable hours (flat-rate hours) are recorded, the top layer "tag" is peeled off by the ASM/SC then pasted on the back of the repair order (figure 15-7). This tells the manager how much will be paid to the technician, assuming the technician is paid by flat rate, to perform all of the work on the repair order.

Punch times on the back side of the RO can indicate the amount of time to complete each operation on the repair order. (See additional discussions for Multiple Operations on a Repair Order and Efficiency.) Also on the back of the repair order will be the technician's comments and specifications or readings taken during the service. The technician, especially for warranty repairs, will write down the repair process followed as well as notes about other issues to point out to the customer such as a problem found during inspection.

At the end of the workday, the ASM/SC will remove the remaining top layer paper and total the flat-rate hours generated for the technician on the second layer (figure 15-8 and 15-9). The ASM/SC may also calculate the technician's efficiency for the day. The second layer will be given to management for analysis and the third layer will be given to the technician as feedback about his or her efficiency for the day (figure 15-9).

ADDITIONAL DISCUSSION

Multiple Operations on a Repair Order and Efficiency

The time ticket is used to signify the start and end of *all* of the work found on a repair order (RO) such as RO #1234. However, a repair order, particularly a warranty repair order, may have several operations on it. For example, there may be three operations, a headlight to install, a door window that won't roll down, and a piece of trim that is loose. When a technician starts an operation on a repair order, such as a replacing the headlight, the technician will "punch on" the repair order and then "punch off" the repair order when the installation is finished. He will immediately "punch on" the RO again to signify the start of the next operation, such as repair of the door window. When finished with the diagnosis or repair, he then will "punch off." Of course, this sequence is repeated for the final repair to fix a loose piece of trim.

The times that each operation requires is for management and perhaps the warranty clerk to analyze. They must determine whether too much time was spent on an operation relative to what is paid by warranty or charged to the customer.

When the amount of time spent on an operation is too long relative to the amount charged, there is an efficiency problem. This can cause a loss to the business because the technician's pay (assume an hourly paid technician for illustration purposes) can be higher than the amount of money charged or paid. When the amount of time spent on an operation is too short relative to the amount charged or paid (say two hours for a job that paid four hours and was done twice as fast as the job paid or 200% efficiency), there might be a question about the quality of work. In either case, the ASM or Service Manager may want to investigate. Too much time spent on a repair may mean the operation's flat-rate time is too low to earn enough profit. Too little time spent on a repair may mean the technician discovered a short cut that is both safe and increases efficiency. So the investigation by the service manager may uncover positive as well as negative findings that should result in improvements for the business.

Shop Production Computer Programs

The process of assigning the technician work and then recording when he or she started and finished it has advanced with technology. A "paperless" service facility will have a computer in each bay (figure 15-10) or a tablet assigned to each technician that is networked to the ASM/SC. Repair orders are sent to the technician once he or she has finished the previous job. The technician will enter into the computer when he or she starts and finishes each job (or operation of each job). Automatically the computer can calculate efficiency, keep track of the flat-rate hours generated for each RO, and record that information into various databases for management. This system even allows the technician to record his or her comments or readings to the repair order file and send messages to the ASM/SC's computer. The technician also has access to past repair orders to review a customer's history and the computer is networked to obtain technical information or even contact a technical hotline with a question.

Influences of Effectiveness and Efficiency on a Productive Work Environment

When a facility is not effective and efficient because jobs do not go "right," productivity declines. When productivity declines, profits turn into losses.

FIGURE 15-10 A dealer service bay with a networked computer that will automatically process repair orders (paperless system) for the technician assigned to the bay.

The work environment will likely become negative or even hostile. Eventually, the environment and losses threaten the future existence of the facility.

Therefore, the first requirement for a positive environment is for a facility to become more effective and efficient. If productivity and profits improve, worker security will improve and morale typically improves as well. To promote effectiveness and efficiency, ASM/SCs must monitor the activities and operations that can influence them. Throughout this book, different methods to monitor operations and measure productivity have been discussed, for example, the use of repair order tracking sheets, the flat-rate system, customer satisfaction surveys, keeping a record of comebacks, customer service scripts with active delivery, and so on.

Improving Effectiveness

When an ASM/SC has a concern about the quality of work produced by the employees, the solution is for everyone to do his or her job correctly. This means adhering to company procedures, using equipment properly, and following factory repair processes. Of course, in the case of educated and skilled employees, their effectiveness depends upon their knowledge and ability to do their job, especially the technicians. This, in turn, means the facility must hire good people. Recruitment, selection, induction, development, and appraisal (evaluation) are part of the second book the authors wrote on *Managing Automotive Businesses*. The assumption here is that the facility has hired employees who can do their job correctly. Therefore, the challenge is to verify that all work is being done effectively.

The most obvious evidence of an ineffective performance is when a customer's automobile has not been properly serviced. Worse it has been damaged from an improper repair at the facility. When a repair has to return for a second repair attempt this is called a "comeback" at many service facilities. Comebacks represent the most serious problem caused by ineffective performances. There may be a number of reasons for a comeback and the ASM/SC must thoroughly review the possibilities. For example, an employee may not have done a job correctly or perhaps was not certified (qualified) or experienced enough to perform the job and made an error. On the other hand, as discussed in the previous chapters, the comeback may be due to the failure of a part placed on the car. In this case, a parts warranty should cover the costs. A comeback could even be due to the customer refusing to have recommended work done at the time of a repair and hopefully, the customer did not refuse the work because the ASM/SC did not present the problem properly to the customer.

Improperly presenting a problem to the customer is another effectiveness concern that can occur when a facility does not identify or gain permission from a customer to perform all of the services needed. To have an automobile leave a facility with a problem that was not found represents an ineffective performance on the part of the technician and the reason must be identified by management. To have a customer leave a facility with an automobile that has a problem found by the technician but not properly called to the customer's attention for consideration also represents an ineffective performance on the part of the ASM/SC.

From the customers' standpoint, the examination of the automobile is important because the inspection adds value to their visit. Actually, many customers seek out and have their maintenance work done at service facilities where a qualified technician can do more than just the maintenance requested: they can inspect the automobile, as opposed to a facility where the maintenance work is conducted by a worker who does not have a technical background. Because this inspection does not cost the customer anything and will likely exceed his or her expectations, it is referred to as a **value-added service** (see Figure 15-11). Therefore, when customers have their maintenance done at a service facility, a technician using a check sheet presented in chapter 4 should conduct the value-added inspection. Customers, of course, hope that nothing wrong will be found with their automobile but, if there is a problem, they typically appreciate the information that will help them keep their automobile in proper working condition.

Value-added service inspections require a qualified (effective) technician to look over each automobile carefully. Technicians who do not have the ability to examine automobiles carefully and assist the ASM/SC in generating value-added service that may result in additional sales must be trained to do so. Technicians who want to perform only the work listed on the repair order and will not look over the automobile should be used as parts installers.

When ASM/SCs recognize the technicians who they can rely on to properly conduct the inspections to identify potential problems that need repair, they should select them to form a team. The technicians and ASM/SC should set procedures so that the ASM/SC is immediately informed of any problems discovered along with the recommendation of the technician. The ASM/SC must then inform the customer as soon as possible about the problem, refer to the repair categories presented in chapter 9

FIGURE 15-11 During a customer's routine service this technician noticed something that could be a potential problem. The technician takes a moment to inspect the issue further. This is an example of a value-added service that was possible because a qualified technician was assigned to the job.

to describe the severity of the problem, and then use the feature-benefit method of selling also presented in chapter 9. If the customer approves the repair, the repair order tracking sheet for the third phase will be followed. When the customer declines the repair, the technician's notations should be clear so the ASM/SC can write a concise statement to be included on the final invoice to warn the customer of the problem found.

Service Consultant/ASM "Up Sales" Effectiveness

The repair order tracking sheet has data to determine the up sale effectiveness of the SC/ASM in the Third Phase. This is done by comparing the Additional Hours Recommended in Row 18 to the Hours Approved in Row 22. This can be done for each job or can be examined over a period time, such as the end of the day, week, or month. For example, consider the five sales on Friday's sheet in Figure 15-12. The total of the Additional Hours Recommended is 5.6 hours in the Second Phase (Row 18) and the Hours Approved in the Third Phase (Row 22) is 4.5 hours. The closing percentage of the Sales Approved compared to the Sales Approved is 80.4%. Presented another way, the SC/ASM closes on four out of five technician recommendations. (Calculation: 4.5 hours approved/5.6 hours possible × 100 = 80.4%; 80% out of 100% is the same as closing four out of five recommendations.) Whether this is good or not depends on comparing this to other data, perhaps against other SC/ASM percentages, other shops SC/ASM percentages, or even past data of the same service consultant to determine improvement over time.

Within the Repair Order Tracking Sheet is data about the hours recommended by the technicians to the SC/ASM to sell. Examination of Additional Recommendations in Row 18 and the technician's identity in

	A	B	C	D	E	F
1	SERVICE CONSULTANT	PS = Parts Specialist		SC= Service Consultant		
2	REPAIR ORDER	RO= Repair Order		Tech=Technician		
3	TRACKING SHEET			TL = Team Leader		
4	**FIRST PHASE - check in**	TODAY IS:	Friday			
5	Customer Last Name	Fosko	Pyle	Dame	Byrne	Stewart
6	RO #					
7	Vehicle year					
8	Vehicle Make					
9	Vehicle Model					
10	Time Promised or Waiting					
11	Initial hours sold	0.9	1.2	0.5	0.3	1.5
12	Time SC gives RO to TL and PS					
13						
14	**SECOND PHASE - initial work**					
15	Text message from Tech: TIME started job					
16	TECH Name	Bill	Bob	Sally	Juan	Kevin
17	TIME Tech reports in person to SC with RO					
18	Additional Work (Hours) recommended?	0	0.5	3	1.1	1
19						
20	**THIRD PHASE - additional work**					
21	TIME SC got Customer Approval					
22	Hours approved (change time promised)	0	0	2.6	1.1	0.8
23	Time SC gave updated RO to Tech (or TL)					
24	Time SC placed part order with PS					
25	Estimated delivery time of ordered parts					
26	TIME PS texts (SC+TECH) pick up parts					
27	Text from Tech to SC additional work started					
28						
29	**FOURTH PHASE - Check Out**					
30	Time Tech reports in person to SC with RO					
31	TIME PS supplies parts receipts to SC					
32	TIME final INVOICE has been completed					
33	TIME customer notified that vehicle done					
34	TIME vehicle Picked up					
35	SC was able to provide Active Delivery?					
36	Thank you note / survey sent to customer?					

FIGURE 15-12 This Repair Order Tracking Sheet has data that can be analyzed for SC/ASM effectiveness.

	A	B	C	D	E	F
1	SERVICE CONSULTANT	PS = Parts Specialist		SC = Service Consultant		
2	REPAIR ORDER	RO = Repair Order		Tech = Technician		
3	TRACKING SHEET			TL = Team Leader		
4	**FIRST PHASE – check in**	TODAY IS:	Friday			
5	Customer Last Name	Fosko				
6	RO #					
7	Vehicle year					
8	Vehicle Make					
9	Vehicle Model					
10	Time Promised or Waiting					
11	Initial hours sold					
12	Time SC gives RO to TL and PS					
13						
14	**SECOND PHASE – initial work**					
15	Text message from Tech: TIME started job					
16	TECH Name	Bill				
17	TIME Tech reports in person to SC with RO	1:15pm				
18	Additional Work (Hours) recommended?					
19						
20	**THIRD PHASE – additional work**					
21	TIME SC got Customer Approval	1:50pm				
22	Hours approved (change time promised)					
23	Time SC gave updated RO to Tech (or TL)					
24	Time SC placed part order with PS					
25	Estimated delivery time of ordered parts					
26	TIME PS texts (SC+TECH) pick up parts					
27	Text from Tech to SC additional work started					

FIGURE 15-13 This Repair Order Tracking Sheet shows that it took the service consultant 35 minutes to do his or her job and contact the customer for approval.

Row 16 can help determine which technicians are participating in value-added service and providing recommendations to the SC/ASM. Among the five technicians, it can be seen that Bill did not present any up-sale recommendations. If this trend is found to be common, it may require investigation by management or additional analysis from the sheet's data. For example, Bill may have been working on a new vehicle with low miles and it did not need anything. Therefore, not recommending any additional work is probably appropriate. However, Bill may not be participating for other reasons that management must determine why.

Service Consultant Efficiency

Assuming the service consultant obtained approval for the repair and will enter the third phase, the time the service consultant got approval would be entered into Row 21. A measure of efficiency (figure 15-13) on the part of the service consultant would be that the time between Row 17 (time the technician reports to the SC with the RO and additional work needed) and Row 21 (the time it took to look up the labor times, calculate the price, and obtain the approval). If the service consultant has done a good job at obtaining all of the customer's contact information and has efficiently looked up the labor times and calculated the parts prices, then communicating the information to the customer, should take minimal time. For a single additional finding, such as a brake lining replacement, this might take less than half an hour from start to finish.

Technical Training

Training technicians is one means to increase effectiveness. When technicians obtain training, they can learn to do a wider variety of jobs with fewer mistakes, and learn technical details that will help them diagnose problems more accurately. Technical training can take many forms: one is in-house training where one technician shadows another

---

CAREER PROFILE

Technical Training and the Job of the Trainer

For some in the automotive industry, they will become technical trainers and developers of training programs. How someone becomes involved in technical training typically relies on his or her presentation skills and the ability to use technical writing skills to help create training lessons. Also a trainer has to have good self-management skills that can be transferred to the classroom. A trainer must have a solid understanding of technical concepts and the ability to relate them to automotive systems with a willingness to keep up with current technology. The way training is delivered to students varies but the formats commonly used are face to face (classroom), "video" training, or by web-based format.

Depending on the company, the qualifications to become a technical trainer vary. Factory trainers for the manufacturers often require industry work experience, industry certifications, and college degree(s) (AAS and/or B.S. degree). Other companies may rely mainly on industry experience and various industry certifications. In a few cases a person may deliver technical content as an instructor at a secondary school or college. Often these jobs require a state teaching license obtained through a university or other special certifications as well as an array of degrees that may include those beyond a B.S. degree. The exact requirements vary by state and institution. For those who are technical trainers, curriculum developers, and automotive educators, the job can be very rewarding.

technician to learn "how to do something." The training may be an inexperienced technician learning from an experienced technician, such as a new person to the industry. In-house training can also be a technician learning from another technician a new skill, such as an experienced technician learning a new skill to rebuild transmissions from another technician.

To help with theory and operation of a manufacturer's specific system, there are online training programs used to train technicians. Manufacturers commonly require dealership technicians and service consultants to complete training online. This helps provide the knowledge the technician and service consultant need to do their jobs. In addition, there are classes led by instructors who teach technicians knowledge and then provide hands-on training in a lab. Videos found on websites can demonstrate how a repair is performed. However, be careful of the quality of information obtained from website demonstrations—it may contain errors or plagued with safety issues!

Improving Efficiency

Efficiency regards the economical use of resources, especially human resources. The use of employee time is a primary concern for ASM/SCs, especially for the technicians. The improper use of technicians is economically damaging and, therefore, is evidence of inefficiency. For example, assume the ASM/SC must ask someone to pick up a part at a local dealership. Assume the consultant could either send a highly skilled technician or a lower-skilled and lower-paid maintenance technician. Although the skilled technician can check out the part before bringing it back to the facility, the maintenance technician should always be sent. If a skilled technician is sent and a customer comes in for a diagnosis and repair, a sale could be lost. If the sale is not lost, the customer will have to wait until the skilled technician returns. This is an inefficient use of a resource, which is the technician's time.

A general rule is that skilled technicians should never be given work maintenance technicians can perform. Of course, if a facility has only maintenance work, the skilled technicians should be used to get the work out the door. While assigning the right work to the proper technician is important, the idea is to keep the technicians "turning wrenches." When the wrench "stops spinning," the shop stops making money. This means the SC/ASM must schedule enough work, encourage technicians to perform "value-added" inspection for up sales, and promote attractive service specials. All of these efforts help to keep the shop productive with the "right kind of work" to earn a profit.

Finally, the SC/ASM must constantly monitor work as it goes through the three systems (Customer Service, Shop Production, and Business Operations) and they must pivot from one system to another as new information becomes available. Acting as a pivot to keep each system accurate is critical to efficiency. When a SC/ASM loses an RO, is unable to find car keys, forgets to call a customer, or allows parts invoices to pile up without processing them onto the invoice; costly delays occur at several points within each system. They can hamper the flow needed between systems and make the service facility look bad in the eyes of a customer who may be watching. Learning an owner's systems and, if few are present, at least thinking in terms of systems, is complex and critical to efficiency.

Monitoring Shop Volume

Volume of customers served is an important measure to understand. Its application is helpful as discussed later for scheduling work and setting improvement targets for technicians, teams, and the shop as a whole. The basic idea is to compare the volume of the shop in terms of the number of customers served to the amount of time spent to service the vehicles. For example, a fast lube shop with a couple bays might have a busy day with 75 customers with each requiring less than a half hour per vehicle to service. A business that overhauls transmissions may serve only six customers per week with each requiring over 10 hours each. Another example is to compare the customer volume to the flat rate labor hours produced. In this case the ASM/SC should total the flat-rate hours produced by the technician, team, or shop and divide it by the number of vehicles repaired (or the number of ROs if there is one RO per vehicle). For example, 40 flat-rate hours and 20 cars = 2 hours per vehicle (40 hours/20 cars = 2 car/hr).

The hours per vehicle calculation can be employed over a short period of time (daily) or weekly, or longer. The longer the period, assume a the less daily work fluctuations will influence the average. This calculation can also be used for each technician as well as for a team, or a shop as a whole. Application would include comparing the numbers to each other for identification of improvements for individual technicians, teams, or the shop as a whole. For example shop had a 1.1 hours per vehicle ratio and management wanted to implement a "value-added" inspection program coupled with appropriate specials to increase the shop by 0.3 hours to 1.4 hours from 1.1 hours. This ratio would be monitored over time to see if there are improvements over time and connected to more sales and hopefully greater profits. Similarly, this ratio can be used to examine individual employee performance and formulate plans for improvement with targets and perhaps incentives.

Another application of the hours per vehicle calculation would be to help an ASM/SC fill empty time in a daily schedule. For example, assume there are four hours of time in tomorrow's schedule that does not have any work. If the average hours of work conducted per vehicle is 1.5 hours (2 cars × 1.5 hrs per car = 3 hours), it is likely that two jobs with unknown repair needs can be scheduled. An ASM/SC perhaps could schedule three vehicles but the risk is going over the four hours of time available (3 cars × 1.5 hrs per car = 4.5 hours). Of course, if the repair needs are known, such as a brake job that will require four hours, then this method is not needed—just schedule the job and don't take in any more work for tomorrow.

Employee Input for Improvement

Employee feedback and suggestions for the improvement of efficiency and effectiveness should be encouraged. Each employee's financial future rides on service facility performance to keep his or her job. Each employee, therefore, has a vested interest in making the business more effective and efficient. Some companies believe that employee feedback is so important that they formally survey employees on a regular basis and conduct interviews regularly. In some of these cases, groups of employees are brought together to identify ways the business can be improved.

The purpose of employee surveys, interviews, and group meetings is to identify current strengths or weaknesses, emerging opportunities, and potential problems. This is important because the immediate as well as the long-term survival of the service facility depends on ideas that can be implemented as a plan to continuously improve the business. For example, when a problem is identified, employee participation to help solve the problem can be beneficial. In addition, the employees often become more sensitive to any activities that can hinder the effective and efficient operations of the business. Not only might they advise management of the problems they see but they can also often come up with solutions to fix them.

Customer Expectations

A positive work environment is also dependent upon meeting the expectations of customers. As stressed in the previous chapters, when customers are satisfied, they are more likely to return for additional service and perhaps recommend the facility to their friends. Their satisfaction is important because facilities depend on repeat customers, as well as a steady stream of new customers, to maintain profitability. One of the best sources of new customers is from the referrals of current customers.

While it can only be assumed that satisfied customers will actually return for future business, it is known that if they are unsatisfied, they will probably not return. Therefore, identification of unsatisfied customers can be more important to determine what made them dissatisfied and how they might suggest the business improve its operations. When a dissatisfied customer provides constructive insight into the service facility's problems, it should be investigated by management. Possible changes should be made so others are not dissatisfied.

Identifying Completion Expectations of Customers

Unfortunately, the personal expectations of customers are hard to determine because of their different needs and backgrounds. As a result, ASM/SCs must attempt to predict customer expectations while they are determining what has to be done to the automobile. This, obviously, is difficult since many customers do not know how automobiles work, what expectations are reasonable, and are skeptical about service facility recommendations. Further, the expectations of customers who see their automobile as simply a means of transportation will likely differ from those who have a personal attachment to it.

ASM/SCs, therefore, must interview the customers to gain an idea of their knowledge and perceptions and to clarify their expectations. Then they must assure them that either their expectations will be met or that they are not realistic. At the same time, ASM/SCs must instill confidence in them that the service recommended will solve their problems and that they will be properly and promptly performed.

After working with the customer, the ASM/SC gives the work order to the technician as described in the earlier chapters. At the same time; however, the ASM/SC should communicate the information gathered about the customer to the service team. In some cases, this additional bit of information can be important to the team's attainment of their common goal, which is to meet customer expectations. For example, one customer at Renrag Auto Repair had his oil changed regularly. Because of his extensive travels, he would have his wife call on the afternoon he would come in. He would always arrive at 4:30 and wanted a specific type and brand of oil, which was not kept on inventory. As a result, a case of this oil was stored in the supply room specifically for this customer. When he came in, extra effort was always put forth to have his car serviced at the time of his arrival. While the oil sales alone might not justify the extra effort, the repairs on his automobile and his referrals resulted in additional profits.

Encouraging Efforts to Exceed Expectations

Long-term success cannot be achieved by a slick program, gimmick, or expensive improvements. It ultimately comes down to the effectiveness and efficiency of the technicians, ASM/SC, manager, and owner as well as their attitude about meeting customer expectations. This positive attitude can translate into a positive environment or culture, which is often what really counts when it comes to customer satisfaction. No one likes going into a business where the personnel are negative or unfriendly, no matter how good they are at doing their job.

Positive attitudes, a positive work environment, effectiveness, and efficiency cannot be bought. In other words, money will not buy better employee attitudes or even better performance. Rather, when employees prove they can do the job and that they care about meeting or exceeding customer expectations, reward bonuses for improved performance can be considered. A company policy to grant bonuses, however, is a complex topic because it is a board/owner prerogative.

Therefore, because bonuses may be subject to company policy, ASM/SCs must use other methods to encourage employees to meet goals and conduct work that meets or exceeds customer expectations. When expectations are met or exceeded, they must be recognized via a compliment or letter to the manager and/or owner about the employee's extra efforts. At the same time, when expectations are not met, they too must be informally recognized with a confidential discussion with the employee and/or formally recognized with a memo. This relatively simple practice can have an effect on the work environment. Too often, employees do not care because they do not get feedback that their performance matters.

The Follow Through

After a repair or maintenance has been completed, ASM/SCs must follow through and make sure all services were provided, recommendations for additional services were reported to the customers, and the correct procedures for preparing the automobile for delivery were conducted (such as ensuring the vehicle is clean and is free of shop materials). To assist the ASM/SC and technician in preparing the automobile for this active delivery, a "check off" sheet should be attached to every repair order (discussed in chapter 9). These sheets, when properly designed and used, should guarantee that customer expectations have been met.

At the time of delivery, the ASM/SC must also translate any of the technician's comments on the repair order into words the customer can understand. As suggested earlier, this should be done to support the environment by instilling customer confidence that the recommended service solved their problem and was properly performed.

A careful translation of the technician's comment into words the customer can relate will help the customer understand what was done to fix the problem. Another possibility is to show the customer the old parts and to offer to give them to the customer provided there is not a core charge or it is not leaking fluids. At Renrag Auto Repair, when old parts were offered, most customers did not take them home.

If a customer would desire to have more specific information on a repair, then the technician's comments should be shown to the customer. In some cases the technician may be asked to explain the repair to the customer. At Renrag Auto Repair, customers were quite impressed when the technician, who actually performed the repair, explained the cause and cure in person. This personal touch also shows customers that a team of employees were looking after their best interests.

What to Say to Avoid Problems

Meeting customer expectations does not mean that ASM/SCs should comply with every request. There are times when a customer request must be denied; however, the denial must be justified. To arrive at a "yes" or "no" answer, there are five questions ASM/SCs should ask themselves. These questions are demonstrated in the following example.

Fred brings in a new chemical additive that according to the chemical manufacturer will do great things for his transmission. He requests that the technician pour this mystery liquid into his automatic transmission when the technician services it. Before agreeing to the customer's request, the ASM/SC should ask these five basic questions.

Will pouring the mystery liquid into the transmission-

1. *Help the Customer?*—Yes. It is nice to be agreeable. Pouring in the liquid might help the customer because he will not have to pour it in himself.
2. *Promote Sales?*—No. The business did not sell him the liquid and the act of pouring it in does not mean the business will generate any additional sales. After all, do people go to a restaurant with their own raw hamburger and ask them to cook it? Probably not, and even if they did, the restaurant probably would not cook it because they have no idea whether the meat is good quality and was handled properly. Therefore, it just makes good sense for the restaurant to use its own meat for the hamburger. The restaurant managers will know that their beef will taste good and will not make customers sick. In the same way, the service facility spends a great deal of time to secure high-quality parts and products for their customers at a reasonable price.
3. *Protect the Company from Risk?*—No. It is unknown what this liquid is and whether it can be used in the transmission. The service facility could be at risk for pouring in the mystery liquid without a written waiver from the customer stating that if it damages the transmission, the business is not responsible. Of course, even with the waiver, the court may view the business as the "automotive experts." In other words, if the liquid caused any damage, the service facility experts should have known better. Thus, the business may still be liable.
4. *Assure Health and Safety?*—No. If there is an accident and the technician spills the liquid on his skin or gets it in his eyes, no one will know how to help him. Also, if he accidentally spills the liquid on the automobile's paint, damage could result. The liquid could even be explosive if handled incorrectly or could damage the transmission's internal parts once inside. Therefore, whether the liquid is safe for the company's technician to handle or use inside the transmission is unknown.
5. *Make a Fair and Honest Profit?*—No. The business will not profit from pouring the liquid into the transmission.

A "no" response to any one of these five questions indicates that the request should be denied. Since there are so many "no" responses in the example, it would be best to use the one that is the most logical to defend. In this case, declining the request based on "Promoting Sales" or "Making a Fair and Honest Profit" would not be relevant to the interests of the customer. In fact, they may be offended that their "favor" is rejected. "Assuring health and safety" would be a stronger response, but using the "protect the company from risk" answer would likely be the easiest to defend. The ASM/SC could deny the request because it is against company policy to pour products into the transmission that are not recommended by the automobile manufacturer.

Review Questions

Multiple Choice

1. A technician turns in a repair order that recommends replacement of the CV boot with no further description. Which of these should the ASM/SC do next?
   A. Estimate replacement of the complete axle.
   B. Verify parts availability.
   C. Determine the reason for the repair.
   D. Check the vehicle repair history
2. When a customer is picking up his or her vehicle: ASM/SC A says that it is important to take the time to explain the work performed in as much detail as the customer requires. ASM/SC B says that if the customer asks questions it indicates he or she does not trust the shop/dealership. Who is correct?
   A. A only
   B. B only
   C. both A and B
   D. neither A nor B
3. A customer calls for an ASM/SC who is already working with a customer. Which of these should the ASM/SC do?
   A. Take the customer's name and number and promise a call back.
   B. Attempt to help the customer.
   C. Place the customer on hold until the consultant is available.
   D. Transfer the call to the owner/service manager.
4. An ASM/SC has prepared an estimate from a technician's diagnosis. Before providing the customer with the estimate which of these should the ASM/SC perform first?
   A. Agree on a completion time with the technician.
   B. Verify availability of necessary parts.
   C. Perform a thorough test drive.
   D. Identify additional maintenance needs.
5. A fellow ASM/SC is upset with one of the shop's technicians. Which of these should the ASM/SC do?
   A. Encourage his fellow consultant to talk with the technician.
   B. Offer to work with that technician until the situation blows over.
   C. Speak to the technician himself.
   D. Alert the service manager immediately.
6. A customer has come to pick up his or her vehicle when the service department is very busy. Which of these is the best way to handle the situation?
   A. Direct the customer to the cashier/cash them out.
   B. Advise the customer that you are very busy.
   C. Review the work performed and the invoice with the customer.
   D. Ask them to come back when it is quieter.

Short Answer Questions

1. How can an ASM/SC create a positive work environment and how will this help productivity?
2. What are completion expectations?
3. Define effectiveness and efficiency.
4. Explain methods to measure effectiveness.
5. Explain how time tickets are used to measure efficiency.
6. From a manager's perspective how are efficiency and effectiveness related to productivity?
7. How are customer expectations related to effectiveness and efficiency?
8. What are the basic requirements a service facility needs to establish and maintain a positive work environment?

9. List the 5 criteria used to deny a customer's request. Elaborate the answer by thinking of a customer request that might be denied for 1 or more the criteria listed then explain in detail why it would be denied.

Activity

Activity Problem #1: Calculate the technicians' efficiency based on working an eight-hour day:

Example: Ted earned 10 hours and worked 8; his efficiency is 125% (10/8=0.125 × 100 = 125%)

Calculate Eduardo's efficiency: He earned 9.9 hours today

Calculate Jenny's efficiency: She earned 8.1 hours today:

Calculate Frank's efficiency: He had three jobs on his time ticket today for 1.8, 2.5, and 3.5 hours. Total his time ticket hours and calculate his efficiency for today:

Analysis Question: Of the three technicians who is the most efficient?

Discussion Question: Higher efficiency is often considered better than lower efficiency. However, consider whether it is always "good."

(Hint: consider quality of work, comebacks, and other issues such as safety in your answer.)

Activity Problem #2: Calculate the shop efficiency using the same method as activity one, for each week. There are five technicians who each worked eight hours a day for five days for a total of 200 hours for the week.

Calculate efficiency for week #1: the shop earned 189 hours

Calculate efficiency for week #2: the shop earned 228 hours

Calculate efficiency for week #3: the shop earned 200 hours

Analysis Question: There is not a continuous upward trend in efficiency over the past three weeks. Discuss what factors might contribute to fluctuations in efficiency.

(Hint: Think in terms of the [economic and shop] environment, workflow, the technicians, and jobs that shop performs. Also consider in your answer external factors management cannot control, such as poor weather, a road to the shop that has been closed, and so forth.)

Activity Problem #3: Calculate the shop's average hours per vehicle each week:

Example: This week the shop had 190 hours and worked on 95 cars; the average hours per vehicle was 2 hour per vehicle (190 hours/95 cars = 2 hours per car).

Calculate the average hours per vehicle for week #1: 88 cars and 189 hours

Calculate average hours per vehicle for week #2: 228 hours and 101 cars

Calculate average hours per vehicle for week #3: 200 hours and 96 cars

Analysis Question: What is the range of average hours per car for weeks 1 to 3?

(Hint: This answer requires two numbers, the highest and lowest average hours per vehicle answer.)

Application Question: The ASM finds that she has 10 unsold hours for tomorrow. How many cars might she try to schedule for tomorrow?

(Hint: Look at the range calculated in the analysis questions. Divide the unsold hours for tomorrow into the highest number and lowest number, then the middle number. Consider your three answers and decide how many cars might be scheduled. Remember to round your number down to a whole number because you do not want to schedule too many cars and disappoint customers.)

Activity Problem #4: The shop's average hours per vehicle are 1.2. Calculate each technician's average hour per vehicle:

Example: Ryan earned nine hours today and worked on four cars, his average hours per vehicle is: 2.25 hour (9 hours/4 cars = 2.25 hour per vehicle).

Calculate Betty's average hours per vehicle: She earned nine hours today and worked on three cars:

Calculate Renaldo's average hours per vehicle: He earned 42 flat-rate hours this week and worked on 35 cars this week:

Calculate Carlo's average hours per vehicle: He earned 20 flat-rate hours and worked on 40 cars (this was over two days).

The Analysis questions #1 to #4 and the Application Question are based on Activity Problem #4

Analysis Question #1: Compare each technician's hours to the shop average. For each technician, which one is above, below, or equal to the shop average of 1.2 hours per vehicle and how many hours are each ahead or behind the shop average?

Analysis Question #2: From a management perspective, why might Carlos be below the average?

(Theorize in terms of the work he was assigned such as fast lube area, attitude or ability to inspect vehicles for concerns, as well as any other factors that management might explore such as only two days' data is available.)

Analysis Question #3: From a management perspective, why might Betty be above the average?

(Think in terms of a single day's data and its accuracy for the analysis. Also consider what efficiency techniques a technician might use to be above the average. Finally, consider negative problems related to being over such as "oversell.")

Analysis Question #4: From a management perspective, consider Renaldo's hours and the length of time the data was collected. Should the analysis be done for short duration (for each job or at the end of each day like efficiency) or is it more useful to analyze this type of data over a longer period of time? Explain your answer.

Application Question: Explain what other information a service manager might want to know to help explain technician performance that is beyond a "calculated number".

(Hint: Consider the type of work assigned to the technician, the technician's ability/certifications, the bay's equipment such as lift, as well as other factors that might cause a single technician's efficiency to be higher or lower than the shop average).

CHAPTER 16

OWNERSHIP OF A SERVICE FACILITY FROM START-UP TO EXPANSION

OBJECTIVES

Upon reading this chapter, you should be able to:

- Understand the steps an owner will go through to establish a service facility.
- Understand problems encountered during the establishment of a service facility business and possible solutions.
- Distinguish between owner investment and return on capital invested: profit and loss.
- Critique owner and management decisions relative to the strategic business plan.
- Analyze the financial planning steps from the strategic business plan to create budgets and objectives that will obtain the outcomes necessary to stay in business.

CAREER FOCUS

You want to own your own business, and you know the transition from a technician or a manager to a business owner is tough. You have the money to buy the business as well as a solid financial plan of how you will pay your personal bills while the business is starting up. You know there will be long days, few, if any, free weekends, and you are prepared to lose your money until you get established. You are also ready to handle many problems that will cause sleepless nights. Regardless, you want to be your own boss and know that you want to serve the motoring public first and foremost. You know that your idea to fix the customers' car right the first time and for the money they agreed to pay is the most important principle that comes *before* profit is considered. Without these principles, customers will not return to do business with you tomorrow.

Keenly you are aware you need profit today to stay in business so you can serve your customers tomorrow. You know that this comes when you deliver both "effective service" (satisfied customers) that is performed "efficiently" (shop time and resources are not wasted). When you provide effective and efficient service, you feel that owning your own business can be rewarding and the money you earn is measure of success not the ends in itself. Therefore, you plan to work hard every day to serve enough customers to make a profit. You realize, as the owner, it is tough to patiently serve customers and improve the business over many years to finally be successful. Further, you know that some people choose to appreciate how complicated it is to own a business. Therefore they want to remain an employee rather than become an owner . . . that you can respect.

Introduction

Past chapters covered the role of the service consultant, ASM, and manager duties, as well as the job of the parts specialist. This final chapter is dedicated to service facility ownership and specifically the steps necessary to start an automotive repair business. For details associated with operation and expansion of a service facility, which is often the duty of the service manager, the next book in the series *Managing Automotive Businesses: Strategic Planning, Personnel, and Finance* should be consulted. The purpose of this chapter is to discuss the basics of service facility ownership and operation relative to investment, start-up activities, and capital needs.

Start-up of a service facility takes planning and as the saying goes, "If you fail to plan, you plan to fail." This book does not provide all of the details necessary to make sufficient plans to start a business. When it comes to planning, this chapter provides general steps. Details that expand the steps can be researched through a variety of means; such as business planning seminars, business consultants, and seeking professional opinions from attorneys and accountants among others with business experience.

The "Owner" as Investor and Worker

An owner of a service facility is someone who has obtained money to start-up or purchase an existing service facility. When the owner invests money in the business, he or she expects a return. When an owner works in the business, perhaps as a technician or manager, he or she expects to be paid a wage that is the "going rate" as compared to a similar job at a similar business. For example, Jill starts a service facility with $75,000 cash she saved and plans to work as the lead technician for $15 per hour in addition to her ownership duties. She may approach the business as hoping to obtain $7,500 per year on her money invested (10% yield on her money) and at 40 hours per week over 52 weeks, a wage of $31,200 as a technician. She may view her ownership duties as "management" and calculates she will spend another 15 hours per week taking care of business problems, customer concerns, paperwork, and business systems. She may hope to earn another $20 per hour doing these duties or $300 per week ($15,600 per year).

If Jill makes the right business decisions and establishes a customer base with enough customers, she may earn her expected wage as a technician plus that of an owner/manager for a total of $46,800 per year for 55 hours of work per week. If there is enough money left over after paying her wages and other bills, she may get the $7,500 at the "end of the year" on the money she invested. When there is not enough money left over, she will get less than the $7,500. When there is not enough money to cover her wages and other bills, she will have a "loss" and will have to get the money to make up the loss. Jill might have some of the $75,000 left over from starting the business to cover the loss. If she doesn't have any of the $75,000 left over, she will have to borrow the money or put more of her own money into the business to pay the shortfall. In some cases, getting more money is not possible and Jill will not be able to pay herself the entire expected wage for being the manager or technician.

The $7,500 is an important number because many entrepreneurs do not have $75,000 cash to start the business. In this case, Jill would have borrowed the $75,000 from perhaps a bank and the $7,500 represents the total yearly principal and interest payments (this is a "rough" number assuming a 7.5% interest rate and 20-year payoff). A business is bankrupt, meaning it must close and sell all of the assets to try and pay back the bank, when the business does not have enough money to pay its bills that include the loan payment of $7,500 (payments are usually monthly and a yearly figure was used for illustration purposes). Bankruptcy for businesses, especially a business with a significant bank loan, is not an uncommon occurrence. Therefore, starting a new business requires caution, industry experience, good advice, planning, capital, and a little luck.

On the other hand, Jill may do very well. She may make more money than she planned and will earn a profit beyond the $7,500 that she planned to earn on her invested money. How well she does depends on many factors and an understanding that some weeks she may do well and earn a profit, others she may have a loss. She earns profit by having more profitable weeks than weeks where she loses money. The independent automotive repair business can be volatile with a good week followed by a bad week. Why this happens is due to a number of factors that include the type of work performed and how well the facility "does the job." It depends on where the facility is located, the local economy, and whether customers have money to pay for automotive repairs. Other factors include local weather conditions, advertising and promotional specials as well as how the company charges/treats the customer among a host of other factors. When Jill has extra profit, she would be smart to save it to cover the bad weeks, buy equipment and tools she needs, and pay for unexpected problems such as broken equipment or building repairs.

Unlike Jill in the past example, an owner may not work in the business. Some partnerships have investors who invest money and do not work in the business. These investors would hope to get a return on what they invested when the business has a profit. When the business has a loss, these investors may have to put more money into the business to cover the loss. When the business has earned exceptional profit, the investors may share in splitting up the extra money earned. It is best if investors get what they expect on the money invested so they do not request to sell their share of the business to someone else. Worse, the investor may want to close an unprofitable business, sell all of the assets to recover as much money as possible, and invest the money he or she gets from the sale into a different business venture. This is often referred to as "reducing the investor's loss." However, when investors get the profit they expect and more, they will keep their money in the business and when the time is right, they might even invest more money into the business to expand it.

The Seven Steps for Service Facility Start-Up

There are seven basic steps for service facility start-up. The amount of time each step takes an owner to complete depends on the amount of research needed. Regardless of when a service facility opens for business, it can take three to five years to reach investor profit targets, if not longer in some circumstances. To shorten the time to reach profitability, some investors prefer to buy established businesses. Established businesses already have gone through the seven-step process and therefore purchasing them requires the new owners to assume the roles of the previous owners if they have the proper expertise and experience. Another means to shorten the seven steps is to purchase a franchise. A franchise will help new owners go through the process of start-up and this will be discussed later in this chapter. The seven steps for start-up and the related questions for the owner to answer are as follows:

1. Strategic Plan of the Business
    a. Questions to answer: What will the business do to make profit? Specifically, what repairs will the business perform and on what type of vehicles? After thoughtful and honest deliberation, the owner or owners should be able to prepare a realistic mission statement. Then the owner must look to the future

and determine what the business should achieve (goals). The goals are to be followed by the identification of objectives needed to achieve the goals and accomplish the mission.
2. Tactical Plan of the Business
   a. Questions to answer: What resources are needed to carry out the mission? Specifically what equipment, building type, business location, service tools, technical training, and information systems are required to satisfy the customer?
3. Business Structure
   a. Question to answer: What name will the business use? Further research is needed to determine whether to operate as a proprietorship, partnership, corporation, or LLC.
4. Financial Operations and Licenses
   a. Question to answer: What documents must be obtained and filed to set up banking accounts, credit card services, tax accounts, and to legally sell services to customers?
5. Operational Plan of the Business
   a. Questions to answer: What contracts must the business owners sign for equipment and building leases, insurance, advertising, phone service, and utilities? In addition what licenses, permits, and facility improvements are required before offering services to customers? Finally, this research leads to how much money do the investors need to start the business and then determine whether or not any bank financing will be required?
6. Marketing Plan of the Business
   a. Question to answer: When the business does not have an established customer base, what is required to get customers to patronize the service facility? The reality is that a start-up business does not have a customer base and it takes time to get enough customers to earn profit.
7. Get to Work
   a. Question to answer: What methods can help an owner determine whether the service facility is making money? Several tracking mechanisms are commonly used at service facilities and selection depends on the type of repair business and the owner's needs.

Step 1: Strategic Plan of the Business

To be in the automotive repair business, the owner must determine what business is needed for the region served. The owner must also consider whether he or she is interested in what the business will sell to the customer and whether he or she has the expertise to run the business. As the owner reflects on these considerations, his or her thoughts must be based on reality and not lead by emotion or fleeting thoughts of fast money and easy work.

Therefore, the process of developing a strategic plan requires self-reflection of the owner's skills and interest, honesty with oneself in terms of personal strengths and limitations, as well as the desire to thoroughly complete the research needed to enter the market. For example, if the owner is not interested in high-tech automotive repair and wants to be in the maintenance and light automotive repair business, then that will affect the answers to the next six steps. For example, a start-up transmission overhaul business will have a different plan than a general automotive repair business.

Furthermore, a business that performs maintenance and light repair may be of interest to the owner; this business may not be the best choice for certain regions. To extend the example, a count of how many other businesses offer the same services in the region can be conducted. The research must determine whether the businesses that perform the same type of work are "busy" or not. When the choice is to start-up a similar business where some of the competitors "aren't busy," it will likely take longer to establish a similar business. In fact, the owner may run out of money before a profit is reached. In reality, it is difficult to enter any market and gain "enough" customers to make a profit unless the owner has an "edge" over the competition. An example of the data a prospective owner might collect to analyze a market may include:

- Car counts within X miles (or X minutes driving distance) of a proposed business location
- Traffic counts in front of the service facility
- Whether the business's customers mostly live near the service facility (serve neighborhoods and the bedroom community) or work near the service facility (serve customers while they work at factories, stores, offices, hospitals, or colleges).
- Demographic information of people in the area (such as age)
- Number of competitors in the same region and their volume of business
- Demand for the proposed services to be provided in the region

All of these factors among others are used to analyze a business opportunity and determine if an "edge" over the competition is possible. The data is not hard to obtain and available online through several census services and market research companies as well as city government offices. However, discussion of how to analyze the data for a market study is beyond the scope of this book. Additional research or expertise should be sought to help determine the need for a certain type of business in a region and the "edge" that is needed to prosper.

The Business Mission Statement

Once the research has been completed and the decision has been made to start a certain type of automotive service business, the owner must write down in a sentence stating what the business will do to make money. The sentence should include the type of repairs and vehicles that will be serviced as covered in chapter 1 of this textbook. This sentence is often referred to as a mission statement and it can be as simple or complex as the owner wishes; however, it should be as short as possible. For example, a short mission statement might include the types of vehicle repairs to be performed and the intent to make a profit from satisfied customers, such as:

> Jill's Import Repairs is in business to service VW, BMW, and Mercedes brakes and steering/suspension so a profit is earned and each customer is satisfied.

A more detailed specific statement might include more information on the customers and employees such as:

Jill's Import Repairs' Mission Statement:
- Establish a positive and long-term relationship with customers who own a VW, BMW, and Mercedes vehicle.
- Provide each customer with professional advice about brake, steering, or suspension work by employing a highly trained workforce.
- Perform quality work for each customer at a price to assure a planned profit is earned so Jill's Import Repairs can serve tomorrow's customers.

A mission statement is helpful to make sure that all employees know what the business is supposed to do and that it is expected to make money. When an opportunity arises, such as adding the servicing of Honda vehicles, the owner and manager can discuss the change relative to the current mission statement. If a change is desired, the mission statement can be changed. However when the mission is changed then a review of each of the subsequent business start-up steps is required to assure the business is prepared to offer the additional services to customers. When an opportunity requires a drastic change, such as a new venture to rent moving trucks, then a new business with its own mission statement might be considered. The seven-step process would then be repeated for a new business, such as Jill's Truck Rental.

The Business Goal and Objectives

After the mission statement, the owner must next determine what the business should achieve (goals) and what objectives need to be met to achieve them. Goals are specific to the owner's wishes. For a start-up service facility a goal might be to "stay in business" and a reputation to "serve customers who feel our advice is forthright, honest, and helpful in making decisions about their vehicle problems," to "employ a competent workforce," and perhaps even to "serve customers who believe our work is a good value for the price we charge." Regardless, owners should determine what they want their business to be in the future. Do they want to be the best, the biggest, or the most respected among other goals?

Below is an example of established service facility's goals. The goals for this business have evolved from perhaps the basic needs of a start-up service facility to include greater ambitions.

The visionary goals of ALPHA Motors Dealership Service Department of Riverside are to:

1. Grow profits through a marketing plan that will not just increase sales but also increase the volume of customers who come to the business.
2. Fire up customers through a plan that will entice them to buy additional services not because of high pressure sales techniques but because of the competence of the staff working on their vehicle.
3. Fund the future with a financing plan that will allow the owner to pay for future expansion with current profits and not additional capital infusion.
4. Create a work plan that focuses both on customer and employee satisfaction.

Once the goals are established, the owner would create his or her objectives to achieve the goals. As part of the process the owner may consider customer surveys about customer satisfaction. There might be training criteria established for employees. The management may create policies and procedures to assure each customer's invoice meets certain criteria, such as following labor time standards and part's markup requirements. Below is an example of the measurable objectives used by an established business to achieve the goals of the service facility. It should be noted that the financial objectives are the result of profit center analysis that is covered in the next textbook in the series *Managing Automotive Businesses: Strategic Planning, Personnel, and Finance*:

The objectives of ALPHA Motors Dealership Service Department of Riverside are:

1. To control costs and grow sales by 1% per month over each of the next nine months. Profits achieved are expected to be 15% per month in service and 35% in parts.
   a. Increase the work schedule by 10 cars per week by reducing unexpected emergency capacity (chapter 10).
   b. Increase the customer base by three new customers per week with additional promotion efforts (chapter 12).
2. Close additional sales and achieve a flat-rate hour per vehicle average above 1.1 hours per vehicle.
   a. Make customers aware of technician concerns using Must, Should, Could format (chapter 8).
   b. Use the first and second opportunity to promote sales to make customers aware of service specials and vehicle maintenance requirements (see chapter 8).

3. Achieve profit targets established by the owner so that within five years, three additional bays can be added to the service department.
4. Survey customers:
   a. Obtain 95% positive feedback for the question: "Do you feel the service advice you were provided was forthright, honest, and helpful in making decisions about your vehicle problem?"
   b. Obtain a 4.0 average on a 5-point scale for the question "Rate on a scale of 1 to 5 with 5 being the highest; 'The quality of work I received was good value for the price I paid.'"
   c. Study human resource initiatives (benefits, working conditions, pay rates, coworker/manager relationships) and implement those that will attract and retain qualified personnel to serve customers.

Step 2: Tactical Plan of the Business

The tactical plan is the owner's study of what assets are needed to conduct business. The assets needed include:

- Tools, shop equipment, information systems, electronic equipment, and business supplies or equipment
  - All of the tools required to serve the customer from the write-up process through the repair process to the final invoice and filing. By examining the mission of the business, the owner can begin to assemble the list of assets needed to conduct business. Naturally, the tools required for a transmission overhaul business are different from those needed for a general automotive repair shop.
- The facility requirement:
  - Location and square footage of space required for shop bays, office, waiting room, and storage. Ceiling height of the shop area for lifts and electrical service, number of parking spots, and bathrooms among other needs.
- Human resources
  - Number of people (if any) to run the business (technicians, service consultants, parts specialist, managers, clerical personnel) and experience needed by each as well as additional training needed so they can do the job required. Days and hours of operation should be established as this will dictate personnel requirements.

Planning can be done up to a certain point; then the needs of the business will depend on other factors such as government regulations, anticipated demand for services, and the number of customers that need to be served to break even (no profit or loss). For example, garages that offer Pennsylvania State Inspection are required to have inspection bays that are 12 feet wide by 22 feet long provided they use an approved headlight aimer.

Furthermore, local governments often have ordinances on minimum number of parking spaces for certain businesses as well as sign size limitations, curb cut length for access to the property, and so forth. When an owner buys a franchise, they will have additional requirements for the owner to consider, such as traffic count minimums on the road in front of the facility and building design requirements. So, the requirements of the business (tactical plan) will depend on many factors, some beyond the owner's control. The owner must know what is needed to conduct business and recognize that a mobile repair business's (drive to the customer and service the car on site, figure 16-1) government requirements, type of facility, and other assets needed are very different than a tire dealer and repair business (figure 16-2).

FIGURE 16-1 Mobile Automotive Repair Business.

FIGURE 16-2 Tire dealer and repair facility.

Step 3: The Business Structure

In the simplest example possible, JOE SMITH can be in business right now and it would be a proprietorship "entity" (one-owner business) as compared to a partnership which has is two more owners. Joe would keep records of the money charged to customers (income) and keep records of the money spent to deliver the service to the customer (called expenses, cost of labor, or cost of parts—see chapter 1 for more detail). Joe would have customers write checks to him and deposit them into his personal checking account. He would pay any business bills out of his checking account. The difference between what he earns (income) and pays out (expenses) would be a profit or loss (called net profit or net loss). Joe would pay federal, social security, state, and local taxes on the net income earned. Joe would pay sales tax (covered later) on the sale of labor and parts to the customer. Joe's social security number would be used on government documents (tax returns for example) to identify the business and he would have to submit his taxes (forms and money) by the due dates.

If Joe wished to be called JOE'S GARAGE or some other name, he could file a fictitious name with his state government. As long as the fictitious name was not already used by a corporation or protected by a federal trademark or copyright, he could DBA (Do Business As) or TA (Trade As), Joe Smith DBA JOE'S GARAGE. With the proper state paperwork, he could set up a checking account as JOE'S GARAGE and use it to identify his business for tax purposes and other business activities.

This discussion becomes more complicated when risk is considered. Specifically what if something goes wrong and a customer's car is damaged by Joe in an accident. Worse, what if, because of the work Joe performed, the customer wrecked the car? NOW WHAT? A business liability insurance policy is something Joe would buy and it would pay a claim Joe submits. However, it is possible that the damages are more than the insurance limits and they will not pay the entire claim. Joe, the owner, could lose all of the business assets because the injured person(s) can be awarded Joe's assets by a court. If the business assets are not enough to make the injured person(s) "whole" (Whole is a term meaning the injured are able to return to their life in no worse condition than before the accident), the court could award the injured person Joe's personal assets (savings, car, home, hobby items, depending on state laws). Considering the owner might lose everything he or she worked for, the owner might want a business entity that will protect the owner's personal assets in case of a problem.

The two most common business entities that can help limit owner liability are the Corporation (IRS Subchapter S-Corporation compared to the C-Corporation are not differentiated in this book) and an LLC (Limited Liability Company). Either can be created with the help of an attorney and with advice the owner obtains from an accountant. Different business structures and entities are discussed in more details in the next textbook in the series: *Managing Automotive Businesses: Strategic Planning, Personnel, and Finance*. For this step, the entity that is most appropriate (proprietorship, partnerships, LLC, S- or C-Corporation) depends on a number of factors based on state law, tax benefits, protection of personal assets, and size of the company among other considerations. Typically, an accountant and attorney are consulted before start-up so the necessary legal and tax-related documents to establish the business entity are submitted to the government.

CASE STUDY

A customer who went to a garage had her car's brakes inspected. The technician did not tighten the lug nuts to the rims tight enough. The tire fell off after the customer drove a few miles. This business owner had been in business for almost 40 years as a proprietorship without an incident and had insurance. Unfortunately, the customer was injured badly when her car hit a telephone pole and her leg and hip were broken. Her medical bills, time away from work, ability to return to work full time, and damage to the vehicle were $835,000 and greater than the insurance company's policy limit of $500,000. The business owner was underinsured. Not only would a lawsuit take his business assets but it would lay claim to his personal assets he had accumulated after 40 years of working.

Business Structure Tax Consequences

There are tax ramifications for different business structures and an accountant can be consulted to determine which one is most advantageous for an owner based on his or her personal finances. This is important because many owners plan to work for the business (subject to social security tax as well as federal, state, and local taxes) and also be an investor (profit is subject to federal and some state and local taxes). Separation of the owner's wages from his or her investment money for the purposes of paying taxes follows different rules depending on the entity structure chosen and state tax law. Depending on the legal and accounting advice obtained, the establishment of the business can be more complicated when the owner holds part of the business assets, such as real estate and lifts, in a separate company.

Therefore, Jill may have two businesses, Jill's Auto Repair Incorporated which is in the business to fix cars, and the assets include tools and shop equipment. Her other business is Jill's Rental Real Estate LLC, which owns the building where Jill's Auto Repair Inc. operates as well as items attached to the building such as in-ground lifts. In this example, Jill's Auto Repair Inc. rents the building and equipment from Jill's Rental Real Estate LLC. The reason some might find this advantageous is twofold; first is so that one business can be sold separate from the other such as selling the automotive repair to one buyer and keeping or selling the building and its rental business to another owner. Therefore, as separate businesses, the cost to buy each is lower so more buyers can be obtained. Many owners think like investors and want the eventual end of the business to result in its sale rather than to close it (in a few cases owner wishes to pass it on as inheritance but that is outside this discussion). A second reason is if something goes wrong in the automotive repair business and it has to close or declare bankruptcy, the building and perhaps some major assets, such as lifts, are retained by the owner(s) in the real estate business. The reasons for an automotive repair business to become bankrupt include a poor local economy, legal action, or mistakes created by the owner that might stem from his or her poor health or personal problems. So understanding business risk and how different business entities can protect assets and provide tax advantages requires legal and accounting advice that is unique to each owner's situation.

Step 4: Financial Operations and Licenses

After step 3 the owner should have the business name and entity paperwork from the state and must next set up the financial operations of the business. To start, the owner needs an identification number for the government to track the business. As stated in step 3, a proprietor would use his or her Social Security number. For partnerships and other types of business entities (LLC, Corporation), the owner(s) would file the forms for an Employer Identification Number (EIN) from the Federal Government. With a federal identification number, the first account the owner will set up is a checking account in the business name. After the checking account is set up, the owner can now write out checks for the various expenses incurred as a result of setting up the business.

While a checking account will allow the owner to deposit customer cash and checks, it will not allow the owner to accept credit cards. Since many customers pay by credit cards, the owner needs to look into credit card agreements and make application. Since the fees vary from bank to bank, the owner must shop for the best deal. Local banks, membership in an organization such as the Chamber of Commerce, or even automotive-related trade groups often have programs for merchants to accept various credit cards from customers. Specifically, credit card agreements are different in terms of the rates, fees, and credit card equipment charges. The owner must be prepared to pay the credit card company processing fees, equipment charges, and as much as 4% of a total sale on each job paid by credit card (this is called a discount rate). In the automotive repair business, accepting credit cards is necessary because customers may not have the money to pay an unexpected auto repair bill. Since it is better to get paid than not to get paid, most owners consider credit card fees a necessary expense of "doing business."

ADDITIONAL DISCUSSION

Cash Flow and a "Line of Credit"

Owners do not know what the future holds for the business but do understand that "running out of cash to pay bills and wages" means the business will have to close. When this happens, they must get more money. Some owners have money in reserves (savings) held by the business to withdraw when big bills or unexpected expenses occur. Others do not have cash in reserve and the owner or investors must deposit more investment capital into the business. A few owners will be forced to use less advantageous means such as borrowing money on a personal loan or business credit card with high interest rates to pay bills. It is suggested that before starting the business the owner should know how to get more money, called capital, to keep the business open when tough times occur. Not only must the owner know where to get more capital but also how much is available. As the owner approaches banks to establish accounts, the owner should inquire about a financial product called a "line of credit." A line of credit will loan the owner money when the owner requests it, up to a maximum amount. When the line of credit is used, the owner pays interest on the money used. A line of credit can be established anytime but it is best if the owner obtains it when starting the business. In some cases, an owner may have to pledge a personal asset such as his or her home as collateral to qualify for the line of credit. The owner must understand that if the line of credit payment is not paid each month, the owner will lose his or her home. Each method to obtain more capital has risk and cost that the owner must know and understand well in advance.

Sales Tax

After establishing the "banking end of the business," the owner must file for a license to collect state sales tax. The sale tax license will permit the owner to buy parts that will be resold to the customer. The parts are bought from the part's store without paying sales tax and when resold, the tax on both the cost of the part plus the business markup is collected from the customer. For example, a part is bought without paying sales tax for $10. The service facility marks it up by $15 for a total charge to the customer of $25. Sales tax is collected on $25 from the customer. To set up an account at a part's store, the owner must fill out an application in addition to completing a state form with all of his or her state's tax license information. The owner can only defer the sales tax payment (pay the sales tax later) on items to be resold to the customer and may not use the license to buy any item that will not be resold to a customer such as tools used in the business. Furthermore, the business must charge the customer sales tax on labor charges, sublet repair charges (services done by other garages such as towing), shop supplies, and any other items the customer buys from the owner's business. Depending on state laws, when the owner sells the parts and collects the sales tax from the customers, he or she may be required to deposit the money in a separate account for payment of taxes (not the business checking account). The separate account assures the owner will have the money available to pay the tax by the state's due date. Payment to the state is usually monthly or quarterly.

When Joe, the owner, does work for another business, such as performing an alignment for another automotive repair business, such as Jill's Auto Repair (this is referred to as a sublet repair), Joe will give Jill a state sales tax exempt form to fill out. The state sales tax exempt form will provide Joe with Jill's sales tax number and other business information. If the state wants to know "why Joe did not collect tax from Jill," he must provide the state official with the paperwork Jill filled out. Whether Joe provides Jill credit terms to pay the bill later is between Joe and Jill. For example, when Joe filled out the parts store application, Joe will be allowed to buy up to a maximum of "X" dollars of parts during the month then pay for them at the start of the next month. This courtesy is not extended to all business owners because the business that extends the credit is at risk of not getting paid.

Other Taxes and Licenses

In addition to sales tax, a business might be required to submit additional "local" sales tax, a business privilege tax, federal and state payroll taxes, as well as array of other government taxes. Filing the proper forms to set up the different tax accounts and submission of the tax on time with correctly filled out forms can be challenging. A business owner usually obtains advice and may contract with an accountant or other professional such as a payroll company to perform these functions. Often it is less expensive to pay someone else to do this work so the owner can stay focused on the mission of the business: fix cars for customers and grow the business so it is profitable.

Depending on the work the shop performs and the location of the business, there may be additional licenses to sell certain service or products to customer. For example, to purchase refrigerant for air conditioning work requires an EPA section 609 license. Parts stores require that the license be presented when purchasing the refrigerant in addition to filling out related EPA forms. Not only must the person purchasing refrigerant have the EPA license, the technician working on the vehicle's A/C system must have the license in addition to the business owning the proper equipment. (EPA inspectors will ask to see "approved" equipment upon visiting the service facility.) Some states require vehicle owners to obtain an emission inspection (decentralized) and state safety inspections (figure 16-3). Depending on the state, to perform the inspection for the customer, the service facility needs to be licensed as well as the technicians (figure 16-4). In a few states, the state may regulate the entire automotive repair industry; such as California's Bureau of Automotive Repair (BAR). Anytime the government requires a business to have a license to conduct business, the owner must be prepared for state authorities to visit: planned or unplanned. Failure to pass a government inspection or supply proper

FIGURE 16-3 Licensed technician applies a state safety inspection sticker to a customer's car.

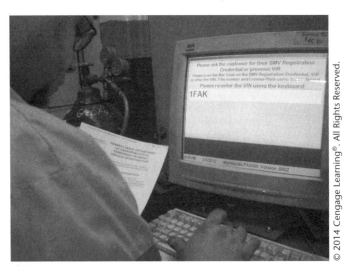

FIGURE 16-4 Licensed technician enters emission inspection information into a computer that is connected to the state before testing the vehicle's emissions (state-administered federal program under the EPA Clean Air Act). Technicians must be licensed to issue emission inspections.

documentation can result in the suspension the state issued license and/or fines. The end result, the service facility cannot make sales to the customer and the owner is out of business entirely or is prohibited from selling specific services. This generally will result in lost profits for the business.

Step 5: Operational Plan of the Business

Next the owner must determine what must be done to obtain a physical location for the equipment needed to operate. To start this process the owner will review the tactical plan from step 2 to determine what tools, equipment, and physical location are required to conduct business. Unlike previous steps, this step can become capital intensive with the owner paying lots of money as well as signing business contracts to get what is needed to open the business. The money needed will likely come from personal savings, and the contracts signed may be complicated requiring legal assistance to review the terms and conditions. Failure to understand the contract can result in unexpected surprises with undesired results when problems arise.

Borrowing the Capital Required to Start the Business

When an owner does not have enough money saved, he or she must borrow it by obtaining bank financing. Borrowing money has two parts: an interest portion which is the "rent" that is paid each month for the privilege of using the money borrowed and a second part which requires the owner to pay back the principal of the loan. The principal of the loan is the amount of money originally borrowed. Often an owner will approach a bank for a loan. If the owner has a solid business plan with good documentation and significant personal collateral to pay or "cover" the loan ("cover" means the owner's personal assets are worth as much if not more than the amount asked for in a loan), the bank will provide the best interest rates and other terms. If the owner has few assets but a solid business plan with good documentation, he or she may go to the government for a Small Business Administration or SBA "backed" loan. In this case, the SBA essentially guarantees the loan principal to the bank in case the business owner does not succeed.

Adjustment of the Tactical Plan

The tactical plan discussed in step 2 may require adjustment when the cost of purchasing the equipment and tools needed is greater than the owner's savings and amount the owner can borrow. To reduce costs, a business owner may need to consider the elimination of equipment features or buy less expensive models than originally desired. The owner might consider the purchase of used equipment and hope the cost of upgrades, repairs, and maintenance will allow him or her to earn profit until better equipment can be purchased.

In addition to equipment, the owner will need to buy inventory such as oil, chemicals, and other materials. This can be costly and the owner must resist "buying too much." For example, the fresh oil tank may hold 250 gallons but realistically the business will use only 100 gallons per month before the next oil delivery. Therefore, it is prudent to buy a little more than needed for the month rather than "filling it to the top" allowing the owner to have the capital to buy other required items.

The plan should be to buy what is needed until the next delivery. This concept is applied whether it is oil or spray cleaners. Caution to such a plan is required as there is a fine line between keeping inventory to a minimum and wasting expensive technician time. For example, a decision not to carry light bulbs because the parts store will deliver the bulb when needed may be penny wise but dollar foolish. This means that the owner might save $100 by not buying an assortment of light bulbs that cost a couple dollars per bulb but results in wasted technician time until a light bulb arrives to finish the job. The savings of a couple dollars results in lost technician time that can be over $20. In the end the owner must balance the needs of the shop because he or she will pay more by having a technician wait for a delivery of a low-cost part than to carry the item in inventory. In general, if the item is used frequently or semi-frequently and the cost per unit is under X dollars, the owner may want to carry a small quantity. If it is used infrequently or higher cost than X, the owner will likely want to get it when needed rather than carry it in stock. There are more details to this concept that are covered in the next book in the series *Managing Automotive Businesses: Strategic Planning, Personnel, and Finance*.

Equipment Leases

Business owners often decide to lease equipment as a means to obtain the assets needed to operate the business. Leases allow the service facility owner, called a "lessee," to use a specific asset, such as a wheel balancer, in exchange for a regular monthly payment. Each month the service facility owner pays a "rental charge" to use the equipment to the "lessor" (equipment owner) until the lease term ends. A lease term is often several years long and the total cost of all of the monthly payments is much higher than the cost to buy the equipment in the first place. Therefore, a lease allows the business to use the equipment to make money, and instead of buying it and using a lot of money, they can make monthly payments. The idea is to avoid spending lots of money up front to buy equipment that the business may not have available to spend because it has other financial needs.

Unfortunately, equipment leases often are not available to a new start-up business owner because a lessor requires a minimum amount of time in business to qualify. Time in business along with a strong financial position (cash in the bank, strong net profit each month, and a history of paying loans on time) allows a business to get the best lease terms and rates. When a business does not have a strong financial history, a lease can be very expensive with rates as high as if they borrowed the money from a credit card company to buy the equipment. As part of the lease conditions, the service facility may be able to buy the leased asset after the term ends for very little cost. However, these leases may be considered capital leases by the IRS with unique tax ramifications that should be handled by an accountant. When leases are not available to the start-up business owner, then the owner will need to get the money perhaps through a larger bank loan to buy all of the assets needed.

Stick to the Tactical Plan with Money Left Over

During step 5, the owner must stick to the tactical plan and buy what is needed to start the business with as much money left over as possible from a business loan or personal savings. The reason is because lots of money is needed for other expenses before the owner generates the first dollar of income. Even once income begins to be earned, the owner will likely not have enough money to pay the bills each month as he or she starts the business. One of the biggest expenses incurred by the owner each month is the lease on the service facility building. The lease on a building is considered a commercial lease and is not like residential leases to rent an apartment. These leases are a business contract that could be for a length of five years and hundreds of thousands of dollars in total rent due over that time period. If the owner closes the business before the lease ends, the building landlord can still hold the owner financially responsible for the money still owed on the contract. Beware that some loans and contracts, such as the building lease, require the owner to sign and pledge personally to pay money owed on the contract and might even require a spouse's signature and pledge as well. Commercial leases can be complex with the owner responsible for all building maintenance, building insurance, and property taxes (called a triple net lease). Assistance from an attorney experienced in commercial lease negotiations is recommended. Often negotiated are options to buy the building from the landlord in the future, automatic renewals of the lease after the term expires, fixed rent increases over time, pre-move-in repairs or upgrades, as well as early rent discounts that can help improve start-up cash flow.

Contracting for Services

Once the owner has made arrangements to obtain the hard assets needed (building, equipment, tools), next services must be contracted for a building security system, sign installation, utility hookup, equipment installation, business insurance, phone, Internet, website services, advertising, and an array of unexpected building expenses. Unexpected expenses can include furnace maintenance, installation of additional lighting, building repairs, shop air line installation or replacement, painting, repair of garage doors, bathroom repairs, replacement of showroom carpet and ceiling tiles, outside landscaping and parking lot lines among a number of other issues.

Building Permits and Business Licenses

Before moving into the building the owner must have or be in the process of obtaining permits or licenses in addition to complying with other regulations. Examples of these include:

- Building occupancy permit
- City, Township, Borough codes inspection of the building or property for safety and mechanical systems violations

- Local zoning approvals or variance (permission if the building is located in a zone that is not approved for automotive repair)
- EPA phase 1 or higher environmental property report performed by an approved contractor (often a requirement of a bank, insurance carrier, or investors before leasing a property for automotive repair or purchase of real estate)
- Floor drain inspection (or removal thereof depending on state regulation) in addition to impact studies if required by a government agency before starting the business (new buildings and a few existing "change of use" buildings may require state or local government agency "impact study reports")
- Holding tank inspection such as belowground lifts, floor drain tanks, waste and fresh fluid holding tanks (usually if underground), as well as compressed air tanks as required by insurance or other government regulations. Below ground gasoline tanks, often found on properties that were gas stations at one time, requires special consideration and research as to applicable regulations.
- "Government-regulated" equipment compliance such as A/C equipment
- State official inspection for emission and/or state safety inspection licenses approval
- Insurance carrier inspection for hazards that might result in a claim
- Other permits, licenses, inspections, compliance depending on regulations to conduct business in a specific region

Some permit, license, and compliance processes require that the owner to be prepared to host a variety of visits. Visitors include local, state, and possibly federal officials as well as insurance company representatives in addition to surprise return visits or unexpected visitors who are vague in what they require. The surprise visitors can be legitimate government officials, nosey neighbors, and even legitimate covert operations (mystery customers) that are charged with oversight of the service facility for the government or a franchise. For all visitors, the owner should be present and take time to work with the visitors' request. The owner must understand compliance requirements, show the visitor what he or she wants to see, and learn from the experience when there is an opportunity. If the visit appears to be beyond the scope of a regulatory requirement, such as showing the visitor where the company safe is located, then questioning the visitor politely should be done before possibly denying the request.

Depending on the region, competition may be fierce and existing business owners in similar businesses may not appreciate the newcomer. In this case, some compliance issues may be generated by complaints that lead to investigations. For example, a complaint to the State Attorney General's Office about the wording of the repair order disclosure the customer must sign will require a state official to call or stop by for a copy of a repair order. They must review the statement(s) the customer must sign to determine if it meets state guidelines. The owner must give a repair order copy to the requesting office and if it is not correct, the inspector will inform the owner of the infraction. These tactics are usually just a nuisance and hopefully unfounded by the investigator. When a problem is found, the investigation can actually help make the business better. Don't worry much about it. The owner may need to dispose of 1,000 invoices before using them but the business must comply with the regulation and keep moving on with the "setup plan."

Thinking through step 5 is very important because this research leads to how much money the owners really need to start the business and how much bank financing is required. If enough money is not available, it is possible to run out of money before opening for business. Even once the business is open for business, a profit will not come for weeks, months, or even years. Bills must be paid and cash from the business loan is needed to keep the business open until profit arrives. This is the risky part because it is unknown how much money the business will lose before it is profitable. Many business owners do not take a wage (much less get a return on the money invested initially) because the business is just not making enough to pay the bills. In some cases, owners must continue to put more money in the business while they continue to work at the shop so the business remains open until profit hopefully arrives in the future.

Step 6: Marketing Plan of the Business

A start-up business does not have a sales record or know what the expenses will be, so it is conjecture whether the business can make enough money to be profitable. Losses are expected for a period of time (sometime months). The main reason is because a start-up business does not have an established customer base. Without a customer base, the business must continually attract new customers. Internet, coupons, newspaper, phonebook, flyers, mailings, social networks, and word of mouth can help get new customers. However, it will take time to establish a customer base that will return over and over again to have the business work on their vehicle. Some businesses will see a customer only once in a great while (such as transmission overhaul). Others will see customers regularly such as a fast lube operation. The requirement to get and keep customers rests not only with advertising but also with the service consultant and technicians. Advertising will get customers in the door, the service consultant will keep the customers at the service facility, and the technician must do high-quality work to keep customers coming back again. So it is a joint effort. A great service consultant with either a poor service system in place to process work or poorly trained or inexperienced technicians will serve the customer once but never again as the customer searches for a better shop. A service consultant with a poor attitude but a good service system to process work and good technicians will likely not keep the customer long enough for the technicians to get a chance to work on the customer's car the first time.

Establishing a Customer Base

The reality is that a start-up business does not have a customer base and in the beginning everyone must work especially hard to get and keep the customers who patronize the business. It takes time to establish a customer base and get enough customers to earn a profit. In creating a marketing plan, the owner must work with a variety of mediums and spend enough money to promote the shop. To get customers in the door a website and ads in a phonebook and newspaper, and on the radio will get the word out to a large area (although some are probably too far away to be a customer). Flyers and mailings are effective at targeting customers who live near the shop or who either work or live in the area and will use the business' services. Social networks and word of mouth are also effective but can take time. Thus, to create a customer base takes a variety of medium products, time, and a success record of satisfying customers with quality repairs at a fair price. Some may argue that a new business should offer "inexpensive repairs" and big coupon savings to bring in customers. Unfortunately, this tactic can bring in money (too often at a loss and not a profit) but more importantly moves the business away from establishing a profitable customer base base as these customers come merely for the "inexpensive specials" and do not return once prices normalize. It is best in the beginning to focus on the customer so each one gets what he or she paid for and more. Care by each member in the repair process is the key to a good customer review that will be passed on from the customer to others.

Maintaining the Customer Base

As the customer base is established, the owner must remember to constantly maintain the relationship. Email, text messages, twitter, Facebook, newsletters, and reminders among other communication sources help keep the customer informed of the owner's desire to serve the customer. Customers will forget the service facility over time and attention to remind them of their next service due as well as other services and specials offered by the service facility is required. Even once established, new customers are needed because some will leave because they moved, retired, bought a new or different brand of car, got a new job out of the area, or died. Therefore, a business must constantly recruit new customers and funds must be set aside to attract them as well as to market to existing customers. For a start-up business, of course, more money is to be spent on attracting new customers and as the business grows, money is to be spent on marketing to existing customers.

Cross Promotion

Another way to attract customers is to identify other businesses that have customers who might need the services offered by the automotive repair business. For example, a mobile automotive repair business may find that a pizza shop with delivery pizza is a good match to cross promote. It reasons that customers who have their pizza delivered may also want their vehicle serviced at home and therefore the two customer bases may overlap. In this situation, the owner of the pizza shop would obtain coupons from the owner of the repair shop. The coupons might be $6 off a service. As pizzas are delivered with the flyer attached to the top of the pizza box, the customer receives a $6 "extra value" for an automotive repair for what is the cost of a $12 pizza. (Basically, a "thank-you" savings brought to the customer compliments of the Pizza Shop owner.) By color-coding the flyers, each pizza delivery person can be given $3 off vehicle services for every coupon redeemed by the customer. This promotes cooperation of the pizza delivery employees to help with the effort (in practice they begin to hand the flyers out not only with the pizza but also to others they see which all leads to more business). As the owner tracks the flyers redeemed, the end result is the mobile automotive repair business grows with each pizza delivery because of the connection of the common customer bases. While this example uses a pizza shop and mobile automotive repair, the concept begins the process of thinking about a customer base differently. Specifically, the owner needs to think about:

- Who are the business's target customers and where are they located?
- Is there another business that shares the same target customer?
- Can a cross promotion (win-win) effort be established with another business or employees of a local college, factory, or hospital who might share the same customers in a nearby location as the service facility?
- If a cross promotion partner can be identified, what is an appropriate way to entice the customers to use the service facility?

    - Discount cards for the employees that are given out by the Human Resources department
    - Donation of X dollars to a hospital charity or college scholarship initiative if Y number of employees/students buys a service over a certain period of time.
    - Cross promotion flyers that help another business sell more products, such as pizzas, while the service facility gets the new customers it needs.

There are many other promotional efforts that can be considered such as grand openings, special event offers, and even target marketing toward a specific demographic group. For example, a service facility may target professionals, such as dentists, an office complex, or neighborhood that seem to like the business services offered. The scope of marketing and promotion to extend the customer base and promote the services, expertise, price, and location of the service facility is limitless and is as much of an art as it is a science with principles to follow. The idea is that to be in business the owner needs to think customers, get them in the door, and keep them coming back.

Step 7: Get to Work

The mission of the business directs the owner and employees to perform a certain type of work to earn a profit. The owner must establish what standards will be implemented to obtain the results desired. In establishing the standards, documentation is required to track the results. If there are too many documents, then the owner will get too much information that may not be useful in measuring what is needed and time can be wasted. Too little documentation and the shop's mission may not be carried out. Often the owner really wants to know how much to charge in terms of labor rate for the expenses that the shop incurs. This answer is much more complex than can appear in a single chapter and the next textbook in the series *Managing Automotive Businesses: Strategic Planning, Personnel, and Finance* has an entire section dedicated to an analysis of sub-business operations (profit centers) and the income/volume needed to cover expenses and costs (labor or parts). Markup as well as more advanced analysis techniques is discussed in more detail.

For this chapter, a start-up business has different concerns because there are usually just too few customers to break even (no profit or loss) as compared to an established business where more advanced financial concepts apply. Specifically, there are three basic items to track in the beginning.

1. Talk to each customer about what he or she thought of the services. This works until the business has more customers than can be spoken to personally and then a survey method should be used to determine their satisfaction on various topics.
2. Technician inspection sheets that will make sure each vehicle is checked over for the items needed to inform the customer of what might be wrong with the vehicle. These should be reviewed by the owner initially and filed for future reference when a customer returns for service. These sheets can also protect the service facility from claims of poor quality work by the customer.
3. Tracking of each technician's billable hours for services performed and then combining all of the technicians to track the shop. Regardless of the technicians' pay system the shop needs to generate enough billable hours to pay the bills. It is often useful to track the billable hours on a chart relative to a breakeven line to determine whether (especially at start-up) enough work is being done (see Figure 16-5).

Is the Business Making Money?

After an owner is assured the customers are satisfied, the technicians are checking over the vehicles, and the hours generated by each technician are known, a review of the profit of the shop and each technician can be performed. To illustrate this procedure, the following example will examine a single technician's performance; however, this example can be expanded to examine an entire shop of technicians if desired. Therefore,

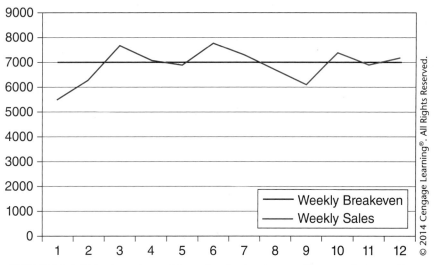

FIGURE 16-5 This chart shows 12 weeks of sales data against a straight weekly breakeven line of $7,000 per week. The chart shows that the business had thousands of dollars of losses in the first couple weeks. Sales fluctuated above and below the breakeven line during the quarter. Examination of the numbers that made up the chart found the business will end the quarter near its breakeven (week 12).

assume an hourly technician is paid $780 (for illustration purposes assume this includes the business's share of taxes) and his or her production is 15 billable hours for the week at a labor rate of $50 per hour. The shop earned $750 from the technician's labor (15 hours × $50/hour) and paid out $780 for loss of $30 on the wages. After tracking the billable hours of the technician, the owner meets with the technician to determine the problem and the owner makes some changes as a result. The next week the technician has a production of 20 hours. The shop earns $1,000 and pays out $780 for a gross profit of $220 (about 22%). The shop however must still cover other expenses and every technician must carry their fair share so a profit is earned. Generally for every $10 the shop brings in, the labor cost must not be more than $4 (40%) and so a gross profit of least $6 (60%) is leftover to pay expenses. Expenses are paid from gross profit and by national statistics for an "average" service facility will use about $4.5 of the $10 sale (45%) to pay bills. After expenses are paid, if there has been enough volume of work, the owner should be able to keep a net profit of $1.5 (15%) of the $10 sale.

For the technician who was just studied, a gross profit was earned from the technician's work. However, not enough was earned (22% gross profit instead of the 60% required) to pay the bills and earn a profit. So what is wrong? It could be a variety of issues from not having enough work to keep the technicians busy to assigning wrong work to them. The technician's skills to do the work quickly could be a problem or even a business system with flaws in processing the work. Regardless, on a $780 pay out in wages, the technician must typically "bring in" *at least* 2.5 times their wage or $1950 (39 billable hours at the $50 stated labor rate – NOT COUNTING PARTS SALES) to carry the expenses of the shop so a profit can be earned. Profit naturally depends on business expenses, also called overhead, and this example can assume only enough customers patronized the service facility.

CASE STUDY

An owner proclaimed: "I need three pieces of information to know whether I am in trouble: how much money is in my checking account (he knew this relative to what he had to pay out each month in wages and other expenses)? what is in the invoice basket to be picked up and paid by customers (what money will be coming in yet)? and finally what does our schedule look like in the coming days (are we busy or not—how much more money will be coming in)?" Based on that information he knew whether he would make money or have to invest more money into the business to keep it going. He went on to explain that it was tenuous each week; he tried not to get too upset when he had to put money into the business. He also did not get too excited when he had more than enough profit and advised to avoid a spending spree or making rash decisions about expansion of the business or improvements. In the end it *should* all work out given he spent a lot of time repeatedly doing analysis and other planning not unlike the steps provided in this chapter. Over time, he found it gets better and he got used to the pressures and better able to manage them. However in the beginning, he didn't sleep a lot of nights and was up very early many days and weekends to start work because he was worried about going bankrupt.

Volume of Business

For start-up shops and during different business times (such as winter), each job may make the required profit margins but there may not be enough work for a technician to produce 39 hours (extended from the previous example). Furthermore, while the "going labor rate" in the area may be $50 per hour, it could be too low for the start-up business to earn a profit if the technicians cannot reach their billable hour goals. Switching the technician to flat rate (paid for the work they complete) may help this situation but the owner must make sure he or she does not violate federal minimum wage laws. The "volume of business and labor rate charge" analysis can be complex to solve and as such the next book in the series *Managing Automotive Businesses; Strategic Planning, Personnel, and Finance* addresses additional concepts that are helpful to figure out the answer to what can be a difficult problem.

Job-by-Job Analysis

While understanding the pay of each technician relative to what each "brings in" is important, the owner for a start-up business will want to make sure that money is made on each job performed. It is possible for the shop as a whole to make money but some jobs may make more money than others. This is information that the owner needs to know! Using the same method as the technician weekly wage to gross profit method, the owner will want to review each repair order. For each repair order the labor sale to the customer must be 2.5 times higher than the cost of the wages paid. In the next textbook in the series, the 2.5 multiplier equates to a 60% gross profit margin using the mass retailer markup calculation. Approximately 60% gross profit margin is reported by national statistics for profitable shops.

To illustrate how the 2.5 multiplier works, a Honda head gasket was sold to the customer for $350 (7 hours as per the labor time guide × $50 labor rate). The technician punched on the job on Thursday at 3:00 p.m. and worked until 5:00 p.m. (two hours) then finished the job the next day at noon (four hours; he started at 8:00 a.m.). For six hours of work, the hourly paid technician earns $17 per hour or $102. $102 paid to the technician means the job had to be sold for at least $255 ($102 × 2.5 = 255). However, the $102 labor cost sold for $350 and this is _more_ than the 2.5 multiplier. So the shop made $95 more than expected ($350 – $255 = $95) because of the technician's fast work. Provided the job does not "come back" in the future, the shop owner might declare—SCHEDULE MORE HONDA HEAD GASKETS AND GIVE THEM TO THAT TECHNICIAN TO PERFORM!

The owner's conclusion and decision might be different if the technician was performing a General Motor's head gasket that was sold to the customer for $300 (6 hours labor time guide time × $50 per hour) and the hourly technician was paid $18 per hour and took 10 hours to do the job at a cost of $180. Based on what the technician was paid, the work did not earn the required 2.5 multiplier ($180 paid × 2.5 = $450 sale price instead of what was charged which was $300). The shop made $120 in gross profit ($300 sale less $180 cost = $120 gross profit) but this was not enough to pay the bills and earn a profit; the shop needed $180 in gross profit. The owner upon inquiry may find that the technician had not worked on this type of vehicle before and it was a "learning experience"; therefore, he or she will do better next time. The owner might also find that the particular engine design has many difficulties in the repair process and it should not be scheduled in the future or at least not for the amount of money charged.

When the service facility wants to do certain types of jobs but the technicians cannot perform the work very well, then training is needed. When they can do the work but not very quickly, practice is needed. In the meantime, the owner will have to accept the loss of profit until the jobs can be done profitably. If training and practice is not an option, then the owner must hire an experienced technician or review the business strategic plan for the type of work performed by the business. The outcome of an owner's review can be limitless as can the possible solutions. This job-by-job analysis merely helps the owner pinpoint problems to address in making sure each job earn enough profit.

Parts Sales, profits areas, and low margin services

Parts sold to the customer must also make a profit. A 2.5 multiplier can be used for parts as well; however, the problem is that a shop can "price themselves out of work" by trying to sell a $1,000 part for $2,500. Discussion of this is covered in the next textbook in the series *Managing Automotive Businesses: Strategic Planning, Personnel, and Finance*. For the point of this chapter, parts that typically cost between $10 and $150 might use the 2.5 multiplier. So when calculating profit there is the labor area profit and a separate parts area profit. Each area (parts and labor) must make enough profit. However, it is possible that some services might not make the profit needed because they sell other work or keep the customer coming back. For example, a service facility may charge a customer $10 for a state safety inspection but spend an hour of a $12 technician's time examining the car for a loss of $2 on each inspection. The reason service facilities like to perform state safety inspections is because they often require three out of five customer cars to have higher-paying steering, suspension or brake work that can earn more than the 2.5 multiplier. So the inspections are not "really a loss" when looking at the profitable work they help generate. Thus, the reason why many shops charge little money for an inspection or even in some states check brakes for free is because they are trying to get other work that is profitable. In addition a shop might sell a $50 tire for $54 making only $4 on the tire plus removal, mounting, and balancing of maybe $9 on a technician cost of $3 (a little more than a 2.5 multiplier at best). The reason for the deep discount is tires are a very competitive business. If the shop does not try to be competitive, the end result is the customer may go to another shop to get the tires before they come to "your" shop for other work. Unfortunately, while at the "other" shop buying tires, the customer will likely buy other services, such as brakes, and then there might not be a reason to come to "your" shop next. So it is best to try and serve the customer if, in fact, the mission of the business is general automotive repair. Decisions and examples will, of course, be different if the mission of the shop is something unrelated to these examples, such as transmission overhaul.

Purchase of an Established Business

The setup and operation of a new business can be challenging. Some investors choose to invest in established businesses with an established customer base, a proven business system, experienced employees, and an income/expense history that is verifiable through tax returns. The price an owner of an established business may want to sell the business could be higher than a start-up. The exact price depends on the assets that are included in the sale as well as the work the owner performs and the cash flow generated. Some commercial realtors will present a Pro forma to prospective buyers that compare the business sale price to the end of the year earnings. Investors can calculate the "yield" or return their investment will earn (the correct term from a "finance perspective" is yield but return is often used to express what the prospective buyer will get for the money invested). The return is expressed as a percentage and an investment of $500,000 that produces a net income of $50,000 is a 10% "return" or stated another way, it will take 10 years for the investor to "earn back" the $500,000 invested based on past financial results. The previous owner may also earn a wage in the business that is in addition to the "return." If an investor chooses to perform those duties, he or she can expect to earn a similar wage. If the investor plans to hire someone to do the job, then he or she can hopefully expect to obtain a similar investment return as the previous owners did. Whether the sale price is fair or not requires more analysis by studying what other repair businesses earn and other investments that are available with less risk and similar returns.

In some cases prospective owners see opportunities for improvement and plan to execute changes that will enable the business to earn more net income (or cut losses in a business going bankrupt) so greater returns are produced. Some businesses are for sale for far less than what they "could" sell for because the current owners are having financial or business system problems among other issues. Whether an investor should buy a business sold at a discount requires a very complex evaluation that is beyond the scope of this book. Hired consultants and background information on different evaluation methods discussed in the *Managing Automotive Businesses: Strategic Planning, Personnel, and Finance* should be examined. In addition a business might be sold at a "premium" which is a price that is well over the asset values and value of the cash flow (various methods of analysis not covered in this book). The investor needs to determine whether the price is worth the money that the business will "likely" produce in net income. For example investing $500,000 to get a $10,000 net income is a 2% return ($10,000/$500,000 = 0.02 × 100 = 2%) or 50 years to get the money invested back ($500,000/$10,000 per year = 50 years) through net income earnings (many assumptions are made in this short example). To justify a very high price an investor needs to examine many factors outside of this chapter's discussion such as the value of the land included in the sale (where the business is located) is potentially worth more over time than the business value itself. Therefore, the investor is speculating on the increased land value in the future.

Once a deal has been arranged to buy the business, the new owner will get the money (savings, bank loan, or previous owner financing) to pay the previous owner. The transfer of the business assets to the new owner may be as simple as the previous owner signing over the stock (corporation) to a more complex re-titling (providing sales receipt) for each asset bought by the new owner (proprietorship). The previous owner may also agree to provide support to the new owner through training, consulting, or letters to customers for a period of time. The new owner must also arrange to take responsibility for utility, phone, and other contracts of the previous owner before "closing." Closing in a traditional sense is when both sides get together to sign papers and exchange money.

Franchises

Some investors/owners feel that franchises are a preferred method of investing in a business. There are two types of franchises, a product franchise and system franchise. A product franchise sells products to the franchisee (figure 16-6). The franchisee follows franchisor rules (terms and conditions) and sells the products to customers so each will make money. Examples of a product franchises are new car dealers as well as some tool dealers such as Snap-on, MATCO, and MAC tools dealers (see Figure 16-6).

A system franchise sells a business system to the franchisee. As part of the franchise agreement the franchisor allows the franchisee to use of their name (see Figure16-7) and provides various types of support such as advertising, training, technical assistance, group buying power, as well as oversight to assure the business system

FIGURE 16-6 The business vehicles for two types of product franchises. The owner of the MAC tool franchise business bought the right to sell the franchisor's tools to technicians in a certain territory. The owner of the tow truck is a new car dealer franchisee that bought the right to sell the franchisor's vehicles, in this case General Motors. As part of the franchise agreement the business owner also agreed to fix the cars and sell General Motor's parts.

FIGURE 16-7 The owner of this business bought a system franchise.

is followed. Examples of franchises that might be purchased by investors include often recognized names such as Midas, Cottman Transmission, Aamco Transmissions, Maaco, Meineke, and Jiffy Lube among others.

For investors/owners, franchises have an appeal because they provide a recognized name, training, and guidance in setting up or buying an existing franchise. In exchange, the franchisee contractually agrees to certain terms and conditions such as paying for national and local advertising, a franchise fee (typically a percentage of gross sales), and adhering to procedures of the franchise. Adhering to the franchise system is required and a franchisor may "shop" (send in a mystery shopper to check the quality of services sold) and inspect the franchisee regularly to assure compliance; otherwise there are contractual consequences against the franchisee.

To understand the franchise offering, the franchisor will provide disclosure documents (figure 16-8) and ample time to consult with an attorney, accountant, and review of the terms of the contract. Due diligence should include visits to other franchisees, hiring of consultants deemed necessary by the investor/owner, as well as thought and careful reading all documents. If an investor/owner chooses to invest in a franchise, he or she will pay an upfront fee as well as sign necessary contracts. A franchise representative will then be assigned to assist the franchisee that will include training sessions and setup assistance. It is hoped that the help and guidance of the franchise will improve success and profits in the future "over and above" the fees paid to the franchise.

FIGURE 16-8 Investors examines franchise disclosure documents and form questions for their legal representative.

Review Questions
Multiple Choice

1. An owner who also works as a technician and serves as manager expects compensation for which of the following?
   A. Technician wages
   B. Return on money invested
   C. Manager's salary
   D. All of the above
2. Which is not a section of the strategic business plan?
   A. Mission
   B. Criteria
   C. Objectives
   D. Goals
3. A mission statement is for whom?
   A. Owner
   B. Employees
   C. Management
   D. All of the above
4. To reduce owner risk of loss of personal assets the business should:
   A. Buy a warranty contract
   B. Form as a corporation or LLC
   C. Sell Insurance
   D. All of the above
5. An EIN is *not* commonly used for:
   A. Proprietorship
   B. Partnership/LLC
   C. Tax documents
   D. Corporate identification
6. A state sale tax license allows the service facility to:
   A. Buy tools for the shop without paying sales tax
   B. Perform work for customer without charging the customer sales tax
   C. Have work done by a contractor and not pay sale tax until charged to the customer
   D. Mark up the parts sold to the customer but pay tax only on cost of the part
7. After start up, when should a profit be expected?
   A. Immediately
   B. Within days of opening
   C. By the end of the first year there will be significant profits
   D. It is difficult to predict when a business will earn a profit

8. A business's monthly profit based on a 20-day work schedule occurs when?
   A. After 5 days the business earns a profit
   B. After 10 days the business will earn a profit
   C. After 15 days the business will earn a profit
   D. It depends on expenses but a profit will likely be the final few days of a 20-day schedule
9. Which repair did not make the 60% required profit margin on labor (2.5 multiplier)?
   A. Labor sale $7.5 and technician pay $3
   B. Labor sale $126 and technician pay $48
   C. Labor sale $44 and technician pay $17
   D. Labor sale $52 and technician pay $22
10. Which yield produces a return on capital in five years or less?
    A. $700 yield per year on $7000 invested
    B. $38,000 yield per year on $228,000 invested
    C. $72,000 yield per year on $345,000 invested
    D. $61,000 yield per year on $312,000 invested
11. Return on capital in five years is a yield of
    A. 15%
    B. 20%
    C. 25%
    D. 30%

Short Answer Questions

1. Explain what happens in terms of the owner's money when a business has a loss.
2. What does an owner's return on capital investment mean?
3. Summarize each of the seven steps to start up a business.
4. If an owner runs out of "initial capital" what are the options to try to remain open?
5. Why would an owner need or want a business loan?
6. What is the role of the SBA in the business loan process?
7. What factors determine whether a business will get the best lease terms?
8. How is the tactical plan connected to the purchase of equipment?
9. Do "cheap" repairs keep customers returning? Explain why?
10. What can a service facility do to keep customers returning?
11. What advantages might a franchise provide to an investor (owner)?

Activity

Activity 1: Find an automotive-related franchise and obtain as much information as possible. Report what the franchisor provides to the franchisee and what must be paid (up-front costs, franchise fees, and minimum advertising expense requirements). Determine how this might help start up a business.

Example for Activity 2:

Part 1 of 7: Dave has gone into business for himself. His overhead is $4,200 per month and his labor rate is $42 per hour, a reasonable rate for his rural market. Neglecting parts profit, how many hours does he need to generate per month to break even (no profit or loss)?

Solution: $4,200 overhead /$42 per hour = 100 hours per month or about 25 hours per week (assuming 4 weeks in a month)

Part 2 of 7: If Dave earns 120 hours per month, how much does he take home (neglecting parts profit)?

Solution: 120 hours − 100 hours = 20 hours × $42 / hour = $840 take home pay.

Part 3 of 7: Dave plans that parts sales will be about 1/2 (50%) of the labor sales. Therefore, his expected parts sales are?

Solution: 120 hours × 42 per hour = $5,040/2 = $2,520 in parts sales.

Part 4 of 7: Dave plans to mark parts up 2× over cost. Therefore his profit on parts is?

Solution: $2,520 parts sales/2 part mark up = $1,260 in parts cost. Part sales $2520 - Part cost $1260 = Parts profit $1260.

Part 5 of 7: How much does the parts profit help in reducing the number of breakeven hours required (use step 1 overhead)?

Solution: $4,200 overhead − $1,260 parts profit = $2,940/$42 per hour = 70 hours per month or about 17.5 hours per week.

Part 6 of 7: How much did Dave take home?

Solution: 120 hours earned − 70 hours breakeven = 50 hours × $42 per hour = $2,100

Part 7 of 7: Additional understanding.

Dave doesn't keep track of his flat-rate hours but rather knows how much he brought in each day. If his labor sales of *money collected* each day this week was $140, $160, $280, $190, $310 how many hours did he make *and* did he make breakeven compared to *Part* 1? Finally, did he make any money for himself and what can he hope to take home?

Solution: $140, $160, $280, $190, $310 = $1,080/$42 = 25.7 hours − 25 hours (part 1 answer) = 0.7; Yes he made breakeven but not very much money for himself (0.7 × $42 = $29.4). He can only hope that his parts sales Labor Sales $1080 / 2 (see part 3) = $540 Part Sale / 2 (mark up see part 4) = $270 part cost. $540 Part sale - $270 Part cost = $270 Part profit help him earn a little more profit for the week.

Activity 2:

Part 1: You go into business for yourself. You have one bay and your overhead in this city market is $9,000 per month. Your labor rate is $120 per hour, a reasonable rate for this market. Neglecting parts profit how many hours are needed to reach breakeven (no profit or loss)?

Part 2: If you earn 100 hours per month, how much do you take home (neglecting parts profit)?

Part 3: You plan that parts sales will be 35% of the labor sale. Therefore your expected parts sales are?

Part 4: You plan to mark parts up 2X over cost. Therefore your profit on parts is?

Part 5: How much does the parts profit help in reducing the number of breakeven hours required (use step 1 overhead)?

Part 6: How much did you take home?

Part 7: Additional understanding.

You didn't keep track of your flat rate hours but rather you know how much you brought in each day. If your labor sales of *money collected* each day this week was $500, $350, $550, $300, $400, how many hours did you make *and* did you make breakeven compared to *Part* 1? Finally, did you make any money for yourself and what can you hope it will be?

GLOSSARY

Active delivery—when a service consultant personally delivers an automobile to a customer to discuss matters of importance, to check the condition of the automobile, and to determine customer satisfaction.

Automobile Service Facility—a business that performs automobile maintenance, repair, and diagnosis services.

Bumper-to-bumper warranty—a contract that covers all of the components in an automobile (see Warranty contract).

Business Liability insurance—an insurance policy that covers a customer who may be accidentally hurt while at the facility.

Business Operations System—part of a service system that oversees the business transactions of the service facility (also see the Service System, Shop Production System, Customer Service System).

Campaign repairs—see Recall/campaign repairs.

Chain service facility—an automobile service facility that is one of several facilities owned by a corporation, such as Pep Boys and Sears, Roebuck, & Co.

Chargeback—occurs when the dealer was "paid" too much for the repair and the manufacturer will reduce a future "payment" to the dealer.

Closing the sale—when a customer gives an approval for a service by signing a repair order.

Comebacks or second attempts—occur when a customer had a repair made to an automobile and must return to have the same repair made again.

Commercial Auto Coverage—coverage for vehicles owned and operated by service facilities.

Company policies—statements that indicate how the owners want to conduct business and legal regulations that direct the way business and services must be conducted.

Customer Service System—part of a service system that directs how the service consultant will interact with the customer (also see the Service System, Shop Production System, Business Operations System)

Cores (part core)—a used part that has been replaced and must be returned to the supplier for a reduction in the cost of the new part.

Corporation—a business owned by one or more people who invest in the business by purchasing shares of stock.

Cross training—when workers are taught how to perform one or more jobs in a business.

Customer automobile inventory sheet—a form used to record customer automobiles located in the building and on the property of the facility.

Effectiveness—when work is done correctly and quality outcomes are achieved.

Efficiency—when resources are used properly and output is maximized.

Emission warranty—a federal requirement that new automobile manufacturers must guarantee the repair of their automobile emission components for a stated period of time.

Extended warranty—purchased by owners of new and used automobiles to provide warranty coverage (sometimes with a deductible) for identified components in an automobile for a given period of time or number of miles that the automobile is driven from the date of purchase.

Feature-benefit selling—a sales approach that explains the service to be provided (feature) and its advantage (benefit).

Flat-rate objective—the number of flat-rate hours a technician hopes to earn in a week.

Flat-rate or flat-rate pay system—Technicians who are allotted (and paid for) and paid for a specified amount of time to perform a job are considered flat-rate technicians and are paid on the flat-rate pay system.

Fleet service facility—a service facility that limits its services to vehicles owned by a company or government body.

Flow diagram—a drawing that illustrates the processing of work or an activity through a business or an organization.

Formatted system—when an operations manual is followed to encourage consistency in performances of an automotive repair facility.

Franchise—a business granted the use of a nationally recognized name in return for a fee and a percentage of sales.

Garage Keepers insurance—liability insurance that covers damages to customers' automobiles while at a facility.

Gross profit—the balance left over from gross sales after subtracting technician salaries and the cost of the parts used to repair customer automobiles.

Gross sales—the total amount of money received from customer sales.

Invoice—final bill (see Vendor invoice).

Job description—a list or description of the job tasks to be performed by a person in a position.

Job duties—details of job tasks.

Job tasks—major work assignments conducted by a person in a specific job.

Lead technician—a technician who assists with the coordination of the work in a shop, in assigning jobs to other technicians, in quality control, in monitoring the condition and use of equipment, and with other supervisory duties assigned by management.

Lemon law—a law that requires automobile manufacturers, and, in some states, the automobile dealers, to buy back an automobile when it is not properly repaired.

Maintenance contract—awarded to or purchased by a customer at the time an automobile is purchased and pays for specified maintenance services for a set period of time from the date of purchase or for a set number of miles an automobile is driven.

Markup—the difference between the amount a facility pays for a part and the amount it is sold to a customer.

Mechanic's lien—state laws that permit a service facility to hold a customer's automobile until payment for a service is received.

Net loss—a negative balance after expenses and business taxes are deducted from the gross profit.

Net profit—a positive balance after expenses and business taxes are deducted from the gross profit.

New automobile manufacturer warranty—a contract awarded to the owners of an automobile that provide for its repair at no charge for a predetermined length of time since its initial purchase or for a predetermined number of miles the automobile is driven.

Newsletters—a mode of direct advertising. It contains articles of interest and useful information about automobiles.

Open business environment—forces outside the control of a service facility that affect its sales potential.

Operational environment—features or forces that have an influence on the daily business activities of a service facility.

Operational procedures and regulations—rules based on company policy.

Operation manual—directs the way work is to be conducted and processed in each work area.

Organizational diagram—presents the relationship between the different positions at an automobile service facility.

Organizational structure—represents the managerial chain of command and the relationships among its positions.

Overselling—when customers are sold a service their automobile does not need.

Partnership—a business owned by two or more people.

Petty cash fund—a fund that makes cash available for the purchase of items below a specified amount of money.

Policy check—a check given to customers for the purchase of goods or services from the service facility.

Pre-priced maintenance menu—a chart presenting different maintenances suggested by a manufacturer and the charges for each.

Product-specific service facility—a service facility that diagnoses, repairs, and performs maintenance on specific makes and models of automobiles, such as a facility that services only Volkswagens.

Property Insurance—an insurance policy that covers damages if a building is damaged in a storm, fire, or other accident.

Proprietorship—a business owned by one person.

Recall/campaign repairs—when a manufacturer requests the owners of a specific year, make, and model of an automobile to take it to a service facility for repair at no charge to them.

Repair categories—a list used to indicate the relative importance of a suggested repair.

Repair Order Tracking Sheet—used in the Shop Production System for the service consultant to keep track of a customer's repair progress (also see the Shop Production System).

Sales promotions—the primary objective is to keep regular customers and attract new ones.

Service System—a process with procedures that help a business function efficiently. An automotive service facility service system contains separate systems for customer service, shop production, and business operations (also see the Customer Service System, Shop Production System, and Business Operations System).

Seamless system—a process that is not disrupted when work is passed from one person or one station to another.

Service—the maintenance, repair, and diagnosis of an automobile.

Shop management system—the procedures, documents, files, and computer used to prepare, store, and retrieve customer information, repair orders, automobiles serviced, flat rate hours, parts markups, parts inventory, and vendor information.

Shop Production System—part of a service system that directs how the repair orders are processed (also see the Service System, Customer Service System, and Business Operations System).

Specialty service facility—services specific makes and models of new and/or used automobiles, such as those sold by a dealership.

Specialty team—a team whose repairs, maintenance, or diagnostic work assignments are limited, such as a team that performs only maintenance work.

Stockholders—people who invest in a corporation by purchasing shares of stock in the business.

Sublet sales—when a facility sends a customer's automobile to another facility for service, has the automobile returned when completed, pays the other facility for the service, and charges the customer for the service conducted by the other facility.

System-specific service facility—a service facility that repairs and maintains one automobile system, such as transmissions or brakes.

Tactical environment—factors that influence the provision and supply of resources needed to conduct business activities.

Technical service bulletin (TSB)—an announcements put out by automobile manufacturers to dealerships and subscribers of computer repair information systems to announce automobile operational concerns and how to repair them.

Technician's hardcopy—a thicker, cardboard-like copy of a repair order given to the technician.

Time tickets—a paper tracking system used by some shops.

Up-selling—when a customer selects higher-quality parts and/or labor as opposed to less expensive parts and/or labor.

Value-added service—a service received by a customer at no extra cost.

Vehicle identification number (VIN)—a seventeen-digit number assigned to an automobile when manufactured.

Vendor invoice—a bill from another business for parts, equipment, or services purchased.

Vendors—a company that sells parts, goods, and services to a facility.

Warranty contract—a contract that requires a third party to pay for all or most of a customer's repair invoice when the vehicle requires a repair. Terms in a warranty contract will specify the parts covered, time period, and mileage limits of the coverage.

Warranty coverage period—a predetermined time period, within which an automobile can be repaired "free of charge."

Workflow—the processing of work from the initial contact with customers to the return of their automobile.

Workman's compensation—an insurance policy that covers employees injuries while at work.

INDEX

A

AAA. *See* American Automobile Association (AAA); Approved Auto Repair
Accidents, customers, 98–99
Active delivery, 182–183, 218–219, 254, 273–274
  actual delivery, 218–219
  care and cleanliness of automobiles, 218
  customer follow-ups, 263–264
  future appointment, 182
  needed services, 179
  positive points, 182–183
  pre- and post-inspection, customer automobiles, 218–219
  workflow process, 218–219
Activity
  additional work phase, repair process, 247–249
  check-in phase, repair process, 243–246
  check out phase, repair process, 249–251
  initial work phase, repair process, 246–247
Additional work phase, repair process, 226–227, 236–241
  activity, 247–249
  anticipated parts delivery delays, 238
  catching problems with tracking sheet, 239
  processing the additional work, 239
  service consultant's role, 236–238
  team leader's role, 238
  work schedule, 238
Advertising, 265–268, 321–322
  balanced advertising plan, 272–273
  internet search engines advertising, 268
  internet "website" advertising, 266
  social network advertising, 266–268

Advertising messages, 268–272
  customer specials, 269–272
  newsletters, 268
Air-conditioning systems, automobile, 14
  Clean Air Act Section 609 and, 14
ALPHA Motors Dealership Service Department of Riverside, 358–359
Alternative transportation, 152–153
American Automobile Association (AAA), 153
Answering customer questions, 137–138
Appointment, customer, 76–77
Appointment book, 208, 209
Appointments schedule, 208–210
  time required for, 208–209
  workflow process, 208–210
Approved Auto Repair (AAA), 20–21
  facilities requirement, 20
  technicians, 20–21
ASE. *See* National Institute for Automotive Service Excellence
ASE certification, 4
Assistant service manager (ASM). *See also* Service consultant
  differences, examination, 284
  example of, job demands, 287
  implied expectations, 296–298
  management authority and responsibility, 290–293
  positions within three service facility groups, 285–287
  real and implied manager responsibility, 296–298
  role as team member, 299
  "staff exempt" employee, 285, 286
  team and lateral support group management, 298–300
  types of, 284–288
  workplace responsibilities with OSHA insight, 293–296
Audits, warranty claim, 104–105
Automobile, examination of, 336–337
Automobile dealership, 9–11

  chain of command, 9–11
  described, 9
  flowchart, 10
Automobile service facility
  classification of, 3–5
  computers and, 78–85
  federal regulations influencing, 13–18
  history of services performed, 70–71
  objective of, 3
  ownership, 5–11
  service consultant requirements, 3
  state guidelines for, 18–19
  types of
    chain, 9
    corporate owned, 6–8
    dealership, 9–11
    fleet service, 11–12
    franchise, 8–9
    privately owned, 5–6
    product-specific, 4–5
Automobile service history, 74–76
  computer database, 74–76
  customer approval, 74
  invoices, 74
  repair orders, 74
Automobile systems, 3–4. *See also* National Institute for Automotive Service Excellence (ASE)
Automotive business consultant, 300
Automotive industry jobs, 12–13
  diversity of, 12, 13
  women as professionals in, 13
Automotive repair business, 355–356
Automotive service, historical discussion of, 48–49

B

Banking, 321
Billable hours, 52, 60–61
  calculation of, 52
Bonuses, 345
Bounces, check, 185
Building repair maintenance, 317

Bumper-to-bumper warranty, 100
Bureau of Automotive Repair (BAR), 365
Business
  job-by-job analysis, 376–378
  marketing plan of, 371–373
  money, 374–376
  operational plan, 367–371
    borrowing money, 367
    building permits and business licenses, 369–371
    contracting for services, 369
    equipment leases, 368
    tactical plan, adjustment of, 367–368
  purchase of, 378–380
  requirement, 359
  strategic plan, 355–359
    goal and objectives, 357–359
    mission statement, 356–357
  structure, 361–363
  tactical plan, 359–360
  tax consequences, 362–363
  volume of, 376
Business contracts
  execution of, 307–308
  goods, resale of, 318–320
  maintenance contracts
    for building, 317
    for equipment, 317
  management authority to enter, 306–307
  parts, supply of, 318–320
    uniforms, 318
  sublet sales, 315–316
Business establishment, 378–380
Business goal and objectives, 357–359
Business liability insurance, 309, 361
Business operations policy (BOP). *See* Business liability insurance
Business operations system, 28–29, 35, 51, 122
Business production system, 78
Business structure, 361–363

C

California Air Resource Board (CARB), 14, 192
California's Bureau of Automotive Repair (BAR) regulations, 18, 205

CARB. *See* California Air Resource Board (CARB)
Cashier, 58
Categorizing repairs, 176–178
Chain facility, 9
  chain of command, 9, 10
  definition, 9
  flowchart, 10
  procedures, 41–42
Chargeback, defined, 103
Check bounces, 185
Check cashing service, 185
Check-in phase, repair process, 226–228, 229–232
  activity, 243–246
Check out phase, repair process, 227–228, 241–243
  activity, 249–251
Clean Air Act, 163
  Section 609, 13, 14
Closing a sale, 173
  explaining the needed repair, 180
  feature-benefit selling, 178–181
  opportunities to sell, 181–183
  over the telephone, 173, 174
  payment methods, 185–186
  selling repairs, 174–178
  willingness to help, 173
Code sticker, 199
Comebacks, 89–90, 336–337
  effective handling of, 90
Commercial auto coverage, 310–311
Communication, 114
  customer communication methods, 173–174
  customer-provided vehicle information, 192
  face-to-face, 118–120
  information and warnings, 190–191
  and social network, 120–122
  stages of customer service system, 124
  telephone, 114–118
Communication skills, 30, 65
Company policy, 205–207
  consumer protection laws, 205
  link with operations manual, 205–206
Comparison shopping, 117–118
Computer database, 74–76, 77
  for cost of parts and labor, 154–155
  for warranties and contracts, 98

Computerized appointment schedule, 208–210. *See also* Appointments schedule
Computerized automobile records, 74–76
Computerized service system
  fully, 83–84
    dealerships, manufacturers integration, 85
    drawbacks, 84
  non-computerized, 82
  partially, 82–83
Computer program, 76–77, 78–82
  choosing, 76–77
  shop production, 80–82
    choosing, 80, 82
    operations, 333
Concern, nonessential, 176
Conduct, team leader, 64
Consumer protection laws, 90
  and company policy, 205
  mechanic's lien, 144
  repair order waiver, 137–138
  routine maintenance, 135–137
Contracts
  business, 304–308
  customers under, 324–325
  external and internal, 308
  insurance, 308–311
  maintenance, 317
  uniform, 318
Cores, defined, 319
Corporate owned facility, 6–8
  chain of command, 7
  flowchart, 7
  specialists, 41–42
Corporation, 5, 361. *See also* Corporate owned facility
Coupons, 269–272
Court complaint, 145–147
Courtesy shuttle, 153
Credit card payment, 185
Cross training, 65
CSI score. *See* Customer satisfaction index (CSI) score
Culture, communication and respect, 328
Customer
  retail, 161
  wholesale, 161
Customer amenities
  alternative transportation, 152–153
  comfort, 152

Customer automobile inventory
   sheet, 33
   example of, 36
Customer base, 71
Customer claim
   documentation of, 110
   lemon law requirements, 109–111
   warranty-related, 109
Customer concerns
   cost, 154–155
   time frame, 154–156
Customer contact
   diagnostics charges, 132–133
   disclosure of costs, 135–137
   maintenance requests, 135
   repair authorization, 136–137
   repair requests, 134–135
   and script for diagnosis, 133–134
   verbal authorization, 140
Customer data, historical, 75–76
Customer disagreements, 257
Customer expectations, 51–52, 344–347
   exceeding expectations, 345–346
Customer feedback, 273–274
Customer files, 76–77
   recording problem, 188–189
   vs. paper files, 84
Customer helpline, 92
Customer neglect, warranty contract, 101
Customer personalities, 255–256
Customer problems
   avoiding, 346–347
   examples of, 346–347
   recording, 189–190
Customer questions, 116
Customer records, 70–71
Customer relations, 88–89
Customer satisfaction, 28, 89, 273–277
   psychology of, 275–277
   ratings, 274–275
   surveys, 89
Customer satisfaction index (CSI) score, 258–260, 274–276
Customer service, 52
Customer service scripts, 116–117
Customer service system, 28–29, 50, 78, 123
   communication stages of, 124
   defined, 36
   interactions, 124–127
   non-computerized, 82
      advantages, 82
   overlap of shop production and, 124–125
   overlapping shop production system, 78
   partially computerized, 82–83
Customer shuttle driver, 58
Customer specials, 269–272
   example of, 271
Customers under contract, 324–325.
   See also Contracts
   delivery service, 324–325
   pick up service, 324–325
   repair service, 324–325
   towing service, 324–325
Customer surveys, 273–275
   analyzing information, 274–275
      CSI score, 275
      dealers' scores, 275
   basic questions on, 274
   phone, 274

D

Daily accounting report, 39
   example of, 40
Damage claim and insurance, 145
Date of production, location of, 196–197
   and door sticker information, 196–197
DBA (Do Business As), 361
Dealership service department
   dispatcher system, 62–63
   formal working relationship, 64
   group system, 63–64
   internal users of, 314–315
   lateral support team, 63–64
   obtaining parts from, 314–315
   training, 41
   warranty repairs, 89, 101, 103
Dealing with angry customers, 277–279
Delegating authority, 290–293
Diagnostic charges, 132–133, 136
Dispatcher system, 62–63
   facilities with, 63
   service manager, role of, 63
Door sticker information, 196–197
Duplicating problem, 191–192
Duties, 28–29, 31–35
   overlapping employee, 35–39
   sample lists of, 33–34

E

Economic environment, 184
Effective communication advice, 260–262
Effectiveness, 329
   influences, 335–336
Effectiveness of team, 328–344
Effective service, 352
Efficiency, 329
   influences of, 335–336
Eligibility for warranty/contract work, 97–98
Email and text messages, 121–122
Emission warranty, 102–103
Employee, expectations of, 345
   bonuses, 345
   compliments, 345–346
   exceeding expectations, 345–346
Employee bonding, 313
Employee evaluation, 31
Employee input
   feedback, 344
   group meetings, 344
   suggestions, 344
   surveys, 344
Employer Identification Number (EIN), 363
Environmental Protection Agency (EPA), 163
   automobile air-conditioning system and, 14
   automobile service waste and, 14–15
   vehicle emissions and, 13–14
EPA. *See* Environmental Protection Agency (EPA)
Estimate, 84
   for diagnostic charges, 133–134
   disclosure to customer, 128–129
   documents, 128–141
   example of, 135–136
   options, 129
   paper estimate, 143–144
   preparing, 85–88, 135–137
   repair charges, 128, 130
   sections include in document of, 128–131
   stages of customer interaction, 131–137
   unexpected service problems, 163
   waiver of, 129
Estimate-only forms, 143–144
Estimation, vehicle information and creating, 193

Extended warranty, 100–102
  deductible, 101
  limitations, 101
External contract, 308

F

Face-to-face communication
  greeting customers, 119–120
  greeting repeat customers, 120
Facility, marketing, 262–263
Facility requirement, business, 359
Factory and franchise
    representatives, 257–260
  factory representative's job,
    259–260
  new vehicle manufacturer,
    259–260
  service facility and new car dealer
    programs, 258–259
Fair Labor Standards Act, 288
Feature-benefit method
  inspection forms, 179
    maintenance schedules, 178
    seasonal specials, 178
  maintenance sales, 178–181
Federal job classification, 280–290
Federal Motor Vehicle Safety
    Standards (FMVSS), 18
Federal regulations, 13–18
  automobile air-conditioning
    systems, 14
  OSHA standards, 16–18
  vehicle emission programs, 13–14
  waste materials, automobile
    service, 14–15
Filing system, 76–77
Financial operations and licenses,
    363–367
First-time customers, 71–74
  complementary service, 74
  detailed inspection of vehicle, 74
  historical database, 74
  inspection form, 73
  maintenance inspection form, 73
  safety inspection form, 74
Flat rate, 60
  example of calculations, 60
  facility cost, 189
  technician wage, 189
Flat-rate hours, 210, 214
Flat rate method, 128
Flat rate objectives, 65–66
  based on performance review, 66
  daily labor report for, 66
  definition, 65

Flat rate report, 66
"Flat-rate time," 238
Fleet customers, 91–92
  automatic maintenance, 91
  financial issues with, 92
  overtime, 91
  priority service, 91
  service discounts, 91
  special treatment, 91
Fleet hotline, staff, 93
Fleet maintenance schedule, 91
Fleet manager, role of, 12
Fleet service department, 11–12
Flow diagram, 56
  team members, 57, 58, 59, 61
FMVSS. *See* Federal Motor Vehicle
    Safety Standards
Follow-through
  active delivery, 346
  checkoff sheet, 346
  procedures for, 346
  technician's comments, 346
Formal working relationship, 64
Franchise manual, 286
Franchise ownership, 8–9,
    380–382
  chain of command, 8–9
  corporate owner, rights of, 6–8
  described, 8
  flowchart, 9
  local owner, requirements of, 8
  and national franchise
    corporation, 8
  obtaining parts from, 318–320
  procedures, 41
Full-service facility, 172

G

Garage keepers insurance, 309
General maintenance inspection
    form, 73
General Motors, 199
  code sticker, 199
General Motors Women's Retail
    Network (WRN), 13
  initiatives for women, in
    automotive industry, 13
General supplies, 322–323
  coffee, 323
  first aid, 323
  office, 322
  restroom, 323
"Goodwill" claim, 260
Gross profit, 21
  gross sale and, 22

Gross sale, 21
  gross profit and, 22
Gross vehicle weight rating, 196

H

Hardcopy, repair order, 144, 189
  backup, 71
Hazardous danger, 176
Host, service consultant as,
    152–153
Human resources, 359

I

I-Car (Inter-Industry Conference on
    Auto Collision Repair), 20
IIHS. *See* Insurance Institute for
    Highway Safety (IIHS)
Imminent danger, 176
Imminent malfunction, 176
Implied expectations, 296–298
Implied manager responsibility,
    296–298
Ineffective performance, 336–337
Inefficient performance, 342
Initial work phase, repair process,
    228, 229–236
  activity, 246–247
  communication monitor, 232
  technician's and parts specialist's
    role, 232–236
Inspection
  automobiles, 218–219
  forms, 73–74, 75, 179, 218
  value-added, 337, 338, 342, 343
  vehicle, 74
Inspections and maintenance (I/M)
    programs
  emissions, 163
Insurance, damage claim and, 145
Insurance card, 192
Insurance contracts
  business liability insurance, 309
  commercial auto coverage,
    310–311
  garage keepers insurance, 309
  property insurance, 309
  service facility, 308–311
  workman's compensation, 311
Insurance Institute for Highway
    Safety (IIHS), 97
Internal communication,
    189–202
Internal contract, 308
  and managers, 314–315

Internet search engines advertising, 268
Internet "website" advertising, 266
Invoice, 141–144
    defined, 85
Invoice stage, 142–143

J

Jill's Import Repairs' Mission Statement, 357
Job-by-job analysis, 376–378
Job description, 30–31, 37
Job tasks, 30

L

Labor
    sale price, 156–157
    time book, 155
    times, 155–156, 236, 237
Labor times, 155–156, 236, 237
Lateral support group management, 298–300
Law and legislations, repair orders, 138–140
Leadership responsibilities, 64
Lead technician, role of, 59–60
    responsibilities of, 65
Legal contract, 127
Lemon law, 109–111
    buy back requirement, 109
    judicial process, 110
    state law, 110
Licenses
    building permits and, 369–371
    financial operations and, 363–365
    taxes and, 365–366
Limited liability company (LLC), 361
Line of credit, 364
LLC. See Limited liability company (LLC)
Loan application, 186
Loaner car, 153
Long-term loan, 185

M

Maintenance
    Clean Air Act, 163
    emission inspection, 163
    inspection, 179
    of newer vehicles, 164
    oil change, 165
    of older vehicles, 163

packages, 163
pre-priced maintenance menus, 163–165
preventive, 166–167
for regular customers, 166
related services, 167
schedule, 163
seasonal, 165, 179
specials, 165
inspection, 165
Maintenance contracts, 96–97
for building, 317
for facility equipment, 317
verification of, 106–108
Maintenance sales, 179–181
    feature-benefit selling, 178
Maintenance schedule, 178
Malfunction
    imminent, 176
    nonessential, 176
    potential, 176
Management authority, 290–293
Management responsibility, 290–293
Managers and internal contracts, 313–315
Manager's job description, 296
Managing Automotive Businesses: Strategic Planning, Personnel, and Finance, 51, 142, 284, 352, 358, 374
Manufacturer recommended maintenance
    packages, 164
    specials, 165–166, 178
Marketing plan, of business, 371–373
Markup, 157
    calculation of, 157
Material Safety Data Sheets (MSDS), 16
Mechanic's lien, 144
Mobile automotive repair business, 360
Model year
    location of, 197–198
    vehicle, 197–198
Monitoring repair progress, 223–251
Monitoring shop volume, 343–344
Monthly payments, 186
MSDS. See Material Safety Data Sheets
Multiple teams, 60–62
    specialty team, 61–62
    support staff, 61
    working relationship, 61, 62

N

National franchise corporation, 8
National Highway Traffic Safety Administration (NHTSA), 18, 97
National Institute for Automotive Service Excellence (ASE), 3–4, 20. *See also* Automobile systems
    automatic transmission/transaxle (A2), 3
    brakes (A5), 3
    certification for, collision repair industry, 4
    certification test series, 4
    electrical/electronic (A6), 3
    engine performance (A8), 3
    engine repair (A1), 3
    heating and air conditioning (A7), 3
    manual drive train and axles (A3), 3
    suspension and steering (A4), 3
Negative balance, 21
Net loss, defined, 21
Net profit, defined, 21
New automobile manufacturer warranty, 98
    state's laws, 98
Newsletters, 269
Newspapers, public advertising in, 272–275
NHTSA. *See* National Highway Traffic Safety Administration (NHTSA)
Noncovered service items, 101
Nonessential concern, 176
Nonessential malfunction, 176
Noteworthy items, 177
Number one management rule, 290, 292

O

Occupational Safety and Health Act of 1970 (OSH Act), 295
Occupational Safety and Health Administration (OSHA), 15–17, 295
    fire regulation, 16
    first aid regulation, 16
    Material Safety Data Sheets (MSDS), 16
    regulatory agency, 295

Office supplies
  ordering, 322
  restocking, 322
Open business environments, 183
Operational environment, 183–184
Operational plan, 367–371
  borrowing money, 367
  building permits and business licenses, 369–371
  of business, 367–371
  equipment leases, 368
  tactical plan, adjustment of, 369
Operational procedures and regulations, 206
Operations (shop production system), 122–127
Operations manual, 36, 216–218, 296–297
Optional equipment, 193
Organizational diagram, 2
Organizational structure, 2
OSHA. *See* Occupational Safety and Health Administration (OSHA)
Overcapacity, 215–216
Overhead expenses, 21
Overselling, 172, 256–257
Overtime, 91
Ownership, 352–354
  types of
    corporation, 5
    partnership, 5
    proprietorship, 5

P

Package offers, 134
Paper files, 84
Paper repair order, 144
  technician's hardcopy, 144
Paper tracking system, 329
Partnership, 5
  flowchart, 6
Parts
  availability, 157–161
  costs, 154–157
  identification by manufacturer, 199–200
  markup, 157
  ordering by manufacturer, 194
  related services and parts sales, 167
  sale price, 156–157
  suppliers of, 318–320
  tracking, 167
  unexpected problems, 162

Parts contract
  credit purchase, 319
  purchase of, 318–319
  for resale, 318–320
  restocking, 319
  resupplying arrangements, 319
Parts runner, 58
Parts warranty, 90–91
Patty cash fund, 320–321
Payment contracts, types of maintenance, 96–97
  recall/campaign repair order, 96
  warranty, 96–97
Payment methods
  check cashing service, 185
  credit card, 185
  30-day same-as-cash, 186
  long-term loan, 185
  monthly payments, 186
  personal check, 185
  and selling, 185–186
  short-term credit, 185
Pennsylvania State Safety Inspection Regulations, 19
Pennsylvania Vehicle Safety Inspection Program Effectiveness Study, 19
Performance review, 66
Personal appearance, team leader, 64
Personal check, 185
Personal communications
  greeting customers, 119–120
  preparation repair authorization, 135–137
  presenting the estimate, 135–136
Personal dignity, 65
Phone scripts, 116–117
Phone shoppers, 117–118, 173–174
Policies, company, 205–207
Policy check, 278–280
  example, 279
  issues, 279–280
  use of, 280
Policy manual, 296
"Policy statement," 205
Potential malfunction, 176
Predetermined mileage, 98
Prepaying maintenance, 97
Pre-priced maintenance, 135
Pre-priced package, 134
Printed Internet registration, 192
Privately owned facility, 5–6
  chain of command, 6
  described, 5
  flowchart, 6

Procedures, 35–36
  follow-through, 346
Processing repair order, 37
Product franchise, 9
Production capacity, 203–222
Production date, 196–197
Production option codes, 199
Productive work environment, 335–336
Productivity of team, 53
  based on expansion, 53
  based on increased efficiency, 53, 328–329
Product knowledge, 30
Product-specific facility, 4–5
Professional image, 42–43
  customer reception area, 43
  service consultant's appearance, 42
Promotions, 321–322. *See also* Advertising
Property insurance, 309
Proprietorship, 5
  flowchart, 6
Public advertising, 272–273
Punch time, time ticket tag, 329–335

R

Radio, public advertising in, 272–273
Real manager responsibility, 296–298
Recall and campaign information, 97
Recall/campaign repair order, 97, 106–108
Receptionist, 58
Recording, customer information, 189–192
Registration card, 192
Rental car, 153, 162. *See also* Loaner car
Repair authorization hotline. *See* Fleet hotline
Repair categories, 176–177
  priority levels
    hazardous danger (2), 176
    imminent danger (1), 176
    imminent malfunction (3), 176
    nonessential concern (6), 176
    nonessential malfunction (5), 176
    potential malfunction (4), 176
Repair charges, estimation of, 128
Repair cost, 154–157
  estimation, 130

Repair facilities, 207
Repair order processing, 226
Repair orders (RO), 37–38, 157, 204, 226, 230–232, 239, 243
 approval of, 137–138
 binding contracts, 134
 customer approval of, 86
 distribution of copies, 143–144
 and efficiency, 335
 explanation of form, 53–55
 flag for first visit, 72
 form, 143–144
 legal challenges to, 137–138
 maintenance, 165–166
 nonpayment by customers, 144
 stage, 138–140
 technician's hardcopy, 144, 189
 tracking sheet, 124, 138
 updation of, 140–141
 what and when, 189–190
Repair order tracking sheet, 39, 204, 211–212, 217–218, 226–251, 339, 340
 additional work phase, 226–227, 236–241
 with all phases, 39
 check-in phase, 226–227, 229–232
 check out phase, 226–228, 241–243
 initial work phase, 226–227, 232–236
 overlapping duties, monitoring, 37
 teaching tool, 226
Repairs, selling, 174–178
Repeat customers, 71–72
 identifying, 71
 recruiting, 71
 retaining, 71
Repeat repair policy, 90
Retail customer, 160–161

S

SAE. See Society of Automotive Engineers
Sale price, parts, 157–158
Sales follow-ups, 263–265
 corrective action, 263
 first follow-up, 263–264
 personal recognitions, 265
 primary objective of, 263
 second follow-up, 264
 web-based services, 264–265
Sales promotions, 260–262, 265–268
Sales skills, 30
Sales tax, 364–365

Second attempt (repeat repair), 89
Security systems, 323–324
 alarms, 323
 doors and windows, 323
 service bays, 323
 video monitors, 323
Selling opportunities, 181–183
Selling services, 254–257
 customer personalities and sales, 255–256
 overselling, 256–257
 selling and angry customers, 255
 up-selling, 256–257
"Senior service consultant," 286
Service consultant, 119–120
 definition of, 3
 duties, 28–29, 31–35, 315–320
 exams for, 20
 greeting customers, 119–120, 255
 handling payment contracts, 96
 as host, 152–153
 job description, 30–31, 37
 job tasks, 30–31
 keeping customer informed, 152, 176–177
 limiting factor, 207
 making appointments, 208–210
 meeting customer expectations, 345–346
 overselling, 172, 256–257
 paperwork, 204
 problem solving, 149–150, 173
 recording the problem example of, 190
 responsibilities, 204–205
 role of, 28–43
 scheduling work, 207–216
 script for diagnosis, 133–134
 selling services, 254–257
 sincerity, 173
 as staff nonexempt employee, 293
 telephone skills, 114–118, 173–174
 training, 41–42
 "turn away" work, 207
 up-selling, 256–257
 using appropriate terminology, 192
 work environment, 328–344
Service consultant efficiency, 338–341
Service facilities
 division, 284–285
 "formatted" service system, 284

procedures. See Operations manual
profit centers, 284–285
warranty information, 129–130
Service facility insurance contracts, 308–311
Service facility start-up
 business structure, 361–362
 financial operations and licenses, 363–366
 marketing plan of business, 371–373
 operational plan of business, 367–371
 seven steps for, 354–355
 strategic plan of business, 355–359
 tactical plan of business, 359–360
 work to earn profit, 374–378
Service manager, role of, 7–8
Service reminders, 269
Service system
 customer expectations of, 51–52
 defined, 50, 78, 225–226
 formatted operating, 36–37
 full computerized, 83–84
 drawbacks of, 84–85
 operations of, 78
 overlapping duties, coordination of, 37–38
 purpose of, 51
 service consultant, role of, 50–51
 team approach within, 50–52
 time targets, 230
 types, 28–29, 37, 50–51
Shop equipment maintenance, 317
Shop management system, 74
Shop operations, 30, 122–127, 139. See also Shop production system
 overlap of customer service system and, 124
Shop production operations, 335
Shop production system, 28–29, 36, 50–51, 78, 224, 226–228, 231, 244
 flow chart, 217
 repair order process in, 56
Short-term credit, 185
Shuttle service, 58, 153

Small independent service facility team, 62
    people comprising, 62
SMOG test, 192
Social network advertising, 267–268
Social networking, 266–267
    to communicate with customers, 120–122
    email and text messages, 121–122
Social Security number, 363
Society of Automotive Engineers (SAE), 14
Softcopy, repair order, 138
Solicitations, public advertising, 272–273
Special circumstances, warranty, 99
Special maintenance inspection form, 73
Specialty service facilities duties, 40–42
Specialty team, 61–62
    for automobile maintenance, 61–62
    working relationships, 61
"Staff exempt" employee, 285–290
Staff nonexempt employees, 286–290
State Attorney General's statement, 128–129
State guidelines, 18–19
    on customer approval, 86
    emissions requirements, 163
    on estimate, 85
    on final invoices, 85, 88
    on overselling, 172
    on repair orders, 85
    safety inspection programs, 19
    state issued license, 18
    on verbal authorizations, 140
    on warranties, 89
State's Attorney General's Consumer Protection Department, 127
    inquiry, 145
Stockholders, defined, 5
Sublet sales, 315–316
    gentleman's agreement, 315
    warranty agreement, 316
    written contract, 315
Support staff, 58, 61, 62
System, defined, 36, 78
System-specific service facility, 4

T

TA (Trade As), 361
Tactical environment, 183
Tactical plan, 359–360
Tax consequences, 362–363
Taxes and licenses, 365–366
Team approach
    mission, 52–53
    customer satisfaction, 52
    efficiency, 52
    purpose of, 50
    within a service system, 50–52
Team leader, 64–65
    attentiveness, regarding process weaknesses, 65
    communication, encouraging, 65
    cross-training, 65
    duties, 64–65
    personal example, setting, 64
    responsibility and authority, 292
    team building, 65
Team members, flow diagrams, 57, 58, 59, 61
Team process, 53–56
    division of labor, 55
    team members, 52
Team support group management, 298–299
Team system
    adding members, 58–59
    flowchart, 57
    four-member, 59
    support personnel, 58
    cashier, 58
    customer shuttle driver, 58
    parts runner, 58
    receptionist, 58
    three-member, 56, 58
    two or more teams at facility, 60–62
Technical hotline, 92
    dealership technicians, use of, 92
Technical service bulletin (TSB), 97, 105
Technical training, 341–342
Technician
    certification requirement, 18–19
    limitations, inspection, 178
    problem solving, 191–192
    using appropriate terminology, 192
Technician hard copy. *See* Hardcopy
Telephone communications, 114–118
Telephone customers, 173–174
Telephone solicitation, 272
Telephone techniques
    answering customer questions, 116
    clear, simple language, 115–116
    confirming information, 131–132
    courtesy, 115–116
    focusing on the customer, 115–116
    identifying customer concerns, 132
    practice, 117
    using scripts, 116–117
Television, public advertising in, 272–273
Text messages and email, 121–122
30-day same-as-cash, 186
Time interval limits, 99
Time tickets, 329–335
Time ticket tag, 331
Tire dealer and repair facility, 360
Towing arrangements, 153–154
Tracking parts, 167
Tracking sheet, repair orders, 124
Tracking technician efficiency, 329–335
TSB. *See* Technical service bulletin (TSB)

U

Undercapacity, 215–216
Uniform contract, 318
    cleaning of, 318
    delivery of, 318
Unqualified technician repairs, 293
Unwritten agreements, 312
Up-selling, 256–257
Used car warranty, 100
Utilities, 321
    disruption of, 321
    electric circuit breakers, 321
    shutoff valves, 321

V

Value-added service, 337, 338, 342, 343
VECI. *See* Vehicle Emission Control Information (VECI)
Vehicle, parts availability, 234
Vehicle Emission Control Information (VECI), 198
Vehicle emission programs, 13–14
    centralized, 13
    decentralized, 14
    EPA and, 13–14
    technicians and, 14
Vehicle identification number (VIN), 76, 128
    and estimation, 193
    location of, 194

parts of, 194
  body, 195
  check digit, 194–195
  engine, 195
  make, 195
  manufacturer, 195
  origin, 195
  plant code, 195
  restraint, 195
  sequence number, 195
  vehicle line, 195
  year, 194, 195
Vehicle model year, 197–198
Vehicle records, 70–71
Vendor invoices, 7
Vendors, of goods and parts, 318–319
Verification of warranty, 106–108
VIN. *See* Vehicle identification number (VIN)
Volume of business, 376
Voluntary certification
  AAA, 20–21
  ACDelco Professional Service Center, 21
  American Car Care, 21
  ASE, 20
  Automotive Service Association (ASA), 21
  I-Car, 20
  NAPA Autocare Centers, 21

W

Warranty claim audits, 104–105
Warranty contracts
  comprehensive, 184
  customer neglect, 102
  deductibles, 109
  disclosure to customer, 108
  Internet purchase of, 102
  noncoverage conditions, 101
  payment contract, 96
  payment in credits conditions, 109
    maximum payment provisions, 109
    reimbursement, 109
  procedures, 108
  processing manufacturer warranty, 103–104
  required information, 190
  state laws regarding, 89
  terms, 108
  validation, 99
  verification of, 106–108
Warranty coverage period, 96, 98
Warranty payment, 130
Warranty policies, types of, 98–106
  bumper-to-bumper warranty, 100
  emission warranty, 102–103
  extended warranty, 102
  new automobile warranty contracts, 98–99
Waste, automobile service, 14–15
  EPA and, 14–15
  service consultant, role of, 15
Wholesale customer, 160–161
Women, in automotive industry, 13
Workers' compensation, 311
  conditions for payment of, 320–321
Workflow process, 203–219
  active delivery, 218–219
  definition, 203
  monitoring repair progress. *See* Monitoring repair progress
  operations manual, 216–218
  scheduling work, 207–216
  service consultant paperwork, 204
  service consultants' responsibilities, 204–205
Workman's compensation, 311
Work schedule, 207–216
  appointments, 208–210
  managing, 210–213
  overcapacity, 215–216
  undercapacity, 215–216
  work hours per vehicle, capacity determination, 213–215
WRN. *See* General Motors Women's Retail Network